D1272783

THE LOEB CLASSICAL LIBRARY

FOUNDED BY JAMES LOEB

EDITED BY

G. P. GOOLD

PREVIOUS EDITORS

T. E. PAGE E. CAPPS

W. H. D. ROUSE L. A. POST

E. H. WARMINGTON

HIPPOCRATES

III

LCL 149

HIPPOCRATES

VOLUME III

WITH AN ENGLISH TRANSLATION BY

E. T. WITHINGTON

HARVARD UNIVERSITY PRESS
CAMBRIDGE, MASSACHUSETTS
LONDON, ENGLAND

PA
6156
.A1G
Hippocrates
v.3

First published 1928
Reprinted 1944, 1948, 1959, 1968, 1984, 1999

LOEB CLASSICAL LIBRARY® is a registered trademark of the President and Fellows of Harvard College

ISBN 0-674-99165-6

Printed in Great Britain by St Edmundsbury Press Ltd, Bury St Edmunds, Suffolk, on acid-free paper.
Bound by Hunter & Foulis Ltd, Edinburgh, Scotland.

CONTENTS

41587

TRANSLATOR'S PREFACE

Anutius Foësius on coming (1594) to the surgical section of his *Hippocrates* says that some will find fault with him for editing treatises so fully discussed by many eminent writers: they will call his work futile and superfluous. Some will also cry out upon his notes as fragmentary, superficial and useless. Such fears are more natural in one who looks back not only on Foës himself and his contemporaries, but on the translation of Adams, the great edition of Petrequin, and the labours of Littré and Ermerins, nowhere more complete than when dealing with these treatises; while behind them all loom the thousand pages of Galenic Commentaries and the dim light of the illustrations of Apollonius. He is overwhelmed by his material, and cannot hope to do more that attempt a fairly accurate translation with fragmentary notes condensing the more important discussions of preceding editors.

The recent revolution in surgery due to anaesthetics, asepsis, radiography and other practical and scientific progress tends to put a modern surgeon rather out of touch with the great ancients. It makes him, perhaps, less able to appreciate their achievements, and more conscious of their unavoidable errors. On the other side, recent criti-

cism of the Corpus Hippocraticum relieves him from the necessity of assuming that Hippocrates wrote *Mochlicon,* and therefore of approaching it hat in hand. Its author assumes rather the appearance of a slave surgeon or student to whom his master gave a rather dilapidated copy of *Fractures-Joints* with instructions to summarise everything to do with dislocations, and be quick about it. That the result should have been held in honour for more than twenty centuries is high tribute to the excellence of the original.

The translation was made independently of that by Adams, though some of his expressions were afterwards adopted. The notes and meanings of words are taken more frequently than usual from the Commentaries of Galen, who is surely our highest authority on the subject. The text is mainly that of Petrequin, a conservative scholar who often successfully defends the manuscript readings against rash alterations by Littré and Ermerins. The recent edition by Kühlewein (Teubner, 1902) is doubtless an improvement even upon *Petrequin,* but was not directly available. Some of his emendations are adopted with due acknowledgment, and many of his variants are given in the notes, including all not otherwise attributed. The excessive "Ionicism" of all previous editions has been reduced in accordance with Kühlewein's principles, as in the other volumes.

In treatises so fully discussed by "so many most noble writers in that part of medicine," as Foës has observed, any novel suggestions are likely to be wrong, and the editor is duly conscious of presumption in submitting views of that character as to the

Hippocratic Bench, the astragalus and the origin of Chapters LXXIX-LXXXI on joints.

PREFACE

The whole of this volume has been entrusted to Dr. E. T. Withington, of Balliol College. Only a trained surgeon can explain the surgical treatises of the Hippocratic Collection.

The fourth (and last) volume will contain *Aphorisms*, *Humours*, *Nature of Man*, *Regimen in Health* I–III, and *Dreams.* The text of all these works has to be worked out from the manuscripts themselves, as Littré's text is here very imperfect.

W. H. S. J.

GENERAL INTRODUCTION

When Marcus Aurelius Severinus gave the title *De efficaci Medicina* to his work on surgery he probably expected to annoy the professors of what was then considered a much higher branch of the healing art, but when he goes on to say that surgery is obviously a strenuous, potent and vital method of treatment, few who have been actively or passively concerned with broken bones, dislocated joints or bleeding wounds will venture to disagree with him. He was doubtless also thinking of Celsus, who had long before declared that the part of *medicine* which cures by hand has a more directly obvious effect than any other.[1] He adds that this is also the oldest part of medicine and, indeed, it must have been recognised from the dawn of reason that, in such common emergencies as those just mentioned, something has to be done, primarily with the hand, and that anyone who can do it quickly, effectively and without causing extreme pain is, for a time at least, "worth many other men."

So says Homer[2] of the army surgeon, and both he and his hearers were well qualified to judge. As a great authority puts it, "Homer was not content to recite in general terms the wounds of the warriors as mere casual slashing; he records each stab with

[1] VII. 1.

[2] *Il.* XI. 514.

anatomical precision, describing the path of the weapon and its effects." Condensing slightly Sir Clifford Allbutt's examples—" A spear driven through the buttock pierces the urinary bladder and comes out under the symphysis pubis (5. 65). The rock hurled by Ajax strikes Hector on the breast, he turns faint, pants for health and spits blood (14. 437). An epigastric wound exposes the pericardium (16. 481). Homer explains that, after the spear of Achilles had transfixed Hector's neck, he could still speak because the weapon had missed the trachea (22. 328). Yet more remarkable is the record (8. 83) of the rotatory movement of one of the horses of Nestor, which followed the stab of a spear at the base of the skull (καίριον, a deadly spot)—the weapon had pierced the cerebellum. We may wonder not only at the poet's surgery, but also that his hearers were prepared to comprehend such particulars."[1]

It will perhaps increase the wonder and interest if we contrast the *Iliad* with our mediaeval Romances of chivalry, where there is no end of wounds and violence but an almost complete absence of definiteness or surgical interest. Take the famous fight between Balin and Balan in the *Morte d'Arthur:* the champions first unhorse and stun one another, but spring up and fight desperately for a prolonged period, " wounding each other grievously " all the time. At length, when "all the place was red with their blood," when " they had smitten either other seven other great wounds so that the least of them might have been the death of the mightiest giant in the world," they have to take a good rest, but go

[1] *Classical Review*, 37. 130.

at it again with undiminished vigour for an indefinite time till at last Balin faints. To a Greek, the pathos of the incident would be obscured by its absurdity, while, of course, there is nothing surgical about it. Perhaps the only interesting wound from this point of view is that received by Sir Launcelot when shot by the lady huntress, "so that the broad arrow smote him in the thick of the buttock over the barbs," and even the ministrations of a hermit could not enable him to sit on his horse for weeks. So too in the *Tale of Troy* translated by Caxton, there is as much slaughter as in the *Iliad*. Did not the good knight Hector slay a thousand Greek knights in one day? "He gave Patroclus a stroke upon his head and cleft it in two pieces, and Patroclus fell down dead." He cleft Archylogus in twain "notwithstanding his harness," and repeated this immediately on another Greek; in fact he must evidently have kept it up for hours. But the only surgically interesting case is that where Ulysses "struck King Philumenus in his throat and cut asunder his original vein, and smote him as half dead," especially if "original" means "jugular," for Philumenus is as vigorous as ever soon afterwards. No one would dream of making a table of mortality from these romances, distinguishing the wounds by localities and weapons, as has been done for the 147 wounds described in the *Iliad*, with results fairly corresponding with surgical probability.[1]

The object of this comparison is to show that the Greeks, during what has been called their "middle ages," were a people who, in interest in their bodies,

[1] Frölich, *Die Militärmedizin Homer's*, 1879.

knowledge of the nature and results of injuries, and respect for those skilled in the methods of healing afterwards called Surgery, surpassed all those whom we know at a corresponding stage of civilisation.

When we add to this the frequent sacrifices (which may help to explain their greater anatomical knowledge compared with that of our mediaeval ancestors), the vigorous funeral games, and the probably already widespread custom of gymnastic training, there seems no need to suppose borrowings from older civilisation to explain the rise of surgery in a few centuries to the height at which we find it in the Hippocratic writings. As regards the palaestra, if we may judge from the famous group of "the Wrestlers," and its great frequency, dislocation of the shoulder joint was often deliberately produced, and Hippocrates will tell us that it was part of a good education to know all the ways of putting it in again.

The fact that medical schools first arose on the rim of the Greek world, especially in that part of the Asiatic coast where Ionian joined Dorian and both came in contact with remains of older cultures from Crete and Caria, as well as with strangers from Egypt and the East, may be partly accounted for by such contacts. Materials and methods of bandaging perhaps came from Egypt, and we may possibly find in a Cretan drain-pipe or Egyptian tomb a sample of that most interesting of Hippocratic instruments, the crown trephine;[1] but the special

[1] A large bronze crown trephine has been found at Nineveh, and was evidently worked with a cord like the Hippocratic instrument. Meyer-Steineg Sudhoff, *Geschichte d. Medizin*, 1921, p. 25.

treatment of Fractures and Dislocations which forms the main and most remarkable part of Hippocratic surgery was, we may be fairly sure, developed by the Greeks themselves.

It is, however, only right to cast an admiring glance in passing on what little is visible of the Edwin Smith Papyrus. This dates from the seventeenth century B.C. at latest, and contained a "Book of Surgery and External Medicine," the remaining part of which comprises forty-eight typical cases extending from the top of the head to the thorax and breasts. The description of each case is divided into Examination, Diagnosis, Verdict, Treatment. No less than fourteen cases are declared incurable, and in nine of them no treatment is suggested. In only one case is the use of a charm mentioned. The following is Case 18, a wound of the temple, condensed from Prof. Breasted's version.[1] "You should probe, and if you find the bone whole without a pšn, a thm or a fracture you should say, Treat it with fresh meat the first day and afterwards with ointment and honey."

This remarkable Papyrus indicates that the Egyptians possessed a semi-scientific surgery not much inferior to that of Hippocrates more than a thousand years before his birth. Whether he was indebted to them is another question, but they evidently knew at least two forms of bone injury besides fracture, and it is not impossible that when we are told what "pšn" and "thm" mean, we may get some light on the origin of the Hippocratic term *hedra*.

[1] In *Recueil d'Études Égyptologiques*, Paris, 1922.

The earliest historical Greek practitioner is represented as being most effective as a surgeon. Democedes, coming from Croton, a city famous for its gymnasts, though without instruments, so excelled his colleagues that he became medical officer with large and increasing salaries in Aegina, Athens and Samos successively. Brought as a slave to Susa, and probably again without instruments, he cured King Darius of an injury thus vividly described by a layman—" his foot was twisted, and twisted rather violently, for he got his astragalus dislocated from its joints." The Greek surgeon restored it effectively with little pain, saved the Egyptians, who had failed to do so, from impalement, fed at the king's table, and, if we may trust Herodotus, became a prominent figure in history. But he can hardly have lived to see the birth of Hippocrates, in whose time the most important of the treatises here translated were composed. According to all surviving evidence from antiquity, they were mostly written by him, and though there is now a tendency to believe that Hippocrates, like other great teachers, may have written nothing, we shall, while indicating the different amount of evidence for the genuineness of the various treatises, use "the writer" and "Hippocrates" as synonymous terms.

To show how these works were valued we may quote a paragraph from a high authority on Greek matters, which also introduces us to the remarkable MS. which contains most of them. "The MS. was written in Constantinople about the year A.D. 950, and it begins with a paean of joy over the discovery of the works of this ancient surgeon, Apollonius, with his accurate drawings to show how the various

dislocations should be set. The text was written out. The illustrations were carefully copied. Where the old drawings were blurred and damaged, the copies were left incomplete lest some mistake should be made. Why? Because this ancient surgeon, living about 150 B.C. [75 is more probable], knew how to set dislocated limbs a great deal better than people who lived a thousand years after him. It was a piece of good fortune to them to rediscover his work. And his writing again takes the form of a commentary on the fifth-century Hippocrates. Hippocrates' own writing does not look back. It is consciously progressive and original." [1]

The writer, indeed, though he teaches with authority and confidence, confesses failures and welcomes improvements. His work, especially that on the surgery of the bones, formed the basis for future progress and did not prevent it. There was, in fact, steady progress for five centuries, and ancient surgery reached its culmination about A.D. 100. It began, says Celsus, to have its professors at Alexandria, but the first eminent practitioner whom we know as "the Surgeon" was Meges of Sidon, who practised at Rome shortly before Celsus, and is the source whence he drew much of his surgical knowledge. At the end of the century, Archigenes and Leonidas performed amputation almost in the modern style, while Heliodorus and his follower Antyllus showed themselves capable of doing all a surgeon could do, without the aid of modern discoveries. The former was especially famous for his work on the skull and lower part of the body

[1] Gilbert Murray, *Rise of the Greek Epic*, 1911, p. 24.

(hernia, fistula, stricture), the latter for the ligature of aneurisms and resection of bones, but he follows Heliodorus so closely that we do not know which was the greater or more original. The surgical writings of the earlier Celsus and the much later Paulus are interesting and very similar, but the first was a layman, the second may or may not have performed the operations he portrays; for both are compilers. But when we pass to the Heliodorus-Antyllus fragments we feel a different atmosphere. There is a definiteness and determination in their language which leaves no doubt that they did what they describe. "The ancients refused to undertake a case of this kind, but we shall" etc., is a phrase which recurs. One is convinced that they did what they say and hopes the unfortunate patient had a large dose of mandragora.[1] This state of excellence, however, does not appear to have lasted. Galen tells us that when he came to Rome he found that serious operations were usually handed over to "those called surgeons."[2] Unless Antyllus was among them, none of their names have come down to us, and when, two centuries later, Oribasius made his great "Collections," he had to go back to him and Heliodorus for the best surgery; while for ordinary fractures and dislocations he could find nothing better than Galen's commentaries on the treatises in this volume.

Heliodorus, however, is introduced here not as part of an inadequate outline of Greek surgery, but

[1] They removed the whole arm-bone (humerus) and part of the shoulder-blade, and call resection of "the lower part of the jaw" an easy operation. Oribasius XLIV. 23.

[2] X. 455.

because he will help us to explain some of the Hippocratic apparatus. The reader of this volume will hear a great deal about bandaging, but very little about definite forms of bandaging. In the surgery, says the writer, the kinds of bandages are the simple (circular) sceparnus, simus, the eye, the rhomb and the hemitome or hemirhomb. This contrasts vividly with the 50 bandages of Heliodorus, the 60 of Soranus, and the 90 odd given in the *De Fasciis* ascribed to Galen.

We should gather from Galen's commentary [1] that three were simple and three complex, the first being a true circle (εὔκυκλος) where each turn covers the former, so that there was no "distribution" up or down. The sceparnus, or "adze," was slightly oblique, and the simus, or "snub," very oblique, both being simple spirals. But Heliodorus,[2] an older and perhaps better authority on this point, says the simple bandage was a simple figure-of-eight used to fix a limb to some support, while the circular, which was called "the εὔκυκλος of Hippocrates," was slightly spiral and could be distributed upwards or downwards, being used to close sinuses.[3] The sceparnus was a complex bandage, and commenced as an open figure-of-eight; which agrees with a still older commentator, Asclepiades,[4] who says the Hippocratic sceparnus was a slightly oblique crossed bandage (χιεζόμενος). The simus is more puzzling: *De Fasciis* says it is not a bandage at all, but refers to the shape of parts to which a sceparnus bandage should be applied.[5] Galen says Hippocrates trans-

[1] XVIII(2). 732. [2] *Orib.* XLVIII. 61.
[3] *Ibid.* 64. [4] *In Erotian*, s.v. [5] XVIII(1). 772.

ferred the term from its use for a snub nose, or the sloping curve at the bottom of a hill, to denote a very sloping bandage, whence Petrequin concludes that it may be our favourite "spiral with reverses." But if this form had been known, it is hardly credible that we should not have had some clear account of it, and it seems more likely that it was sloping figure-of-eight.

The complex bandages are described in detail by Heliodorus as "the Hippocratic eye" (ὀφθαλμός), very similar to the existing bandage for one eye, "the Hippocratic rhomb" which covered the top of the head, and the hemirhomb intended for the side of the face or unilateral dislocation of the jaw.

Hippocrates was also fond of a bandage rolled up to the middle from either end and put on obliquely from two heads, and was evidently acquainted with many complex and ornamental forms though he does not approve of them. He had a peculiar method[1] of bandaging fractures with an under and upper layer separated by splints and compresses, the under-bandaging being done according to a rule clearly laid down, but this, says Galen, went out of use, leaving only the technical terms ὑπόδεσις and ὑποδεσμίδες.

Ointments.—The under-bandages and the folded pieces of linen called σπλῆνες (pads or compresses) were usually soaked in some application, the most important being two forms of "cerate," (1) white or liquid, which consisted of wax liquefied in olive oil or oil of roses,[2] supposed to prevent inflammation, while (2) (which was the same with the addition of

[1] *Surgery*, XII. [2] XVIII(2). 365.

some pitch[1]) was used for inflamed or open wounds, and was supposed to have anodyne properties and to favour the production of healthy pus; wine and oil were also used.[2]

Splints.—Of the ordinary splints (νάρθηκες) we know curiously little. The name (like the Latin *ferulae*) implies that they were stalks of an umbelliferous plant.[3] They were put on separately; Celsus[4] tells us they were split (*fissae*) and Paulus[5] that they were wrapped in wool or flax. The nature of the large hollow splint (σωλήν), the *canalis* of Celsus,[6] is not altogether certain, in spite of much description. It is usually taken to be gutter-shaped, but Galen tells us[7] that it went right round the limb, more so than did the box splint (γλωσσόκομον), from which it also differed in being circular outside; it was therefore tubular and cylindrical. But the limb could be put upon it, so it must have been opened, and, indeed, we hear of an opened (ἀνοικτός) solen in the Galenic writings.[8] Perhaps this was a gutter splint, and the only form used in later times, for Paulus, who says the solen was made of earthenware as well as wood, uses σωληνοειδής in a sense which must mean "like a gutter." So also in Soranus (1. 85) a baby's pillow is to be hollowed, σωληνοειδῶς, so as *not* to go right round its head: but Rufus uses the word of the spinal canal, and Dioscorides of a funnel pipe, so it will be prudent to keep to the ambiguous "hollow

[1] XVIII(2). 538.
[2] In the case of club foot the ointment was stiffened with resin.
[3] The giant fennel, light and strong, used by the Bacchants.
[4] VIII. 10. 1. [5] VI. 99. [6] VIII. 10. 5.
[7] XVIII(2). 504. [8] XIV. 795.

splint." The writer's account of more complicated "machines" can only be made clearer by illustrations.[1]

In conclusion we must mention a theory which brings together, and throws light upon, most of these treatises. *Wounds in the Head* has a place by itself, to be considered shortly, the other four have peculiar titles. In *Fractures* the Greek ἄγμος (for κάταγμα) is strange, as observed by Galen. *Joints* clearly means *Reduction of dislocated joints,* and is so given in our oldest MS., but the correction seems too obvious to be correct.[2] Both these treatises have abrupt beginnings, are probably mutilated and certainly in disorder, yet they rank in the first class of "genuine" works of Hippocrates. *In* (or *About*) *a Surgery,* often ambiguously shortened to *Surgery,* but more instructively expanded to *Concerning things done in the Surgery,* is a collection of notes, chiefly on bandaging, and is obviously derived in part from *Fractures,* yet it contains at least one passage requisite to explain a statement in *Fractures.* Lastly the *Mochlicon* (Leverage), usually rendered *Instruments of Reduction,* begins with a chapter on the Nature of Bones, while the rest is almost entirely an abridgment from *Joints.*

The Hippocratic *Corpus* contains a treatise on the *Nature of Bones* which, after a very few remarks on that subject, is occupied by a variety of confused accounts of blood vessels. It is a wreck which has gathered debris from various sources; yet it contains several peculiar words which are quoted in the

[1] See Appendix: Supplementary Note.

[2] Still, the περὶ ἄρθρων of Apollonius and Galen may be an abbreviation; following which example we shall call it "Joints."

Hippocratic Lexicons of Erotian and Galen as being closely connected with *Mochlicon.* The author of *Joints* says he intends to write a treatise on the veins and arteries and other anatomical matters.

This condensed summary may suffice to lead up to the following inferences:—

The Hippocratic part of the *Nature of Bones* originally came after the first chapter of *Mochlicon,* which is really its first chapter. This treatise, thus enlarged, had as Preface our *Surgery,* the whole being an abridgment from an earlier work by the great Hippocrates "for use in the Surgery," which was perhaps its original title (see p. 56). Such a work would be well adapted either for teaching or for refreshing a surgeon's memory.

Of the larger and older work our *Fractures* and *Joints* are important fragments, but there was probably an Introduction (now lost) containing the passage now extant in *Surgery* necessary to explain the later statement in *Fractures.* This earlier work may also have comprised an original treatise by Hippocrates on bones and blood vessels, of which part of our *Nature of Bones* is an abridgment. Both these surgical works got broken up, and assumed something like their present form before reaching the haven of the Alexandrian Library.

Littré has hints of the above theory, but it is more fully worked out by O. Regenbogen,[1] who carries it a step further. The seven books of *Epidemics* were, even before Galen's time, divided into three sections: I and III were universally held to be the oldest and most genuine; II, IV, VI,

[1] *Op. cit.*, infra.

which, as Galen says,[1] are not composed works (συγγράμματα) but memoranda (ὑπομνήματα), were generally supposed to have been compiled by Thessalus, son of Hippocrates, from his father's note-books; V and VII, as Galen remarks,[2] are beyond the range of the Hippocratic spirit (γνώμη), and, we may add, within that of the Macedonian artillery, which indicates a date later than 340 B.C.[3] Galen has his doubts about the single authorship of the middle section, and these are shared by modern critics; but there is no doubt that *Epidemics* II, IV and VI are closely connected with the three works, *Surgery, Bones, Mochlicon,* which we have ventured to call an abridgment, but which, if we had not got a good deal of the original, might aptly be termed memoranda. Not only do whole passages in either set correspond verbally, or almost verbally, but there are peculiar philological similarities; in particular the verb δρᾶν, which, before the rise of drama, was typically Doric, occurs in all six treatises, and a few others belonging to what may be called the middle Hippocratic period, but neither in the earlier nor the later ones. It is not found, for instance, in *Fractures* or *Joints*, nor in *Epidemics* V and VII. Perhaps it is not too fanciful to suggest that after the triumph of Sparta (404 B.C.) these strangers from Cos, who had their surgeries along the northern edge of the Greek world from Perinthus to Crannon, may have remembered that they too might claim to

[1] VII. 890. Cf. also VII. 825, 854. [2] XVII. 579.

[3] Littré tries, not very successfully, to get them all into the fifth century. V. 16 ff. The date of *Epidemics* V, VII, is fixed by the siege of Daton where a patient (94) was wounded by "an arrow from a catapult."

be Dorians and might have expressed the claim by occasional use of a strong Doric word.[1] Anyhow, there seems all the evidence we can expect that *Surgery* and *Mochlicon* formed part of an "abridgment" used in the first half of the fourth century by the practitioners who compiled *Epidemics* II, IV, VI, while *Fractures, Joints* and *Wounds in the Head* belong to the previous generation.[2]

Some little evidence as to the order of these treatises is given by grammarians. They point out that the infinitive used as imperative, characteristic of older Greek, is especially prominent in the Hippocratic *Corpus*. During the fifth century it was being driven out by the imperative and became demoralised in the process. This "depraved" use was shown mainly by the substitution of the accusative for the nominative of the participle to represent the second person imperative.[3] Now, as regards our treatises, "depraved infinitives" occur only in *Surgery* and *Mochlicon*, and are absent from *Fractures* and *Joints*, except those parts of the latter which are interpolated from *Mochlicon*. We thus have further evidence that these chapters are interpolated, and that *Surgery* and *Mochlicon* are not by the author of *Fractures—Joints*.

[1] The popularity of the Athenian dramatists, who use the word frequently, is perhaps a simpler explanation.

[2] Cf. Schulte, *op. cit.*, infra.

[3] "In cases of the second person the subject is in the nominative, but when the infinite is equivalent to the third person of the imperative its subject is in the accusative." Goodwin, *Greek Moods and Tenses*, p. 784.

Manuscripts, Editions and Commentaries

The Hippocratic manuscripts and editions have already been discussed in these volumes by a more competent authority. The chief MSS. of the surgical works are: (1) B (Laurentianus 74. 7) ninth or tenth century, referred to above, and described in detail by Schöne in the preface to his *Apollonius*, (Teubner, 1896); (2) M (Marcianus Venetus 269) eleventh century; (3) V (Vaticanus Graecus 276), twelfth century. M and V, with their progeny, form the basis of all editions up to the last by Kühlewein (Teubner, 1902), in which B is for the first time fully utilised. Unfortunately the whole of *Mochlicon* and the last five chapters of *Wounds in the Head* have been cut out of this oldest MS.

The chief editors have paid marked attention to these treatises, and Petrequin's *Chirurgie d'Hippocrate*[1] —text and translation with very copious notes and appendices, the fruit of thirty years' labour by a practising surgeon—probably represents the most thorough treatment of any ancient medical documents. It is to this work that the present edition is mainly indebted.

Francis Adams translated the treatises in his *Genuine Works of Hippocrates.*[2] He could spare less time and had fewer advantages than Petrequin. The translation, based upon Littré's text, is straightforward and readable, and the notes have special value owing to the author's practical experience in almost Hippocratic circumstances, though they are

[1] Paris, 1877–1878.
[2] Sydenham Society, 1849.

sometimes flatly opposed to the views of the equally experienced Petrequin.

Since the appearance of Schöne's beautiful edition of *Apollonius of Kitium* (Illustrated Commentary on the Hippocratic Treatise on Joints), German scholars have paid much attention to the subject. Schöne himself attempted to show that *Fractures—Joints* at any rate was a genuine work of the great Hippocrates, but was opposed by the eminent scholar Hermann Diels.[1] More recently, three interesting *Theses* on the connections,[2] grammar[3] and style[4] respectively of the surgical treatises have appeared. Their contents are very briefly outlined in the introductions, and will repay study by those interested in the subject.[5]

[1] Diels, *Sitzungsberichte der k.p. Akademie*, 1910, p. 1140 f.
[2] Regenbogen, O., *Symbola Hippocratea*, 1914.
[3] Schulte, E., *Observationes Hippocrateae Grammaticae*, 1914.
[4] Krömer, J., *Questionum Hippocraticarum capita duo*, 1914.
[5] See also Kühlewein, H., *Die chirurgischen Schriften des Hippocrates*, Nordhausen, 1898.

Abbreviations in Notes

B. M. V. = the three chief MSS. noted above.

Erm. Pq. Kw. = the three more recent editors: Ermerins 1856, Petrequin and Kühlewein as above.

HIPPOCRATES

ON WOUNDS IN THE HEAD

INTRODUCTION

No Hippocratic work has attracted more attention than this short treatise. All the prominent Alexandrian medical commentators discussed it, and it is in Erotian's list of genuine works. Galen, of course, wrote a commentary, though only a fragment survives.[1] All ancient writers on the subject from Celsus to Paulus had it before them. At the Renaissance it attracted the attention both of anatomists and surgeons, and continued to do so almost to our own times. Its genuineness has hardly been questioned except by those who doubt whether Hippocrates wrote anything.

This celebrity is perhaps equally due to its excellence and its peculiarities. The former may be seen in its clear descriptions and magisterial language; the writer teaches with authority. The latter are two: its account of the sutures, and its doctrine as to trephining. With regard to the former, we may say that, as modified by Galen to the effect that the H form is the only normal one, it is fairly correct so far as it goes, and that it is much better than the later account of Aristotle—that men have three sutures radiating from a centre and women one, which goes in a circle.[2] The ancients (and Vesalius) accepted this view of

[1] In *Oribasius*, XLVI. 21. [2] *Hist. Anim.* 1. 7.

the sutures, but all surgeons, from the post-Hippocratic age onwards, have been troubled by his rule as to trephining, which may be condensed as follows:—

If the skull is contused or fissured, you should trephine at once, but an open depressed fracture does not usually "come to trephining," and is less dangerous; in short, an injured skull should have a hole made in it if there is not one already.

The Alexandrians, as we gather from Celsus, rejected this: "the ancients," he says (piously leaving Hippocrates unnamed), advised immediate operation, but it is better to use ointments—and wait for symptoms. The vast majority of surgeons have done so, but many have regretfully wondered, after the patient's death, whether the Hippocratic trephining might not have saved a life. "Hippocrates" (as the supposed author of *Epidemics* V. 27) is praised by Celsus, and many others, for confessing that he thought a fissure was a suture and so left a patient untrephined. Symptoms appeared later; he trephined on the fifteenth day, but the patient died on the sixteenth; yet this is just what any later surgeon would have done, even had he recognised the fissure. The reader will find in *Littré* and *Petrequin* extensive quotations from French surgeons, and from our own Percival Pott, on the probability of lives being saved by preventive trephining used as an operation of choice before it is obviously necessary, but the Hippocratic rule is no more likely to be reintroduced than is the use of vigorous venesection, which would also doubtless sometimes save life.

The use of the common word πρίων as a semi-

technical term for a complicated surgical instrument brings us to another noticeable point in the treatise: there seems to be an attempt to establish a medical vocabulary. Eminent theologians have recently settled the controversy on St. Luke's alleged medical language by declaring that the Greeks had none, "the whole assumption of medical language in any ancient writer is a mare's nest," [1] but if the writer of Acts had told us that St. Paul at Lystra got a *hedra* in the region of the *bregma* which penetrated to the *diploe,* they would have been fairly confident that he was a physician who made a rather pedantic use of his medical vocabulary. Here are three simple Greek words which are given such peculiar meanings that they have to be defined and not translated.

The last term had some difficulty in keeping, or recovering, the somewhat unnatural sense [2] here given to it, and probably did so only through the prestige of this little work. *Hedra* could not be saved even by the authority of Hippocrates and his care in defining it. It is that form of skull injury which is left as its mark (or seat) by the weapon, and varies in size and shape accordingly from a prick to a gash, but without depression, "for then it becomes a depressed fracture." It included mainly what are now called "scratch fractures" and, as Galen says, would also comprise an oblique slice—ἀποσκεπαρνισμός. It was too vague to last, and was partly replaced by ἐγκοπή—incision. Its vagueness has made some confusion in the treatise, for though

[1] Jackson and Lake, *Prolegomena to Acts*, II. 355.

[2] *i.e.* the porous bone tissue between the two hard layers of the skull bones.

there is little doubt that Hippocrates intended to describe five forms of skull injury—as is twice asserted by Galen[1]—later scribes by splitting up the *hedra* have tried to make seven, though, strange to say, no MS. mentions a sixth.

Several cases in *Epidemics* V. seem intended as illustrations to this treatise. A patient with contusion of the skull is trephined largely down to the diploe, he gets inflammatory swelling of the face (erysipelas) and is purged: the Hippocratic rules being thus followed, he recovers (V. 16). The patient with fissure (V. 27) is left untrephined till it is too late A girl dies because the trephining was insufficient. She has spasm on the side opposite the injury (V. 28).

These cases are more remarkable because skull injuries have nothing to do with epidemics, and there is no such notice of bodily fractures or dislocations. *Epidemics* V., as we have seen, probably belongs to the third Hippocratic generation, when the rules of the Master, as to the treatment of wounds in the head, may have begun to be called in question.

With regard to the style of the treatise, every reader will be struck by the frequent repetition of the same words and phrases, often unnecessarily. This occurs in another manner and to a less extent in *Fractures* and *Joints*, where we shall discuss it further in considering the probability of a common authorship.

[1] XVIII(2). 672. *Orib.* as above.

ΠΕΡΙ ΤΩΝ ΕΝ ΚΕΦΑΛΗΙ ΤΡΩΜΑΤΩΝ

I. Τῶν ἀνθρώπων αἱ κεφαλαὶ οὐδὲν ὁμοίως σφίσιν αὐταῖς, οὐδὲ αἱ ῥαφαὶ τῆς κεφαλῆς πάντων κατὰ ταὐτὰ πεφύκασιν. ἀλλ' ὅστις μὲν ἔχει ἐκ τοῦ ἔμπροσθεν τῆς κεφαλῆς προβολὴν—ἡ δὲ προβολή ἐστιν αὐτοῦ τοῦ[1] ὀστέου ἔξεχον στρογγύλον παρὰ τὸ ἄλλο—τούτου εἰσὶν αἱ ῥαφαὶ πεφυκυῖαι ἐν τῇ κεφαλῇ ὡς[2] γράμμα τὸ ταῦ, Τ, γράφεται, τὴν μὲν γὰρ βραχυτέρην γραμμὴν ἔχει πρὸ τῆς προβολῆς ἐπικαρσίην πεφυκυῖαν· τὴν δὲ
10 ἑτέρην γραμμὴν ἔχει διὰ μέσης τῆς κεφαλῆς κατὰ μῆκος πεφυκυῖαν ἐς τὸν τραχήλον αἰεί. ὅστις δ' ὄπισθεν τῆς κεφαλῆς τὴν προβολὴν ἔχει, αἱ ῥαφαὶ τούτῳ πεφύκασι τἀναντία ἢ τῷ προτέρῳ· ἡ μὲν γὰρ βραχυτέρη γραμμὴ πρὸ τῆς προβολῆς πέφυκεν ἐπικαρσίη· ἡ δὲ μακροτέρη διὰ μέσης τῆς κεφαλῆς πέφυκε κατὰ μῆκος ἐς τὸ μέτωπον αἰεί. ὅστις δὲ καὶ[3] ἀμφοτέρωθεν τῆς κεφαλῆς προβολὴν ἔχει, ἔκ τε τοῦ ἔμπροσθεν καὶ ἐκ τοῦ ὄπισθεν, τούτῳ αἱ ῥαφαί εἰσιν ὁμοίως πεφυκυῖαι
20 ὡς γράμμα τὸ ἦτα, Η, γράφεται· πέφυκασι δὲ τῶν γραμμέων αἱ μὲν μακραὶ πρὸ τῆς προβολῆς ἑκατέρης ἐπικάρσιαι πεφυκυῖαι· ἡ δὲ βραχεῖη διὰ μέσης τῆς κεφαλῆς κατὰ μῆκος πρὸς ἑκατέρην τελευτῶσα τὴν μακρὴν γραμμήν.[4] ὅστις δὲ μηδὲ

ON WOUNDS IN THE HEAD

I. Men's heads are not alike nor are the sutures of the head disposed the same way in all. When a man has a prominence in the front of his head—the prominence is a rounded outstanding projection of the bone itself—his sutures are disposed in the head as the letter *tau*, T, is written; for he has the shorter line disposed transversely at the base of the prominence; while he has the other line longitudinally disposed through the middle of the head right to the neck. But when a man has the prominence at the back of his head, the sutures in his case have a disposition the reverse of the former, for while the short line is disposed transversely at the prominence, the longer is disposed through the middle of the head longitudinally right to the forehead. He who has a prominence at each end of his head, both front and back, has the sutures disposed in the way the letter *eta*, H, is written, for the long lines have a transverse disposition at either prominence and the short goes through the middle of the head longitudinally, ending each way at the long lines. He who has no

---

[1] So B. Kw. for τὸ τοῦ Pq. The older MSS. BV omit the letters T H X.

[2] ὥσπερ. [3] Omit καί.

[4] τῇσι μακρῇσι γραμμῇσιν.

ἑτέρωθι μηδεμίην προβολὴν ἔχει, οὗτος ἔχει τὰς ῥαφὰς τῆς κεφαλῆς ὡς γράμμα τὸ χῖ, Χ, γράφεται· πέφυκασι δὲ αἱ γραμμαὶ ἡ μὲν ἑτέρη ἐπικαρσίη πρὸς τὸν κρόταφον ἀφήκουσα· ἡ δὲ ἑτέρη κατὰ μῆκος διὰ μέσης τῆς κεφαλῆς.

30 Δίπλοον δ' ἐστὶ τὸ ὀστέον κατὰ μέσην τὴν κεφαλήν· σκληρότατον δὲ καὶ πυκνότατον αὐτοῦ πέφυκεν τό τε ἀνώτατον ᾗ[1] ἡ ὁμοχροίη τοῦ ὀστέου ἡ ὑπὸ τῇ σαρκὶ καὶ τὸ κατώτατον τὸ πρὸς τῇ μήνιγγι ᾗ[1] ἡ ὁμοχροίη τοῦ ὀστέου ἡ κάτω· ἀποχωρέον δὲ ἀπὸ τοῦ ἀνωτάτου ὀστέου καὶ τοῦ κατωτάτου, ἀπὸ τῶν σκληροτάτων καὶ πυκνοτάτων ἐπὶ τὸ μαλθακώτερον καὶ ἧσσον πυκνὸν καὶ ἐπικοιλότερον ἐς τὴν διπλόην αἰεί. ἡ δὲ διπλόη κοιλότατον καὶ μαλθακώτατον καὶ μάλιστα
40 σηραγγῶδές ἐστιν· ἔστι δὲ καὶ πᾶν τὸ ὀστέον τῆς κεφαλῆς, πλὴν κάρτα ὀλίγου τοῦ τε ἀνωτάτου καὶ τοῦ κατωτάτου σπόγγῳ ὅμοιον· καὶ ἔχει τὸ ὀστέον ἐν ἑωυτῷ ὅμοια σαρκία πολλὰ καὶ ὑγρά, καὶ εἴ τις αὐτὰ διατρίβοι τοῖσι δακτύλοισι αἷμα ἂν διαγίνοιτο ἐξ αὐτῶν· ἔνεστι δ' ἐν τῷ ὀστέῳ καὶ
46 φλέβια λεπτότερα καὶ κοιλότερα αἵματος πλέα.

II. Σκληρότητος μὲν οὖν καὶ μαλθακότητος καὶ κοιλότητος[2] ὧδε ἔχει. παχύτητι δὲ καὶ λεπτότητι, οὕτως·[3] συμπάσης τῆς κεφαλῆς τὸ ὀστέον λεπτότατόν ἐστι καὶ ἀσθενέστατον τὸ κατὰ βρέγμα, καὶ σάρκα ὀλιγίστην καὶ λεπτοτάτην ἔχει ἐφ ἑωυτῷ ταύτῃ τῆς κεφαλῆς τὸ ὀστέον, καὶ ὁ ἐγκέφαλος κατὰ τοῦτο τῆς κεφαλῆς πλεῖστος ὕπεστιν. καὶ δὴ ὅτι οὕτω ταῦτα ἔχει, τῶν τε

[1] Kw. omits. [2] So BV. Kw. Pq. has dative throughout. [3] Kw. omits.

prominence at either end has the sutures of his head as the letter *chi*, X, is written: the lines are disposed one transversely coming down to the temple, the other longitudinally through the middle of the head.

The skull is double along the middle of the head, and the hardest and most dense part of it is disposed both uppermost where the smooth surface of the skull comes under the scalp, and lowest where the smooth surface below is towards the membrane.[1] Passing from the uppermost and lowest layers, the hardest and most dense parts, the bone is softer, less dense and more cavernous right into the diploe. The diploe is very cavernous and soft and particularly porous. In fact, the whole bone of the head except a very little of the uppermost and lowest is like sponge, and the bone contains numerous moist fleshy particles like one another and one can get blood out of them by rubbing them with the fingers. There are also rather thin hollow vessels full of blood contained within the bone.

II. Such then is the state of hardness, softness and porosity, but in thickness and thinness of the skull generally, the bone is thinnest and weakest at the bregma,[1] and has the least and thinnest covering of flesh in this part of the head, and there is most underlying brain at this part of the head. It follows from such a state of things that when a man is wounded

[1] Dura mater.

[2] The bregma comprises the front part of the top of the head, where the skull remains longest open.

τρωσίων καὶ τῶν βελέων ἴσων τε ἐόντων κατὰ
10 μέγεθος καὶ ἐλασσόνων, καὶ ὁμοίως τε τρωθεὶς
καὶ ἧσσον, τὸ ὀστέον ταύτῃ τῆς κεφαλῆς φλᾶταί
τε μᾶλλον καὶ ῥήγνυται καὶ ἔσω ἐσφλᾶται, καὶ
θανασιμώτερά ἐστι καὶ χαλεπώτερα ἰητρεύεσθαί τε
καὶ ἐκφυγγάνειν τὸν θάνατον ταύτῃ ἤ που ἄλλοθι
τῆς κεφαλῆς· ἐξίσων τε ἐόντων τῶν τρωμάτων
καὶ ὁμοίως τε τρωθεὶς καὶ ἧσσον, ἀποθνήσκει ὁ
ἄνθρωπος, ὁπόταν καὶ ἄλλως μέλλῃ ἀποθανεῖσθαι
ἐκ τοῦ τρώματος, ἐν ἐλάσσονι χρόνῳ ὁ ταύτῃ ἔχων
τὸ τρῶμα τῆς κεφαλῆς ἤ που ἄλλοθι. ὁ γὰρ
20 ἐγκέφαλος τάχιστά τε καὶ μάλιστα κατὰ τὸ
βρέγμα αἰσθάνεται τῶν κακῶν τῶν γινομένων ἔν
τε τῇ σαρκὶ καὶ τῷ ὀστέῳ· ὑπὸ λεπτοτάτῳ γὰρ
ὀστέῳ ἐστὶ ταύτῃ ὁ ἐγκέφαλος καὶ ὀλιγίστῃ σαρκί,
καὶ ὁ πλεῖστος ἐγκέφαλος ὑπὸ τῷ βρέγματι
κεῖται. τῶν δὲ ἄλλων τὸ κατὰ τοὺς κροτάφους
ἀσθενέστατόν ἐστιν· συμβολή τε γὰρ τῆς κάτω
γνάθου πρὸς τὸ κρανίον, καὶ κίνησις ἔνεστιν ἐν
τῷ κροτάφῳ ἄνω καὶ κάτω ὥσπερ ἄρθρου· καὶ ἡ
ἀκοὴ πλησίον γίνεται αὐτοῦ, καὶ φλὲψ διὰ τοῦ
30 κροτάφου τέταται κοίλη τε καὶ ἰσχυρή. ἰσχυρότερον δ' ἐστι τῆς κεφαλῆς τὸ ὀστέον ἅπαν τὸ
ὄπισθεν τῆς κορυφῆς καὶ τῶν οὐάτων ἢ ἅπαν
τὸ πρόσθεν, καὶ σάρκα πλέονα καὶ βαθυτέρην
ἐφ' ἑωυτῷ ἔχει τοῦτο τὸ ὀστέον. καὶ δὴ τούτων
οὕτως ἐχόντων, ὑπό τε τῶν τρωσίων καὶ τῶν
βελέων ἴσων ἐόντων,[1] καὶ ὁμοίων καὶ μεζόνων καὶ
ὁμοίως τιτρωσκόμενος καὶ μᾶλλον, ταύτῃ τῆς
κεφαλῆς τὸ ὀστέον ἧσσον ῥήγνυται καὶ φλᾶται
ἔσω, κἢν μέλλῃ ὥνθρωπος ἀποθνήσκειν καὶ ἄλλως
40 ἐκ τοῦ τρώματος, ἐν τῷ ὄπισθεν τῆς κεφαλῆς

equally or less, the wounding and weapons being equal or smaller, the bone in this part of the head is more contused or fractured, and fractured and contused with depression, the lesions are more mortal, medical treatment and escape from death more difficult here than in any part of the head. When wounded equally or less, the wounds being alike, the patient, if he is going to die in any case from the wound, dies sooner when he has it in this part of the head than anywhere else; for it is at the bregma that the brain is most quickly and especially sensitive to evils that arise in scalp or skull, since the brain is covered here by thinnest bone and least flesh, and the greatest part of the brain lies under the bregma. Of the other parts, that at the temples is weakest, for the junction of the lower jaw with the cranium is at the temple, and there is an up-and-down movement there as in a joint. Near it is the organ of hearing, and a large and thick blood vessel extends through the temporal region. The whole skull behind the vertex and the ears is stronger than any part in front, and this bone has a fuller and thicker covering of flesh. It follows from such a state of things that when a man is stricken equally or more severely by woundings or weapons which are equal and similar or larger in this part of the head, the bone is less fractured, or contused with depression; and if the man is going to die in any case from the wound, he takes

---

[1] ἀπάντων Pq.

ἔχων τὸ τρῶμα ἐν πλείονι χρόνῳ ἀποθανεῖται· ἐν πλείονι γὰρ χρόνῳ τὸ ὀστέον ἐμπυΐσκεταί τε καὶ διαπυΐσκεται κάτω ἐπὶ τὸν ἐγκέφαλον διὰ τὴν παχύτητα τοῦ ὀστέου, καὶ ἐλάσσων ταύτῃ τῆς κεφαλῆς ὁ εγκέφαλος ὕπεστι, καὶ πλέονες ἐκ φυγγάνουσι τὸν θάνατον τῶν ὄπισθεν τιτρωσκομένων τῆς κεφαλῆς ὡς ἐπὶ τὸ πολὺ ἢ τῶν ἔμπροσθεν. καὶ ἐν χειμῶνι πλείονα χρόνον ζῇ ὥνθρωπος ἢ ἐν θέρει, ὅστις καὶ ἄλλως μέλλει[1]
50 ἀποθανεῖσθαι ἐκ τοῦ τρώματος ὅπου ἂν τῆς
51 κεφαλῆς ἔχων[2] τὸ τρῶμα.

III. Αἱ δὲ ἕδραι τῶν βελέων τῶν ὀξέων κακουφοτέρων, αὐταὶ ἐπὶ σφῶν αὐτέων γινόμεναι ἐν τῷ ὀστέῳ ἄνευ ῥωγμῆς τε καὶ φλάσιος καὶ ἔσω ἐσφλάσιος—αὗται δὲ γίνονται ὁμοίως ἔν τε τῷ ἔμπροσθεν τῆς κεφαλῆς καὶ ἐν τῷ ὄπισθεν—ἐκ τούτων ὁ θάνατος οὐ γίνεται κατά γε δίκην, οὐδ᾽ ἢν γένηται. ῥαφὴ δὲ ἐν ἕλκει φανεῖσα, ὀστέου ψιλωθέντος, πανταχοῦ τῆς κεφαλῆς τοῦ ἕλκεος γενομένου, ἀσθενέστατον γίνεται τῇ τρώσει
10 καὶ τῷ βέλει ἀντέχειν, εἰ τύχοι τὸ βέλος ἐς αὐτὴν τὴν ῥαφὴν στηριχθέν·—πάντων δὲ μάλιστα, ἢν τὸ βέλος[3] ἐν τῷ βρέγματι γενόμενον κατὰ τὸ ἀσθενέστατον τῆς κεφαλῆς—καὶ αἱ ῥαφαὶ εἰ τύχοιεν ἐοῦσαι περὶ τὸ ἕλκος καὶ τὸ βέλος
15 αὐτέων τύχοι τῶν ῥαφῶν.

IV. Τιτρώσκεται δὲ ὀστέον τὸ ἐν τῇ κεφαλῇ τοσούσδε τρόπους· τῶν δὲ τρόπων ἑκάστου πλείονες ἰδέαι γίνονται τοῦ κατήγματος ἐν τῇ τρώσει. ὀστέον ῥήγνυται τιτρωσκόμενον καὶ τῇ ῥωγμῇ[4] ἐν τῷ περιέχοντι ὀστέῳ τὴν ῥωγμήν, ἀνάγκη φλάσιν προσγενέσθαι, ἥνπερ ῥαγῇ· τῶν

longer time dying when he has it in the back of the head. For suppuration of the bone takes longer to come on and penetrate down to the brain because of the thickness of the skull; also there is less brain in this part of the head, and, as a rule, more of those wounded in the hinder part of the head escape death than of those wounded in front. In winter, too, a man lives longer than in summer, if he is going to die from the wound in any case, in whatever part of the head he may have the wound.

III. *Hedrae*[1] of sharp and light weapons, occurring by themselves in the skull without fissure, contusion or contused depression (these happen alike in front and at the back of the head) do not, at any rate by rights, cause death even if it occurs. If a suture appears in the wound when the bone is denuded, wherever the wound may be, the bone makes very weak resistance to lesion or weapon [if the weapon happens to get stuck in the suture itself][2]—most of all if the weapon gets in the bregma, the weakest part of the head—and if, when the sutures happen to be in the region of the wound, the weapon also happens to strike the sutures themselves.

IV. The bone of the head is injured in the following number of modes, and for each mode several forms of fracture occur in the lesion. The bone is fractured when wounded, and the fracture is necessarily complicated by contusion of the bone about it, if it was really fractured. For the very

[1] See Introduction. [2] This seems a superfluous gloss.

---

[1] ὅστις ἂν ἄλλως μέλλῃ.
[2] ἔχῃ Kw.'s conjecture.
[3] ἕλκος Pq. Erm. βέλος Kw. codd.
[4] τῆς ῥωγμῆς Pq.; V omits.

γὰρ βελέων ὅ τι περ ῥήγνυσι τὸ ὀστέον, τὸ αὐτὸ τοῦτο καὶ φλᾷ τὸ ὀστέον ἢ μᾶλλον ἢ ἧσσον, αὐτό τε ἐν ᾧπερ καὶ ῥήγνυσι τὴν ῥωγμὴν καὶ τὰ
10 περιέχοντα ὀστέα τὴν ῥωγμήν· εἷς οὗτος τρόπος. ἰδέαι δὲ ῥωγμέων παντοῖαι γίνονται· καὶ γὰρ λεπτότεραί τε καὶ λεπταὶ πάνυ, ὥστε οὐ καταφανέες γίνονται, ἔστιν αἱ τῶν ῥωγμέων,[1] οὔτε αὐτίκα μετὰ τὴν τρῶσιν, οὔτ' ἐν τῇσιν ἡμέρῃσιν ἐν ᾗσιν ἂν καὶ πόνων ὄφελος γένοιτο τοῦ θανάτου τῷ ἀνθρώπῳ.[2] αἱ δ' αὖ παχύτεραί τε καὶ εὐρύτεραι ῥήγνυνται τῶν ῥωγμέων, ἔνιαι δὲ καὶ πάνυ εὐρέαι. ἔστι δὲ αὐτέων καὶ αἱ μὲν ἐπὶ μακρότερον ῥήγνυνται, αἱ δὲ ἐπὶ βραχύτερον· καὶ
20 αἱ μὲν ἰθύτεραι, αἱ δὲ ἰθεῖαι πάνυ, αἱ δὲ καμπυλώτεραί τε καὶ καμπύλαι· καὶ βαθύτεραί τε ἐς τὸ κάτω καὶ διὰ παντὸς τοῦ ὀστέου [καὶ ἧσσον
23 βαθεῖαι καὶ οὐ διὰ παντὸς τοῦ ὀστέου].[3]

V. Φλασθείη δ' ἂν τὸ ὀστέον μένον ἐν τῇ ἑωυτοῦ φύσει, καὶ ῥωγμὴ τῇ φλάσει οὐκ ἂν προσγένοιτο ἐν τῷ ὀστέῳ οὐδεμία· δεύτερος οὗτος τρόπος. ἰδέαι δὲ τῆς φλάσιος πλείους γίνονται· καὶ γὰρ μᾶλλόν τε καὶ ἧσσον φλᾶται καὶ ἐς βαθύτερόν τε καὶ διὰ παντὸς τοῦ ὀστέου, καὶ ἧσσον ἐς βαθὺ καὶ οὐ διὰ παντὸς τοῦ ὀστέου, καὶ ἐπὶ πλέον τε καὶ ἔλασσον μήκεός τε καὶ πλατύτητος. ἀλλὰ οὐ[4] τούτων τῶν ἰδεῶν
10 οὐδεμίαν ἐστὶν ἰδόντα τοῖσιν ὀφθαλμοῖς γνῶναι ὁποίη τίς ἐστιν τὴν ἰδέην καὶ ὁπόση τις τὸ μέγεθος· οὐδὲ γὰρ εἰ πέφλασται ἐόντων τε πεφλασμένων καὶ τοῦ κακοῦ γεγενημένου γίνεται τοῖσιν ὀφθαλμοῖσιν καταφανὲς ἰδεῖν αὐτίκα μετὰ

[1] ἔστι δ' αἴτιον ῥωγμέων Pq., V.

same part of the weapon which breaks the bone also contuses it more or less; and this happens just at the place where it makes the fracture, and in the bones containing the fracture. This is one mode.[1] As to forms of fracture, all kinds occur, for some are rather small and very small, so as to be not noticeable either immediately after the lesion or in the days during which the patient might be helped in his sufferings and saved from death. Again, some of the fractures are larger and wider, and some very broad. Some are longer, some shorter, rather straight or quite straight, rather curved or bent, going rather deep and right through the bone [and not so deep and not through the bone].[2]

V. The bone may be contused and keep in its place, and the contusion may not be complicated by any fracture of the bone. This is a second mode.[3] There are many forms of contusion; for the bone is more contused or less, to a greater depth, going right through, or less deeply, not going through the bone, and to a greater or smaller extent in length and breadth. Now none of these forms can be distinguished by the eye as to its precise shape and size, for it is not even clear to the eye immediately after the injury whether contusion has taken place, even if the parts are contused and the damage done;

[1] "Fissure fracture." [2] Littré's insertion.
[3] "Contusion."

---

[2] Obscure passage: "help for sufferings may be also help against death." Littré suggests καὶ τοῦ θανάτου.
[3] Added by Littré.
[4] οὐ Kw.; Pq. omits.

τὴν τρῶσιν, ὥσπερ οὐδὲ τῶν ῥωγμέων ἔνιαι ἑκὰς[1]
16 ἐοῦσαί τε καὶ ἐρρωγότος τοῦ ὀστέου.

VI. Ἐσφλᾶται τὸ ὀστέον ἐκ τῆς φύσιος τῆς ἑωυτοῦ ἔσω σὺν ῥωγμῆσιν· ἄλλως γὰρ οὐκ ἂν ἐσφλασθείη· τὸ γὰρ ἐσφλώμενον, ἀπορρηγνύμενόν τε καὶ καταγνύμενον, ἐσφλᾶται ἔσω ἀπὸ τοῦ ἄλλου ὀστέου μένοντος ἐν φύσει τῇ ἑωυτοῦ· καὶ δὴ οὕτω ῥωγμὴ ἂν προσείη τῇ ἐσφλάσει· τρίτος οὗτος τρόπος. ἐσφλᾶται δὲ τὸ ὀστέον πολλὰς ἰδέας· καὶ γὰρ ἐπὶ πλέον τοῦ ὀστέου καὶ ἐπ' ἔλασσον, καὶ μᾶλλόν τε καὶ ἐς βαθύτερον
10 κάτω, καὶ ἧσσον καὶ ἐπιπολαιότερον.

VII. Καὶ ἕδρης γενομένης ἐν τῷ ὀστέῳ βέλεος προσγένοιτο ἂν ῥωγμὴ τῇ ἕδρῃ, τῇ δὲ ῥωγμῇ καὶ φλάσιν προσγενέσθαι ἀναγκαῖόν ἐστι ἢ μᾶλλον ἢ ἧσσον, ἤνπερ καὶ ῥωγμὴ προσγένηται ἔνθαπερ καὶ ἕδρη ἐγένετο καὶ ἡ ῥωγμή, ἐν τῷ ὀστέῳ περιέχοντι τήν τε ἕδρην καὶ τὴν ῥωγμήν· τέταρτος οὗτος τρόπος. καὶ ἕδρη μὲν ἂν γένοιτο φλάσιν ἔχουσα τοῦ ὀστέου περὶ αὐτήν, ῥωγμὴ δὲ οὐκ ἂν προσγένοιτο τῇ ἕδρῃ καὶ τῇ φλάσει ὑπὸ
10 τοῦ βέλεος· [πέμπτος οὗτος τρόπος] [καὶ ἕδρη δὲ τοῦ βελέος γίνεται ἐν τῷ ὀστέῳ· ἕδρη δὲ καλεῖται, ὅταν μένον τὸ ὀστέον ἐν τῇ ἑωυτοῦ φύσει τὸ βέλος στήριξαν ἐς τὸ ὀστέον δῆλον ποιήσῃ ὅπου ἐστήριξεν[2]] ἐν δὲ τῷ τρόπῳ ἑκάστῳ πλείονες ἰδέαι γίνονται καὶ περὶ μὲν φλάσιός τε καὶ ῥωγμῆς, ἢν ἄμφω ταῦτα προσγένηται τῇ ἕδρῃ, καὶ ἢν φλάσις μούνη γένηται, ἤδη πέφρασται ὅτι πολλαὶ

[1] ἐλάσσους Kw.'s suggestion in Hermes XX., but he does not print it.

[2] Kw. puts this passage first, as is done in the translation.

just as some fractures are not visible, being far from the wound,[1] though the bone be broken.

VI. The bone is contused and depressed inwards from its natural position with fractures, for otherwise it would not be depressed. For the depressed bone, broken off and fractured, is crushed inwards away from the rest of the bone, which keeps its place; and of course there will thus be a fracture as well as a contused depression. This is a third mode. Contused depressed fracture has many forms, for it extends over more or less of the skull, is more depressed and deeper, or less so and more superficial.

VII. Again, a weapon *hedra* occurs in the skull. It is called "hedra" when, the bone keeping its natural position, the weapon sticks into it and makes a mark where it stuck.[2] When a weapon *hedra* occurs in the skull, there may be a fracture as well as the *hedra;* and the fracture must necessarily be accompanied by more or less contusion (if a fracture also occurs) where the *hedra* and fracture happened, in the bone containing the *hedra* and fracture. This is a fourth mode. And a *hedra* may occur with contusion of the bone about it, without being accompanied by a fracture in addition to contusion by the weapon. [This is a fifth mode.[3]] Of each mode there are many forms; and as regards contusion and fracture (whether both of them accompany the *hedra,* or contusion only), it has already been declared that there are many forms,

[1] Or, "rather small," Kw.

[2] *Vestigium teli,* "scratch fracture." This passage is obviously out of place in the Greek text.

[3] Pq. omits.

ἰδέαι γίνονται καὶ τῆς φλάσιος καὶ τῆς ῥωγμῆς. ἡ δὲ ἕδρη αὐτὴ ἐφ' ἑωυτῆς γίνεται μακροτέρη καὶ
20 βραχυτέρη ἐοῦσα, καὶ καμπυλωτέρη, καὶ ἰθυτέρη, καὶ κυκλοτερής· καὶ πολλαὶ ἄλλαι ἰδέαι τοῦ τοιούτου τρόπου, ὁποῖον ἄν τι καὶ τὸ σχῆμα[1] τοῦ βέλεος ᾖ· αἱ δὲ αὐταὶ καὶ βαθύτεραι τὸ κάτω καὶ μᾶλλον καὶ ἧσσον, καὶ στενότεραί τε καὶ ἧσσον στεναὶ καὶ εὐρύτεραι, καὶ πάνυ εὐρέαι, ᾖ διακεκόφαται· διακοπὴ δὲ ὁποσητισοῦν γινομένη μήκεός τε καὶ εὐρύτητος ἐν τῷ ὀστέῳ, ἕδρη ἐστίν, ἢν τὰ ἄλλα ὀστέα τὰ περιέχοντα τὴν διακοπὴν μένῃ ἐν τῇ φύσει τῇ ἑωυτῶν, καὶ μὴ συνεσφλᾶται
30 τῇ διακοπῇ ἔσω ἐκ τῆς φύσιος τῆς ἑωυτῶν· οὕτω
31 δὲ ἔσφλασις ἂν εἴη καὶ οὐκ ἔτι ἕδρη.

VIII. Ὀστέον τιτρώσκεται ἄλλῃ τῆς κεφαλῆς ἢ ᾗ τὸ ἕλκος ἔχει ὤνθρωπος καὶ τὸ ὀστέον ἐψιλώθη τῆς σαρκός· πέμπτος[2] οὗτος τρόπος. καὶ ταύτην τὴν συμφορήν, ὁπόταν γένηται, οὐκ ἂν ἔχοις ὠφελῆσαι οὐδέν· οὐδὲ γάρ, εἰ πέπονθε τὸ κακὸν τοῦτο, οὐκ ἔστιν ὅπως χρὴ αὐτὸν ἐξελέγξαντα εἰδέναι, εἰ πέπονθε τὸ κακὸν τοῦτο
8 ὤνθρωπος, οὐδὲ ὅπῃ[3] τῆς κεφαλῆς.

IX. Τούτων τῶν τρόπων τῆς κατήξιος ἐς πρῖσιν ἀφήκει ἥ τε φλάσις ἡ ἀφανὴς ἰδεῖν καὶ ἥν πως τύχῃ φανερὴ γενομένη καὶ ἡ ῥωγμὴ ἣν ἀφανὴς ἰδεῖν καὶ ἤν φανερὴ ᾖ. καὶ ἤν, ἕδρης γενομένης τοῦ βέλεος ἐν τῷ ὀστέῳ, προσγένηται ῥωγμὴ καὶ φλάσις τῇ ἕδρῃ, καὶ ἢν φλάσις μούνη προσγένηται ἄνευ ῥωγμῆς τῇ ἕδρῃ, καὶ αὕτη ἐς πρῖσιν ἀφήκει. τὸ δὲ ἔσω ἐσφλώμενον ὀστέον ἐκ τῆς φύσιος τῆς ἑωυτοῦ ὀλίγα τῶν πολλῶν πρίσιος
10 προσδεῖται· καὶ τὰ μάλιστα ἐσφλασθέντα καὶ

both of the contusion and of the fracture. The *hedra* taken by itself is long or short, rather bent, or straighter, or rounded; and there are many other forms of this mode, according to the shape of the weapon. These same *hedrae* vary in depth and narrowness, and may be rather broad or very broad where there is a cleft; for a cleft in the bone of any size whatsoever as to length and breadth is a *hedra* if the rest of the bone round the cleft keeps its natural place and is not crushed in by the cleft; for this would be a contused depressed fracture, and no longer a *hedra*.

VIII. The skull is wounded in a part of the head other than that in which the patient has the lesion and the bone is denuded of flesh. This is a fifth mode.[1] When this accident occurs, you can do nothing to help; for if the man has suffered this injury, there is no possible way of examining him to make sure that he has suffered it, or whereabouts in the head it is.

IX. Of these modes of "fracture,"[2] contusion, whether invisible or somehow becoming manifest, is a case for trephining, also fissure-fracture, whether invisible or manifest; and if, when there is a weapon *hedra* in the bone, the *hedra* is accompanied by fracture and contusion, or if contusion alone accompanies the *hedra* without fracture, this also is a case for trephining. But as for contused depressed fractures, only a small proportion of them require trephining; and the more the bones are contused,

[1] Seventh Kw., our "contrecoup."
[2] Evidently taken as = injury.

[1] στόμα. [2] ἕβδομος. [3] ὅπου Erm.

μάλιστα καταρραγέντα, ταῦτα πρίσιος ἥκιστα κέχρηται· οὐδὲ ἕδρη αὐτὴ ἐφ' ἑωυτῆς γενομένη ἄτερ ῥωγμῆς καὶ φλάσιος, οὐδὲ αὕτη πρίσιος δεῖται·[1] οὐδ' ἡ διακοπὴ ἢν[2] μεγάλη καὶ εὐρεῖα ᾖ,
15 οὐδ' αὕτη· διακοπὴ γὰρ καὶ ἕδρη τωὐτόν ἐστιν.

X. Πρῶτον δὲ χρὴ τὸν τραυματίην σκοπεῖσθαι, ὅπῃ ἔχει τὸ τρῶμα τῆς κεφαλῆς, εἴτ' ἐν τοῖσιν ἰσχυροτέροισιν εἴτ' ἐν τοῖσιν ἀσθενεστέροισι, καὶ τὰς τρίχας καταμανθάνειν τὰς περὶ τὸ ἕλκος, εἰ διακεκόφαται ὑπὸ τοῦ βέλεος, καὶ εἰ ἔσω ἤϊσαν[3] ἐς τὸ τρῶμα, καὶ ἢν τοῦτο ᾖ, φάναι κινδυνεύειν τὸ ὀστέον ψιλὸν εἶναι τῆς σαρκὸς καὶ ἔχειν τι σῖνος τὸ ὀστέον ὑπὸ τοῦ βέλεος.[4] ταῦτα μὲν οὖν χρὴ ἀπόπροσθεν σκεψάμενον λέξαι, μὴ ἁπτόμενον τοῦ
10 ἀνθρώπου· ἁπτόμενον δ' ἤδη πειρᾶσθαι εἰδέναι σάφα εἴ ἐστι ψιλὸν τὸ ὀστέον τῆς σαρκὸς ἢ οὔ· καὶ ἢν μὲν καταφανὲς ᾖ τοῖσι ὀφθαλμοῖσι τὸ ὀστέον, ψιλόν· εἰ δὲ μή, τῇ μήλῃ σκέπτεσθαι. καὶ ἢν μὲν εὕρῃς ψιλὸν ἐὸν τὸ ὀστέον τῆς σαρκὸς καὶ μὴ ὑγιὲς ἀπὸ τοῦ τρώματος, χρὴ τοῦ ἐν τῷ ὀστέῳ ἐόντος τὴν διάγνωσιν πρῶτα ποιεῖσθαι, ὁρῶντα ὁπόσον τέ ἐστι τὸ κακὸν καὶ τίνος δεῖται ἔργου. χρὴ δὲ καὶ ἐρωτᾶν τὸν τετρωμένον ὅπως ἔπαθε καὶ τίνα τρόπον. ἢν δὲ μὴ καταφανὲς ᾖ
20 τὸ ὀστέον, εἰ ἔχει τι κακὸν[5] ἢ μὴ ἔχει, πολλῷ ἔτι χρὴ μᾶλλον τὴν ἐρώτησιν ποιεῖσθαι, ψιλοῦ τε ἐόντος τοῦ ὀστέου, τὸ τρῶμα ὅπως ἐγένετο καὶ ὅντινα τρόπον· τὰς γὰρ φλάσιας καὶ τὰς ῥωγμὰς τὰς οὐ φαινομένας ἐν τῷ ὀστέῳ, ἐνεούσας δέ, ἐκ τῆς ὑποκρίσιος[6] τοῦ τετρωμένου πρῶτον διαγινώ-

[1] δεῖται—εὐρεῖα Kw. B. [2] οὐδ' ἢν διακοπή. [3] εἴησαν.

depressed and comminuted, the less they require trephining. Nor does a *hedra*, occurring by itself without fracture or contusion, require trephining, and even if the cleft is large and wide, not even then; for cleft and *hedra* are the same.

X. The first thing to look for in the wounded man is whereabouts in the head the wound is, whether in the stronger or weaker part, and to examine the hair about the lesion, whether it has been cut through by the weapon and gone into the wound. If this is so, declare that it is likely that the bone is denuded of flesh and injured in some way by the weapon. One should say this at first inspection, without touching the patient. It is while handling the patient that you should try to make sure whether the bone is denuded of flesh or not. If the bone is visible to the eye, it is bare; if not, examine with the probe. Should you find the bone bare of flesh and injured by the wound, you should first distinguish the nature of the osseous lesion, its extent, and the operation required. And you should also ask the wounded man how he suffered the injury, and of what kind it was. If the bone is not visible so as to show whether it is or is not affected,[1] it is far more necessary than when the bone is bare to make the interrogation as to the origin and nature of the wound. For, in the case of contusions and fractures which do not appear in the bone, though they are there, you should first try to

[1] Reading *νόσημα*.

---

[4] I give Kw.'s order of these sentences.
[5] *νόσημα* B. Kw. [8] *ἀποκρίσιος*.

σκειν πειρᾶσθαι, εἴ τι πέπονθε τούτων τὸ ὀστέον ἢ οὐ πέπονθεν. ἔπειτα δὲ καὶ λόγῳ καὶ ἔργῳ ἐξελέγχειν, πλὴν μηλώσιος. μήλωσις γὰρ οὐκ ἐξελέγχει, εἰ πέπονθέ τι τούτων τῶν κακῶν τὸ
30 ὀστέον, καὶ εἴ τι ἔχει ἐν ἑωυτῷ, ἢ οὐ πέπονθεν· ἀλλ' ἕδρην τε τοῦ βέλεος ἐξελέγχει μήλωσις, καὶ ἢν ἐμφλασθῇ τὸ ὀστέον ἐκ τῆς φύσιος τῆς ἑωυτοῦ, καὶ ἢν ἰσχυρῶς ῥαγῇ τὸ ὀστέον, ἅπερ καὶ τοῖσι
34 ὀφθαλμοῖσι καταφανέα ἐστὶν ὁρῶντα γινώσκειν.[1]

XI. Ῥήγνυται δὲ τὸ ὀστέον τάς τε ἀφανέας ῥωγμὰς καὶ τὰς φανεράς, καὶ φλᾶται τὰς ἀφανέας φλάσιας, καὶ ἐσφλᾶται ἔσω ἐκ τῆς φύσιος τῆς ἑωυτοῦ, μάλιστα ὁπόταν ἕτερος ὑφ' ἑτέρου τιτρωσκόμενος ἐπίτηδες τρῶσαι [2] βουλόμενος ἢ ὁπόταν ἀέκων—καὶ ὁπόταν ἐξ ὑψηλοτέρου γίνηται ἡ βολὴ ἢ ἡ πληγή, ὁποτέρη ἂν ᾖ, μᾶλλον ἢ ὁπόταν ἐξ ἰσοπέδου τοῦ χωρίου, καὶ ἢν περικρατῇ τῇ χειρὶ τὸ βέλος, ἤν τε βάλλῃ ἤν τε τύπτῃ, καὶ
10 ἰσχυρότερος ἐὼν ἀσθενέστερον τιτρώσκῃ. ὁπόσοι δὲ πίπτοντες τιτρώσκονται πρός τε τὸ ὀστέον καὶ αὐτὸ τὸ ὀστέον, ὁ ἀπὸ ὑψηλοτάτου πίπτων καὶ ἐπὶ σκληρότατον καὶ ἀμβλύτατον, τούτῳ κίνδυνος τὸ ὀστέον ῥαγῆναί τε καὶ φλασθῆναι καὶ ἔσω ἐσφλασθῆναι ἐκ τῆς φύσιος τῆς ἑωυτοῦ· τῷ δὲ ἐξ ἰσοπέδου μᾶλλον χωρίου πίπτοντι καὶ ἐπὶ μαλθακώτερον, ἧσσον ταῦτα πάσχει τὸ ὀστέον ἢ οὐκ ἂν πάθοι. ὁπόσα δὲ ἐσπίπτοντα ἐς τὴν κεφαλὴν βέλεα τιτρώσκει πρὸς τὸ ὀστέον καὶ αὐτὸ τὸ
20 ὀστέον, τὸ ἀπὸ ὑψηλοτάτου ἐμπεσὸν καὶ ἥκιστα ἐξ ἰσοπέδου, καὶ σκληρότατόν τε ἅμα καὶ ἀμβλύτατον καὶ βαρύτατον, καὶ ἥκιστα κοῦφον καὶ

[1] Lobeck considers the last two words superfluous, but they are in all MSS.

distinguish by the patient's report whether the skull has or has not suffered in these ways. Then test the matter by reasoning and examination, avoiding the probe; for probing does not prove whether the bone has or has not suffered one of these evils, and what is the result. What probing proves is the existence of a *hedra* or weapon mark, or whether the skull has a contused fracture with depression, or is badly broken, things which are also clearly obvious to ocular inspection.

XI. The skull suffers invisible and visible fractures, invisible and visible contusions, and contused fracture with depression from its natural place, especially when one person is deliberately and wilfully wounded by another, rather than when the wound is unintentional; when the missile or the blow, whichever it be, comes from above rather than from level ground; when the weapon, whether used to throw or strike, is in full control,[1] and when a stronger man wounds a weaker. As to those who are wounded about the skull or in the skull itself by falling, he who falls from a very great height upon something very hard and blunt is likely to get his skull broken or contused, or to have a contused fracture with depression; while if a man falls from more level ground on to something rather soft, his skull suffers less in this way, or not at all. As to missile weapons which wound the parts about the skull or the skull itself, a thing will fracture or contuse the bone in proportion as it falls from a great height rather than the level, and is very hard as well as blunt, and

[1] Adams' "if the instrument be of a powerful nature" seems hardly correct.

---

[2] ἔτρωσεν; Pq. text obscure.

ἥκιστα ὀξὺ καὶ μαλθακόν, τοῦτο ἂν ῥήξειε τὸ ὀστέον καὶ φλάσειεν.

Καὶ μάλιστά γε ταῦτα πάσχειν τὸ ὀστέον κίνδυνος, ὁπόταν ταῦτά τε γίνηται καὶ ἐς ἰθὺ τρωθῇ καὶ κατ' ἀντίον γένηται τὸ ὀστέον τοῦ βέλεος, ἤν τε πληγῇ ἐκ χειρὸς ἤν τε βληθῇ ἤν τέ τι ἐμπέσῃ αὐτῷ καὶ ἢν αὐτὸς καταπεσὼν
30 τρωθῇ καὶ ὁπωσοῦν τρωθεὶς κατ' ἀντίον γενομένου τοῦ ὀστέου τῷ βέλει. τὰ δ' ἐς πλάγιον τοῦ ὀστέου παρασύραντα βέλεα ἧσσον καὶ ῥήγνυσι τὸ ὀστέον καὶ φλᾷ καὶ ἔσω ἐσφλᾷ, κἢν ψιλωθῇ τὸ ὀστέον τῆς σαρκός· ἔνια γὰρ τῶν τρωμάτων τῶν οὕτω τρωθέντων οὐδὲ ψιλοῦται τὸ ὀστέον τῆς σαρκός. τῶν δὲ βελέων ῥήγνυσι μάλιστα τὸ ὀστέον τάς τε φανερὰς ῥωγμὰς καὶ τὰς ἀφανέας καὶ φλᾷ τε καὶ ἐσφλᾷ ἔσω ἐκ τῆς φύσιος τῆς ἑωυτοῦ τὸ ὀστέον
40 τὰ στρογγύλα τε καὶ περιφερέα καὶ ἀρτίστομα, ἀμβλέα τε ἐόντα καὶ βαρέα καὶ σκληρά· καὶ τὴν σάρκα ταῦτα φλᾷ τε καὶ πέπειραν ποιεῖ καὶ κόπτει· καὶ τὰ ἕλκεα γίνεται ὑπὸ τῶν τοιούτων βελέων, ἔς τε πλάγιον καὶ ἐν κύκλῳ ὑπόκοιλα, καὶ διάπυά τε μᾶλλον γίνεται καὶ ὑγρά ἐστιν καὶ ἐπὶ πλέονα χρόνον καθαίρεται· ἀνάγκη γὰρ τὰς σάρκας τὰς φλασθείσας καὶ κοπείσας πῦον γενομένας ἐκτακῆναι. τὰ δὲ βέλεα τὰ προμήκεα ἐπὶ πολὺ λεπτὰ ἐόντα καὶ ὀξέα καὶ κοῦφα, τήν τε
50 σάρκα διατάμνει μᾶλλον ἢ φλᾷ καὶ τὸ ὀστέον ὡσαύτως· καὶ ἕδρην μὲν ἐμποιεῖ αὐτὸ καὶ διακόψαν [1]—διακοπὴ γὰρ καὶ ἕδρη τωὐτόν ἐστι—φλᾷ δὲ οὐ μάλα τὸ ὀστέον τὰ τοιαῦτα βέλεα οὐδὲ ῥήγνυσιν οὐδ' ἐκ τῆς φύσιος ἔσω ἐσφλᾷ.

heavy—in other words, the least light, sharp, and soft.

And the skull is especially likely to suffer this when the wound happens in those circumstances, and is perpendicular, the skull being directly opposed to the weapon, whether the agent be a blow or missile or something falling on the patient, or the patient falling himself, or being wounded in any way whatsoever, so long as the bone is at right angles to the weapon. When weapons graze the skull obliquely, they are less apt to cause fracture, or contusion, or contused fracture with depression, even if the bone is denuded; for in some wounds of this kind the bone is not even denuded of flesh. Those weapons which especially cause visible and invisible fractures, and contuse and crush in the bone out of its natural place, are rounded, smooth-surfaced, blunt, heavy and hard. These contuse the scalp, and pound it to a pulp. The wounds caused by such weapons become undermined both at the side and all round, and more likely to suppurate; they are moist and take long to cleanse, for the crushed and pounded tissue must necessarily become pus and slough away. Elongated weapons being usually slender, sharp and light, cut through the flesh rather than bruise it, and likewise the skull; they make a *hedra* in it and a cleaving[1] (for cleft is the same as *hedra*), but such weapons do not readily contuse the bone or break it, or crush it inwards out of its place.

[1] Or, "It leaves a *hedra* while cleaving."

---

[1] In these words *αὐτὸ* refers to *ὀστέον*, *διακόψαν* to *βέλεα* (*βέλος*). Erm.

Ἀλλὰ χρὴ πρὸς τῇ ὄψει τῇ ἑωυτοῦ, ὅ τι ἂν σοι φαίνηται ἐν τῷ ὀστέῳ, καὶ ἐρώτησιν ποιεῖσθαι πάντων τούτων. τοῦ γὰρ μᾶλλόν τε καὶ ἧσσον τρωθέντος ταῦτά ἐστι σημεῖα, καὶ ἢν ὁ τρωθεὶς καρωθῇ καὶ σκότος περιχυθῇ καὶ ἢν
60 δῖνος ἔχῃ καὶ πέσῃ.

XII. Ὁπόταν δὲ τύχῃ ψιλωθὲν τὸ ὀστέον τῆς σαρκὸς ὑπὸ τοῦ βέλεος, καὶ τύχῃ κατ᾽ αὐτὰς τὰς ῥαφὰς γενόμενον τὸ ἕλκος, χαλεπὸν γίνεται καὶ τὴν ἕδρην τοῦ βέλεος φράσασθαι τὴν ἐν τῷ ἄλλῳ ὀστέῳ φανερὴν γενομένην, εἴτ᾽ ἔνεστιν ἐν τῷ ὀστέῳ εἴτε μὴ ἔνεστιν, καὶ ἢν τύχῃ γενομένη ἡ ἕδρη ἐν αὐτῇσι τῇσι ῥαφῇσιν. συγκλέπτει[1] γὰρ αὐτὴ ἡ ῥαφὴ τρηχυτέρη ἐοῦσα τοῦ ἄλλου ὀστέου, καὶ οὐ διάδηλον ὅ τι τε αὐτοῦ ῥαφή ἐστι καὶ ὅ τι
10 τοῦ βέλεος ἕδρη, ἢν μὴ κάρτα μεγάλη γένηται ἡ ἕδρη. προσγίνεται δὲ καὶ ῥῆξις τῇ ἕδρῃ ὡς ἐπὶ τὸ πολὺ τῇ ἐν τῇσι ῥαφῇσι γινομένῃ,[2] καὶ γίνεται καὶ αὐτὴ ἡ ῥῆξις χαλεπωτέρη φράσασθαι, ἐρρωγότος τοῦ ὀστέου, διὰ τοῦτο ὅτι κατ᾽ αὐτὴν τὴν ῥαφὴν ἡ ῥῆξις γίνεται, ἢν ῥήγνυται, ὡς ἐπὶ τὸ πολύ· ἕτοιμον γὰρ ταύτῃ ῥήγνυσθαι τὸ ὀστέον καὶ διαχαλᾶν διὰ τὴν ἀσθενείην τῆς φύσιος τοῦ ὀστέου ταύτῃ καὶ διὰ τὴν ἀραιότητα, καὶ δὴ ἅτε τῆς ῥαφῆς ἑτοίμης ἐούσης ῥήγνυσθαι καὶ δια-
20 χαλᾶν. τὰ δὲ ἄλλα ὀστέα τὰ περιέχοντα τὴν ῥαφὴν μένει ἀρραγέα, ὅτι ἰσχυρότερά ἐστι τῆς ῥαφῆς. ἡ δὲ ῥῆξις ἡ κατὰ τὴν ῥαφὴν γινομένη καὶ διαχάλασίς ἐστι τῆς ῥαφῆς, καὶ φράσασθαι οὐκ εὐμαρής, οὔτε εἰ[3] ἀπὸ ἕδρης τοῦ βέλεος γενομένης ἐν τῇ ῥαφῇ, ἐπειδὰν ῥαγῇ καὶ διαχαλάσῃ, οὔτε ἢν φλασθέντος τοῦ ὀστέου κατὰ τὰς

Now, besides your own inspection of what you may see in the bone, inquiry should be made into all these things, for they are indications of the greater or less gravity of the wound, also as to whether the patient was stupefied and plunged in darkness, or had vertigo and fell down.

XII. Whenever the skull happens to be laid bare of flesh by the weapon, and the wound happens to occur just at the sutures, it becomes difficult to make an assertion as to the presence or absence of a weapon *hedra* in the bone which would be obvious in another part, especially if the *hedra* happens to come in the sutures themselves. For the suture itself being more uneven than the rest of the skull is deceptive, and it is not very clear which part is suture and which *hedra,* unless the *hedra* is very large. As a rule, too, fracture accompanies the *hedra* when it occurs in the sutures, and the fracture itself is harder to make out—though the bone is broken—for this reason, viz. that when there is a break it comes, as a rule, just in the suture. For the skull here is readily fractured or comes apart owing to the natural weakness of the bone in this place, and because of its porosity. Besides, the suture as such is ready to rupture and come apart, but the bones containing it remain unbroken because they are stronger than the suture. Fracture occurring in a suture includes a giving way of the suture, and it is not easy to make out whether the breaking and coming apart follows a weapon *hedra* occurring in the suture, or whether it is after contusion of the

---

[1] Scaliger's emendation for *συμβλέπει*, confirmed by B. (*συνκλεπτη*).

[2] *αὐτῆσιν . . . γιγνομενῆσι* Pq. [3] *ἥν.*

σάρκας, ῥαγῇ καὶ διαχάλασῃ· ἀλλ᾽ ἔστι χαλεπώτερον φράσασθαι τὴν ἀπὸ τῆς φλάσιος ῥωγμήν. συγκλέπτουσι γὰρ τὴν γνώμην καὶ τὴν ὄψιν τοῦ
30 ἰητροῦ αὗται αἱ ῥαφαὶ ῥωγμοειδέες φαινόμεναι καὶ τρηχύτεραι ἐοῦσαι τοῦ ἄλλου ὀστέου, ὅτι μὴ ἰσχυρῶς διεκόπη καὶ διεχάλασεν· διακοπὴ δὲ καὶ ἕδρη τὠυτόν ἐστιν. ἀλλὰ χρή, εἰ κατὰ τὰς ῥαφὰς τὸ τρῶμα γένοιτο καὶ πρός γε τὸ ὀστέον καὶ ἐς τὸ ὀστέον στηρίξειε τὸ βέλος, προσέχοντα τὸν νόον ἀνευρίσκειν ὅ τι ἂν πεπόνθῃ τὸ ὀστέον. ἀπὸ γὰρ ἴσων τε βελέων τὸ μέγεθος καὶ ὁμοίων καὶ πολλῷ[1] τε ἐλασσόνων, καὶ ὁμοίως τε τρωθεὶς καὶ πολλῷ[2] ἧσσον, πολλῷ μέζον ἐκτήσατο τὸ κακὸν ἐν τῷ
40 ὀστέῳ ὁ ἐς τὰς ῥαφὰς δεξάμενος τὸ βέλος ἢ ὁ μὴ ἐς τὰς ῥαφὰς δεξάμενος. καὶ τούτων τὰ πολλὰ πρίεσθαι δεῖ· ἀλλ᾽ οὐ χρὴ αὐτὰς τὰς ῥαφὰς πρίειν, ἀλλ᾽ ἀποχωρήσαντα ἐν τῷ πλησίον ὀστέῳ
44 τὴν πρῖσιν ποιεῖσθαι, ἢν πρίῃς.

XIII. Περὶ δὲ ἰήσιος τρωσίων τῶν ἐν τῇ κεφαλῇ καὶ ὅπως χρὴ ἐξελέγχειν τὰς πάθας τὰς ἐν τῷ ὀστέῳ γενομένας τὰς μὴ φανεράς, ὧδέ μοι δοκεῖ. ἕλκος ἐν τῇ κεφαλῇ οὐ χρὴ τέγγειν οὐδενί, οὐδὲ οἴνῳ, ἄλλως ἥκιστα·[3] οὐδὲ καταπλάσσειν, οὐδὲ μοτῷ τὴν ἴησιν ποιεῖσθαι, οὐδ᾽ ἐπιδεῖν χρὴ ἕλκος ἐν τῇ κεφαλῇ, ἢν μὴ ἐν τῷ μετώπῳ ᾖ τὸ ἕλκος, ἢ ἐν τῷ ψιλῷ τῶν τριχῶν, ἢ περὶ τὴν ὀφρὺν καὶ τὸν ὀφθαλμόν. ἐνταῦθα δὲ γινόμενα τὰ ἕλκεα κατα-
10 πλάσιος καὶ ἐπιδέσιος μᾶλλον κέχρηται ἤ που

[1] πολλόν.
[2] πολύ.
[3] ἀλλ᾽ ὡς ἥκιστα Pq., but with less support from MSS. or the context.

skull and flesh that it breaks and comes apart. Still, the fracture that follows contusion is harder to make out. For the sutures themselves, having a fracture-like appearance, and being more uneven than the rest of the skull, deceive the mind and eye of the physician, when not violently cleft or gaping —cleft and *hedra* are the same.[1] Now, if the wound is at the sutures, and the weapon penetrated the parts about the bone, and to the bone, you should devote your attention to finding out what injury the bone has suffered. For a person wounded by weapons of equal, similar or much less size to a similar or much less extent suffers far greater mischief in his skull if he receives the weapon at the sutures than when it is not so received, and the majority of these cases require trephining. You should not, however, trephine the sutures themselves, but, leaving an interval, operate on the adjacent part of the bone, if you do trephine.

XIII. The following is my view of the treatment of wounds in the head, and the way to discover affections of the skull which are not manifest. A lesion [2] in the head should not be moistened with anything, not even wine, much less anything else,[3] nor should the treatment include plasters or plugging, nor ought one to bandage a lesion in the head, unless it is on the forehead or in the part devoid of hair, or about the eyebrow or eye. Wounds occurring here are more suited to plasters and bandaging than those

[1] Surely an insertion.

[2] ἕλκος is defined by Galen as "a lesion of continuity in the soft parts." The "wound," therefore, concerns the scalp only.

[3] Or, reading ἀλλ' ὡς ἥκιστα "except the least possible," but the "correction" seems needless.

ἄλλοθι τῆς κεφαλῆς τῆς ἄλλης· περιέχει γὰρ ἡ κεφαλὴ ἡ ἄλλη τὸ μέτωπον πᾶν· ἐκ δὲ τῶν περιεχόντων τὰ ἕλκεα, καὶ ἐν ὅτῳ ἂν ᾖ τὰ ἕλκεα, φλεγμαίνει καὶ ἐπανοιδίσκεται δι' αἵματος ἐπιρροήν. χρὴ δὲ οὐδὲ τὰ ἐν τῷ μετώπῳ διὰ παντὸς τοῦ χρόνου καταπλάσσειν καὶ ἐπιδεῖν, ἀλλ' ἐπειδὰν παύσηται φλεγμαίνοντα, καὶ τὸ οἴδημα καταστῇ παύσασθαι καταπλάσσοντα καὶ ἐπιδέοντα· ἐν δὲ τῇ ἄλλῃ κεφαλῇ ἕλκος οὔτε μοτοῦν
20 χρή, οὔτε καταπλάσσειν οὔτ' ἐπιδεῖν, εἰ μὴ καὶ τομῆς δέοιτο.

Τάμνειν δὲ χρὴ τῶν ἑλκέων τῶν ἐν κεφαλῇ γενομένων, καὶ ἐν τῷ μετώπῳ, ὅπου ἂν τὸ μὲν ὀστέον ψιλὸν ᾖ τῆς σαρκός, καὶ δοκῇ τι σίνος ἔχειν ὑπὸ τοῦ βέλεος, τὰ δὲ ἕλκεα μὴ ἱκανὰ τὸ μέγεθος τοῦ μήκεος καὶ τῆς εὐρύτητος ἐς τὴν σκέψιν τοῦ ὀστέου, εἴ τι πέπονθεν ὑπὸ τοῦ βέλεος κακὸν καὶ ὁποῖόν τι πέπονθε, καὶ ὁπόσον μὲν ἡ σὰρξ πέφλασται καὶ τὸ ὀστέον ἔχει τι σίνος, καὶ
30 δ' αὖτε εἰ ἀσινές τέ ἐστι τὸ ὀστέον ὑπὸ τοῦ βέλεος καὶ μηδὲν πέπονθε κακόν, καὶ ἐς τὴν ἴησιν, ὁποίης τινὸς δεῖται τό τε ἕλκος ἥ τε σὰρξ καὶ ἡ πάθη τοῦ ὀστέου· τὰ δὲ τοιαῦτα τῶν ἑλκέων τομῆς δεῖται. καὶ ὅταν[1] μὲν τὸ ὀστέον ψιλωθῇ τῆς σαρκός, ὑπόκοιλα δὲ ᾖ ἐς πλάγιον ἐπὶ πολὺ ἐπανατάμνειν τὸ κοῖλον, ὅπου μὴ εὐχερὲς τῷ φαρμάκῳ ἀφικέσθαι, ὁποίῳ ἄν τινι χρή· καὶ τὰ κυκλοτερέα τῶν ἑλκέων καὶ ὑπόκοιλα ἐπὶ πολὺ καὶ τὰ τοιαῦτα ἐπανατάμνων τὸν κύκλον διχῇ
40 κατὰ μῆκος, ὡς πέφυκεν ὥνθρωπος, μακρὸν ποιεῖν τὸ ἕλκος.

Τάμνοντι δὲ κεφαλήν, τὰ μὲν ἄλλα τῆς

elsewhere in the head, for the rest of the head surrounds the whole forehead, and it is from the surrounding parts that lesions, wherever they may be, get inflamed and swollen by afflux of blood. Not even on the forehead should you use plasters and bandaging all the time, but when inflammation ceases and the swelling subsides, stop plasters and bandaging. On the rest of the head you should not plug, plaster, or bandage a wound unless incision is also required.

One should incise wounds occurring in the head and forehead where the bone is laid bare and seems to be in some way injured by the weapon, while the wounds are not long and broad enough for inspection of the bone, to see whether it has suffered any harm from the weapon, the nature of the injury and extent of the contusion of the flesh and any lesion of the bone, or, on the other hand, whether the bone is uninjured by the weapon, and has suffered no harm; also, as regards treatment to see what the wound requires, both as regards the flesh and the bone lesion. These are the kinds of wounds that require incision. When the skull is laid bare and there is considerable undermining on one side, open out by incision the hollow part where it is not easy for the suitable remedy to penetrate. In the case of circular wounds which are undermined to a considerable extent, open these out also by a double incision up and down as regards the patient [1] so as to make the wound a long one.

Incisions may be safely made by the surgeon in

[1] *i.e.* at opposite sides of the wound above and below.

---

[1] ἂν μὲν P.

κεφαλῆς ἀσφαλείην ἔχει ταμνόμενα· ὁ δὲ κρόταφος, καὶ ἄνωθεν ἔτι τοῦ κροτάφου, κατὰ τὴν φλέβα τὴν διὰ τοῦ κροτάφου φερομένην, τοῦτο δὲ τὸ χωρίον μὴ τάμνειν, σπασμὸς γὰρ ἐπιλαμβάνει τὸν τμηθέντα· καὶ ἢν μὲν ἐπ' ἀριστερὰ τμηθῇ κροτάφου,[1] τὰ ἐπὶ δεξιὰ ὁ σπασμὸς ἐπιλαμβάνει, ἢν δὲ ἐπὶ τὰ δεξιὰ τμηθῇ κροτάφου, τὰ
50 ἐπ' ἀριστερὰ ὁ σπασμὸς ἐπιλαμβάνει.

XIV. Ὅταν οὖν τάμνῃς ἕλκος ἐν κεφαλῇ ὀστέου εἵνεκα τῆς σαρκὸς ἐψιλωμένου, θέλων εἰδέναι εἴ τι ἔχει τὸ ὀστέον κακὸν ὑπὸ τοῦ βέλεος ἢ καὶ οὐκ ἔχει, τάμνειν χρὴ τὸ μέγεθος τὴν ὠτειλήν,[2] ὁπόση ἂν δοκῇ ἀποχρῆναι. τάμνοντα δὲ χρὴ ἀναστεῖλαι τὴν σάρκα ἀπὸ τοῦ ὀστέου ᾗ πρὸς τῇ μήνιγγι καὶ πρὸς τῷ ὀστέῳ πέφυκεν, ἔπειτα διαμοτῶσαι τὸ ἕλκος πᾶν μοτῷ, ὅστις ἂν εὐρύτατον τὸ ἕλκος παρέξει ἐς τὴν ὑστεραίην σὺν
10 ἐλαχίστῳ πόνῳ· μοτώσαντα δὲ καταπλάσματι χρῆσθαι ὁπόσον ἄν περ χρόνον καὶ τῷ μοτῷ, μάζης ἐκ λεπτῶν ἀλφίτων, ἐν ὄξει δὲ μάσσειν, ἕψειν δὲ καὶ γλίσχρην ποιεῖν ὡς μάλιστα. τῇ δὲ ὑστεραίῃ ἡμέρῃ, ἐπειδὰν ἐξέλῃς τὸν μοτόν, κατιδὼν τὸ ὀστέον ὅ τι πέπονθεν, ἐὰν μή σοι καταφανὴς ᾖ ἡ τρῶσις, ὁποίη τίς ἐστιν ἐν τῷ ὀστέῳ, μηδὲ διαγινώσκῃς εἴ τέ τι ἔχει τὸ ὀστέον κακὸν ἐν ἑωυτῷ, ἢ καὶ οὐκ ἔχει, τὸ δὲ βέλος δοκῇ ἀφικέσθαι ἐς τὸ ὀστέον καὶ σίνασθαι, ἐπιξύειν χρὴ τῷ
20 ξυστῆρι κατὰ βάθος καὶ κατὰ μῆκος τοῦ ἀνθρώπου ὡς πέφυκε, καὶ αὖθις ἐπικάρσιον τὸ ὀστέον τῶν ῥηξίων εἵνεκα τῶν ἀφανέων ἰδεῖν καὶ τῆς

[1] ἐν τῷ . . . κροτάφῳ also below ἐν τῷ ἐπὶ δεξιὰ τμηθῇ κροτάφῳ, Kw.

any other part of the head, but he should not incise the temple, or the part above it in the region traversed by the temporal blood-vessel, for spasm seizes the patient. And if incision of the temple is made on the left, spasm seizes the parts on the right, while if the incision is on the right, spasm seizes the parts on the left.

XIV. When, therefore, you incise a head wound because the bone is denuded, and you want to know whether it has, or has not, suffered any injury from the weapon, the size of the open wound should be such as seems fully sufficient. When operating you should detach the scalp from the skull where it is adherent to the membrane [1] and to the bone. Then plug the whole wound with lint, so that next day it will present the widest possible lesion of continuity with least pain. When plugging use a plaster of dough from fine barley meal to be kept on as long as the lint. Knead it up with vinegar and boil, making it as glutinous as possible. Next day, when you take out the lint, if, on looking to see what the bone has suffered, the nature of the lesion is not clear, and you cannot even see whether the skull has anything wrong with it, yet the weapon seems to have reached and damaged the bone, you should scrape down into it with a raspatory, both up and down as regards the patient, and again transversely so as to get a view of latent fractures and contusion which

[1] Vidius suggests that this refers to the connections between pericranium and dura mater at the sutures. Celsus seems to translate "membranula quae sub cute, calvariam cingit." VIII. 4.

---

[2] τομήν, Kw's conjecture.

φλάσιος εἵνεκα τῆς ἀφανέος τῆς οὐκ ἐσφλωμένης ἔσω ἐκ τῆς φύσιος τῆς κεφαλῆς τοῦ ἄλλου ὀστέου. ἐξελέγχει γὰρ ἡ ξύσις μάλα τὸ κακόν, ἢν μὴ καὶ ἄλλως καταφανέες ἔωσιν αὗται αἱ πάθαι αἱ ἐοῦσαι ἐν τῷ ὀστέῳ [τοῦ βέλεος].[1] καὶ ἢν ἕδρην ἴδῃς ἐν τῷ ὀστέῳ τοῦ βέλεος, ἐπιξύειν χρὴ αὐτήν τε τὴν ἕδρην καὶ τὰ περιέχοντα αὐτὴν ὀστέα, μὴ πολ-
30 λάκις τῇ ἕδρῃ προσγένηται ῥῆξις καὶ φλάσις, ἢ μόνη φλάσις, ἔπειτα λανθάνῃ οὐ καταφανέα ἐόντα.

Ἐπειδὰν δὲ ξύσῃς τὸ ὀστέον τῷ ξυστῆρι, ἢν μὲν δοκῇ ἐς πρῖσιν ἀφήκειν ἡ τρῶσις τοῦ ὀστέου, πρίειν χρή, καὶ τὰς τρεῖς ἡμέρας μὴ ὑπερβάλλειν ἀπρίωτον, ἀλλ᾽ ἐν ταύτῃσι πρίειν, ἄλλως τε καὶ τῆς θερμῆς ὥρης, ἢν ἐξ ἀρχῆς λαμβάνῃς τὸ ἴημα. Ἢν δὲ ὑποπτεύῃς μὲν τὸ ὀστέον ἐρρωγέναι ἢ πεφλάσθαι, ἢ ἀμφότερα ταῦτα, τεκμαιρόμενος ὅτι
40 ἰσχυρῶς τέτρωται ἐκ τῶν λόγων τοῦ τρωματίου, καὶ ὅτι ὑπὸ ἰσχυροτέρου τοῦ τρώσαντος, ἢν ἕτερος ὑφ᾽ ἑτέρου τρωθῇ, καὶ τὸ βέλος ὅτῳ ἐτρώθη, ὅτι τῶν κακούργων βελέων ἦν, ἔπειτα τὸν ἄνθρωπον ὅτι δῖνός τε ἔλαβε καὶ σκότος, καὶ ἐκαρώθη καὶ κατέπεσεν· τούτων δὲ οὕτω γενομένων, ἢν μὴ διαγινώσκῃς εἰ ἔρρωγε τὸ ὀστ᾽ον ἢ πέφλασται, ἢ καὶ ἀμφότερα ταῦτα, μήτε ἄλλως[2] ὁρέων δύνῃ, δεῖ δὴ ἐπὶ τὸ ὀστέον τὸ τηκτὸν τὸ μελάντατον δεύσας,[3] τῷ μέλανι φαρμάκῳ τῷ τηκομένῳ στεῖλαι[4] τὸ ἕλκος, ὑποτείνας ὀθόνιον
50 ἐλαίῳ τέγξας·[5] εἶτα καταπλάσας τῇ μάζῃ ἐπιδῆσαι. τῇ δὲ ὑστεραίῃ ἀπολύσας, ἐκκαθήρας τὸ ἕλκος ἐπιξῦσαι. καὶ ἢν μὴ ᾖ ὑγιές, ἀλλ᾽ ἐρρώγῃ καὶ

[1] Omit B. Kw. [2] ὅλως Pq. [3] δεύσαντα.

is latent because the rest of the bone is not crushed in out of its natural position. For rasping shows up the mischief well, even if these lesions though existing in the bone are not otherwise manifest. And if you see a weapon *hedra* in the bone, you should scrape the *hedra* itself and the bone containing it, in case, as often happens, fissure with contusion or contusion alone accompanies the *hedra*, and not being well marked, is overlooked.

When you scrape the bone with the raspatory, if the skull lesion seems to be a case for trephining, you should operate and not leave the patient untrephined till after the three days, but trephine in this period, especially in the hot season, if you take on the treatment from the first.

Should you suspect the skull to be fractured or contused or both, judging from the patient's account that the blow was severe and inflicted by a stronger person—if he was struck by someone else—and that the instrument with which he was wounded was of a dangerous kind; further, that the man suffered vertigo and loss of sight, was stunned and fell down: in such circumstances if you cannot otherwise distinguish by inspection whether the skull is fractured or contused or even both, then you must drop on the bone the very black solution, anoint the wound with the dissolved black drug, putting linen on it and moisten with oil, and then apply the barley-meal plaster and bandage. Next day, having opened and cleansed the wound, scrape further, and, if it is not sound but fractured and contused,

---

[4] Difficult text. στεῖλαι = *supertegere, inungere.*
[5] τέγξαι.

πεφλασμένον ᾖ, τὸ μὲν ἄλλο ἔσται ὀστέον λευκὸν ἐπιξυόμενον· ἡ δὲ ῥωγμὴ καὶ ἡ φλάσις, κατατακέντος τοῦ φαρμάκου, δεξαμένη τὸ φάρμακον ἐς ἑωυτὴν μέλαν ἐόν, ἔσται μέλαινα ἐν λευκῷ τῷ ὀστέῳ τῷ ἄλλῳ. ἀλλὰ χρὴ αὖθις τὴν ῥωγμὴν ταύτην φανεῖσαν μέλαιναν ἐπιξύειν κατὰ βάθος· καὶ ἢν μὲν ἐπιξύων [τὴν ῥωγμὴν ταύτην φανεῖσαν μέλαι-
60 ναν][1] ἐξέλῃς καὶ ἀφανέα ποιήσῃς, φλάσις μὲν γεγένηται τοῦ ὀστέου ἢ μᾶλλον ἢ ἧσσον, ἥτις περιέρρηξε καὶ τὴν ῥωγμὴν τὴν ἀφανισθεῖσαν ὑπὸ τοῦ ξυστῆρος· ἧσσον δὲ φοβερὸν καὶ ἧσσον ἂν πρῆγμα ἀπ' αὐτῆς γένοιτο ἀφανισθείσης τῆς ῥωγμῆς. ἢν δὲ κατὰ βάθος ᾖ καὶ μὴ ἐθέλῃ ἐξιέναι
66 ἐπιξυομένη, ἀφήκει ἐς πρῖσιν ἡ τοιαύτη συμφορή.

XV. Ἀλλὰ χρὴ πρίσαντα τὰ λοιπὰ ἰητρεύειν τὸ ἕλκος. φυλάσσεσθαι δὲ χρὴ ὅπως μή τι κακὸν ἀπολαύσῃ τὸ ὀστέον ἀπὸ τῆς σαρκός, ἢν κακῶς ἰητρεύηται. ὀστέῳ γὰρ καὶ πεπρισμένῳ καὶ ἄλλως ἀπρίστῳ ἐψιλωμένῳ δέ, καὶ[2] ὑγιεῖ δὲ ἐόντι καὶ ἔχοντί τι σίνος ὑπὸ τοῦ βέλεος, δοκέοντι δὲ ὑγιεῖ εἶναι, κίνδυνός ἐστι μᾶλλον ὑπόπυον γενέσθαι, ἢν καὶ ἄλλως μὴ μέλλῃ, ἢν καὶ ἡ σὰρξ ἡ περιέχουσα τὸ ὀστέον κακῶς
10 θεραπεύηται, καὶ φλεγμαίνῃ τε καὶ περισφίγγηται· πυρετῶδες γὰρ γίνεται καὶ πολλοῦ φλογμοῦ πλέον· καὶ δὴ τὸ ὀστέον ἐκ τῶν περιεχουσῶν σαρκῶν ἐς ἑωυτὸ θέρμην τε καὶ φλογμὸν καὶ ἄραδον ἐμποιεῖ καὶ σφυγμόν, καὶ ὁπόσα περ ἡ σὰρξ ἔχει κακὰ ἐν ἑωυτῇ, καὶ ἐκ τούτων ὧδε[3] ὑπόπυον γίνεται. κακὸν δὲ καὶ ὑγρήν τε εἶναι τὴν σάρκα ἐν τῷ ἕλκει καὶ

[1] Probably a gloss: many codd. and editt. omit.

the rest of the bone will be white after scraping, but the fracture and contusion will have absorbed the dissolved drug and will be black in the white bone. You should again scrape down into this fracture which shows black, and if on further scraping [this fracture which shows black] you clear it away and make it invisible, there has been more or less contusion of the bone, which also produced the fracture now abolished by the raspatory, but it is less formidable and less danger will result from it now the fracture has disappeared. Should it go deep and refuse to disappear when scraped, such an accident is a case for trephining.

XV. After the operation you should use the other treatment requisite for the wound.[1] You should guard against any mischief spreading from the tissues to the skull owing to improper treatment. For when the bone is trephined or otherwise denuded without trephining—whether really sound, or injured in some way by the weapon though apparently sound —there is greater risk of suppuration, even if it would not otherwise occur, if the flesh about the bone receives improper treatment and gets inflamed and strangulated. For a sort of fever occurs in it, and it becomes full of burning heat, and finally the bone draws into itself heat and inflammation from the tissues about it, also irritation and throbbing, and everything bad which the flesh already contains, and so it becomes purulent. It is also bad for the tissues in the wound to be moist and

[1] Vidius: "cetera facienda sunt quae ulceris curatio postulat."

[2] ἀπρίστῳ δέ, καὶ B.Kw.; the rest omit. [3] οὕτως.

μυδῶσαν καὶ ἐπὶ πολλὸν χρόνον καθαίρεσθαι· ἀλλὰ χρὴ διάπυον μὲν ποιῆσαι τὸ ἕλκος ὡς
20 τάχιστα· οὕτω γὰρ ἂν ἥκιστα φλεγμαίνοι τὰ περιέχοντα τὸ ἕλκος καὶ τάχιστ' ἂν καθαρὸν εἴη. ἀνάγκη γὰρ ἔχει τὰς σάρκας τὰς κοπείσας καὶ φλασθείσας ὑπὸ τοῦ βέλεος, ὑποπύους γενομένας, ἐκτακῆναι. ἐπειδὰν δὲ καθαρθῇ, ξηρότερον χρὴ γίνεσθαι τὸ ἕλκος· οὕτω γὰρ ἂν τάχιστα ὑγιὲς γένοιτο, ξηρῆς σαρκὸς βλαστούσης καὶ μὴ ὑγρῆς, καὶ οὕτως οὐκ ἂν ὑπερσαρκήσειε τὸ ἕλκος. ὁ δὲ αὐτὸς λόγος καὶ ὑπὲρ[1] τῆς μήνιγγος τῆς περὶ τὸν ἐγκέφαλον·
30 ἢν γὰρ αὐτίκα ἐκπρίσας τὸ ὀστέον καὶ ἀφελὼν ἀπὸ τῆς μήνιγγος ψιλώσῃς αὐτήν, καθαρὴν χρὴ ποιῆσαι ὡς τάχιστα καὶ ξηρήν, ὡς μὴ ἐπὶ πολὺν χρόνον ὑγρὴ ἐοῦσα μυδῇ τε καὶ ἐξαίρηται·[2] τούτων γὰρ οὕτω γινομένων σαπῆναι αὐτὴν
35 κίνδυνος.

XVI. Ὀστέον δὲ ὅ τι δὴ ἀποστῆναι δεῖ ἀπὸ τοῦ ἄλλου ὀστέου, ἕλκεος ἐν κεφαλῇ γενομένου, ἕδρης τε ἐούσης τοῦ βέλεος ἐν τῷ ὀστέῳ, ἢ ἄλλως ἐπὶ πολὺ ψιλωθέντος τοῦ ὀστέου, ἀφίσταται ἐπὶ πολὺ ἔξαιμον γενόμενον. ἀναξηραίνεται γὰρ τὸ αἷμα ἐκ τοῦ ὀστέου ὑπό τε τοῦ χρόνου καὶ ὑπὸ φαρμάκων τῶν πλείστων. τάχιστα δ' ἂν ἀποσταίη, εἴ τις τὸ ἕλκος ὡς τάχιστα καθήρας ξηραίνοι τὸ λοιπὸν τό τε ἕλκος καὶ τὸ
10 ὀστέον, καὶ τὸ μέζον καὶ τὸ ἧσσον. τὸ γὰρ τάχιστα ἀποξηρανθὲν καὶ ἀποστρακωθὲν τούτῳ μάλιστα ἀφίσταται ἀπὸ τοῦ ἄλλου ὀστέου τοῦ

[1] περί.

macerated, and to take a long time to clean up. You should rather make the wound suppurate as quickly as possible; for thus the parts about it will be least inflamed and it will be most rapidly cleansed; for the tissues that are pounded and contused by the weapon must necessarily become purulent and slough away. When the wound is cleansed it should get rather dry, for so it will soonest become healthy, the growing tissue[1] being dry and not moist, and thus the wound will have no exuberance of flesh. The same principle applies to the membrane covering the brain. For if you trephine at once and by taking away the bone denude this membrane, you should make it clean and dry as soon as possible, lest by being moist a long time it should fungate and swell up, for in such circumstances there is risk of its becoming putrid.

XVI. Any bone which is bound to separate from the rest, when a wound has occurred in the head and there is a weapon *hedra* in the skull, or when the bone is otherwise extensively denuded, usually separates after becoming bloodless, for the blood in the bone is dried up both by time and by most applications. The separation would occur most rapidly if, after cleansing the wound as soon as possible, one should next dry both the wound and the bone whether larger or smaller. For what is soonest dried up and made like a potsherd, thereby most readily separates from the rest of the bone which is full of blood and life, having

[1] Our "granulation tissue."

[2] ἐξερῇται.

ἐναίμου τε καὶ ζῶντος, αὐτὸ ἔξαιμόν τε γενόμενον
14 καὶ ξηρὸν [τῷ ἐναίμῳ καὶ ζῶντι μάλα ἀφίσταται].[1]

XVII. Ὅσα δὲ τῶν ὀστέων ἐσφλᾶται ἔσω ἐκ τῆς φύσιος τῆς ἑωυτῶν, καταρραγέντα ἢ καὶ διακοπέντα πάνυ εὐρέα, ἀκινδυνότερα τὰ τοιαῦτα γίνεται, ἐπὴν ἡ μῆνιγξ ὑγιὴς ᾖ· καὶ τὰ πλέοσι ῥωγμῇσιν ἐσκαταρραγέντα καὶ εὐρυτέρῃσιν ἔτι ἀκινδυνότερα καὶ εὐμαρέστερα ἐς τὴν ἀφαίρεσιν γίνεται. καὶ οὐ χρὴ πρίειν τῶν τοιούτων οὐδέν, οὐδὲ κινδυνεύειν τὰ ὀστέα πειρώμενον ἀφαιρεῖν πρὶν ἢ αὐτόματα ἐπανίῃ· εἰκὸς πρῶτον χαλά-
10 σαντος.[2] ἐπανέρχεται δὲ τῆς σαρκὸς ὑποφυομένης· ὑποφύεται δὲ ἐκ τῆς διπλόης τοῦ ὀστέου καὶ ἐκ τοῦ ὑγιέος, ἢν ἡ ἄνωθεν μοίρη τοῦ ὀστέου μούνη σφακελίσῃ. οὕτω δ' ἂν τάχιστα ἥ τε σὰρξ ὑποφύοιτο καὶ βλαστάνοι καὶ τὰ ὀστέα ἐπανίοι, εἴ τις τὸ ἕλκος ὡς τάχιστα διάπυον ποιήσας καθαρὸν ποιήσηται.[3] καὶ ἢν διὰ παντὸς τοῦ ὀστέου ἄμφω αἱ μοῖραι ἐσφλασθῶσιν ἔσω ἐς τὴν μήνιγγα, ἥ τε ἄνω μοίρη τοῦ ὀστέου καὶ ἡ κάτω, ἰητρεύοντι ὡσαύτως τὸ ἕλκος ὑγιὲς
20 τάχιστα ἔσται, καὶ τὰ ὀστέα τάχιστα ἐπάνεισι
21 τὰ ἐσφλασθέντα ἔσω.

XVIII. Τῶν δὲ παιδίων τὰ ὀστέα καὶ λεπτότερά ἐστι καὶ μαλθακώτερα διὰ τοῦτο, ὅτι ἐναιμότερά ἐστι, καὶ κοῖλα καὶ σηραγγώδεα καὶ οὔτε πυκνὰ οὔτε στερεά. καὶ ὑπὸ τῶν βελέων

[1] Following Kw.'s reading and punctuation of this much controverted passage. Scaliger and others omit the last words.

[2] "This passage is corrupt and depraved in all the examples." Foës.

become itself bloodless and dry [it readily comes away from the vascular and living part].

XVII. Cases of contused fracture of the bones with depression when they are broken up and even comminuted very widely, are less dangerous (than other injuries) if the covering of the brain is unharmed, and where the bones are broken in with many and rather wide fractures they are still less dangerous, and are more readily removed. In such cases you should do no trephining, nor run risk in trying to remove bone fragments before they come up of their own accord: they naturally come up when there is a loosening.[1] Now the fragments come up when the flesh grows from below, and it grows up from the diploë of the skull and its healthy part, if there is necrosis of the upper table of the skull only. Such upgrowth from below and burgeoning of the flesh will take place most rapidly if one brings the wound as soon as possible to suppuration and cleanses it. If the whole bone with both its "tables,"[2] both upper and lower, is contused inwards and depressed into the cerebral membrane, it is by the same treatment that the wound will heal soonest and the bone fragments that are crushed inwards come up most quickly.

XVIII. The (skull) bones of young children are thinner and softer because they contain more blood and are hollow and porous and neither dense nor hard. And when wounded by equal or weaker

[1] "Subsidence of the swelling," Adams, reading *οἴδεος* for *εἰκὸς* as Littré.

[2] Literally "parts."

---

[3] *ποιήσειεν.*

ἴσων τε ἐόντων καὶ ἀσθενεστέρων, καὶ τρωθέντων ὁμοίως τε καὶ ἧσσον, τὸ τοῦ νεωτέρου παιδίου καὶ μᾶλλον καὶ θᾶσσον ὑποπυΐσκεται ἢ τὸ τοῦ πρεσβυτέρου, καὶ ἐν ἐλάσσονι χρόνῳ· καὶ ὅσα ἂν ἄλλως μέλλῃ ἀποθανεῖσθαι ἐκ τοῦ τρώματος,
10 ὁ νεώτερος τοῦ πρεσβυτέρου θᾶσσον ἀπόλλυται.

Ἀλλὰ χρή, ἢν ψιλωθῇ τῆς σαρκὸς τὸ ὀστέον, προσέχοντα τὸν νόον, πειρῆσθαι διαγινώσκειν ὅ τι μὴ ἔστι τοῖσιν ὀφθαλμοῖσιν ἰδεῖν, καὶ γνῶναι εἰ ἔρρωγε τὸ ὀστέον καὶ εἰ πέφλασται, ἢ μοῦνον πέφλασται, καὶ εἰ, ἕδρης γενομένης τοῦ βέλεος, πρόσεστι φλάσις ἢ ῥωγμὴ ἢ ἄμφω ταῦτα. καὶ ἤν τι τούτων πέπονθε τὸ ὀστέον, ἀφεῖναι τοῦ αἵματος τρυπῶντα τὸ ὀστέον σμικρῷ τρυπάνῳ, φυλασσόμενον ἐπ' ὀλίγον· λεπτότερον
20 γὰρ τὸ ὀστέον καὶ ἐπιπολαιότερον τῶν νέων ἢ
21 τῶν πρεσβυτέρων.

XIX. Ὅστις δὲ μέλλει ἐκ τρωμάτων ἐν κεφαλῇ ἀποθνήσκειν, καὶ μὴ δυνατὸν αὐτὸν ὑγιᾶ γενέσθαι μηδὲ σωθῆναι, ἐκ τῶνδε τῶν σημείων χρὴ τὴν διάγνωσιν ποιεῖσθαι τοῦ μέλλοντος ἀποθνήσκειν, καὶ προλέγειν τὸ μέλλον ἔσεσθαι. πάσχει γὰρ τάδε· ὁπόταν τις ὀστέον κατεηγὸς ἢ ἐρρωγὸς ἢ πεφλασμένον, ἢ ὅτῳ γοῦν τρόπῳ κατεηγὸς ἐννοήσας ἁμάρτῃ, καὶ μήτε ξύσῃ μήτε πρίσῃ μήτε δεόμενον, μήτε[1] δὲ ὡς ὑγιέος ὄντος τοῦ
10 ὀστέου, πρὸ τῶν τεσσερακαίδεκα ἡμερέων πυρετὸς ἐπιλήψεται, ὡς ἐπὶ πολὺ ἐν χειμῶνι, ἐν δὲ τῷ θέρει μετὰ τὰς ἑπτὰ ἡμέρας ὁ πυρετὸς ἐπιλαμβάνει. καὶ ἐπειδὰν τοῦτο γένηται, τὸ ἕλκος ἄχροον γίνεται

[1] This fourth μήτε puzzles nearly all the translators. They leave it out. I follow Petrequin. μεθῇ δὲ Litt. Erm.

weapons to a similar or less extent the skull of the younger child suppurates more readily and rapidly than that of the elder and for a shorter period,[1] and when they are going to die in any case from the wound, the younger perishes sooner than the elder.

But if the bone is denuded of flesh you should devote your intelligence to trying to distinguish a thing which cannot be known by inspection—whether there is fracture and contusion of the skull or only contusion, and whether, if there is a weapon *hedra,* it is accompanied by contusion or fracture, or both of these. If the bone is injured in any of these ways, let blood by perforating with a small trepan, keeping a look-out at short intervals,[2] for in young subjects the skull is thinner and more on the surface[3] than in older persons.

XIX. When anyone is going to die from wounds in the head, and it is impossible to make him well or even save his life, the following are the signs from which one should make the diagnosis of approaching death and foretell what is going to happen. He has the following symptoms—when, after recognising that the skull is injured, either broken or contused, or injured in some way, one makes a mistake and neither scrapes nor trephines as though it were not required, yet the bone is not sound, fever as a rule will seize the patient within fourteen days in winter, and in summer just after seven days. When this occurs, the lesion

[1] So Petrequin, avoiding a tautology.
[2] Cf. *θαμινὰ σκοπούμενος*, XXI.
[3] *i.e.* has less depth.

καὶ ἐξ αὐτοῦ ἰχὼρ ῥεῖ σμικρός· καὶ τὸ φλεγμαῖνον ἐκτέθνηκεν ἐξ αὐτοῦ· καὶ βλιχῶδες [1] γίνεται καὶ φαίνεται ὥσπερ τάριχος, χροιὴν πυρρόν, ὑποπέλιον· καὶ τὸ ὀστέον σφακελίζειν τηνικαῦτα ἄρχεται, καὶ γίνεται περκνὸν λεῖον ὄν,[2] τελευταῖον δὲ ἔπωχρον γενόμενον ἢ ἔκλευκον. ὅταν
20 δ' ἤδη ὑπόπυον ᾖ, ἐπὶ τῇ γλώσσῃ φλυκταῖναι γίνονται, καὶ παραφρονέων τελευτᾷ. καὶ σπασμὸς ἐπιλαμβάνει τοὺς πλείστους τὰ ἐπὶ θάτερα τοῦ σώματος· ἢν μὲν ἐν τῷ ἐπ' ἀρίστερα τῆς κεφαλῆς ἔχῃ τὸ ἕλκος, τὰ ἐπὶ δεξιὰ τοῦ σώματος ὁ σπασμὸς λαμβάνει· ἢν δ' ἐν τῷ ἐπὶ δεξιὰ τῆς κεφαλῆς ἔχῃ τὸ ἕλκος, τὰ ἐπ' ἀριστερὰ τοῦ σώματος ὁ σπασμὸς ἐπιλαμβάνει. εἰσὶ δ' οἳ καὶ ἀπόπληκτοι γίνονται, καὶ οὕτως ἀπόλλυνται πρὸ ἑπτὰ ἡμέρων ἐν θέρει ἢ τεσσάρων καὶ δέκα
30 ἐν χειμῶνι· ὁμοίως δὲ τὰ σημεῖα ταῦτα σημαίνει, καὶ ἐν πρεσβυτέρῳ ἐόντι τῷ τρώματι ἢ καὶ ἐν νεωτέρῳ.

Ἀλλὰ χρή, εἰ ἐννοίης τὸν πυρετὸν ἐπιλαμβάνοντα καὶ τῶν ἄλλων τι σημεῖον τούτῳ προσγενόμενον, μὴ διατρίβειν, ἀλλὰ πρίσαντα τὸ ὀστέον πρὸς τὴν μήνιγγα ἢ καταξύσαντα τῷ ξυστῆρι—εὔπριστον [3] δὲ γίνεται καὶ εὔξυστον—ἔπειτα τὰ λοιπὰ οὕτως ἰητρεύειν ὅπως ἂν δοκῇ συμφέρειν,
39 πρὸς τὸ γινόμενον ὁρῶν.

XX. Ὅταν δ' ἐπὶ τρώματι ἐν κεφαλῇ ἀνθρώπου ἢ πεπριωμένου ἢ ἀπριώτου, ἐψιλωμένου δὲ τοῦ ὀστέου, οἴδημα ἐπιγένηται ἐρυθρὸν καὶ ἐρυσιπελατῶδες ἐν τῷ προσώπῳ καὶ ἐν τοῖσιν ὀφθαλμοῖσιν ἀμφοτέροισιν ἢ τῷ ἑτέρῳ, καὶ εἴ τις ἅπτοιτο τοῦ οἰδήματος, ὀδυνῷτο, καὶ πυρετὸς

gets a bad colour and a little ichor flows from it, the inflammation dies completely out of it, it gets macerated and looks like dried fish of a rather livid reddish colour. Necrosis of the bone then sets in, it gets dark coloured instead of white,[1] finally turning yellowish or dead white. When it has become purulent, blebs appear on the tongue and the patient dies delirious. Most cases have spasm of the parts on one side of the body; if the patient has the lesion on the left side of the head, spasm seizes the right side of the body; if he has the lesion on the right side of the head, spasm seizes the left side of the body. Some also become apoplectic and die in this state within seven days in summer and fourteen in winter. These symptoms have the same value both in an older and a younger patient.

If, then, you recognise that fever is seizing upon a patient and that any of these symptoms accompanies it, make no delay but, after trephining the bone down to the membrane, or scraping with the raspatory (for the bone becomes easy to saw or scrape), treat the case in future as may seem best in view of the circumstances.

XX. When in case of a wound in the head, whether the patient has been trephined or not, the bone being denuded, there supervenes a red erysipelatous oedema of the face and one or both eyes and the oedema is painful when touched,

[1] Reading λευκόν. λεῖον Pq. and codd. "without ceasing to be smooth" (?).

---

[1] So Kw. following Erotian and Archigenes. γλισχρῶδες Pq. codd.
[2] λευκὸν ἐόν Kw. etc. [3] καπυρόν.

ἐπιλαμβάνοι[1] καὶ ῥῖγος, τὸ δὲ ἕλκος αὐτό τε[2] ἀπὸ τῆς σαρκὸς καλῶς ἔχοι ἰδέσθαι καὶ τἀπὸ τοῦ ὀστέου, καὶ τὰ περιέχοντα τὸ ἕλκος ἔχοι
10 καλῶς, πλὴν τοῦ οἰδήματος τοῦ ἐν προσώπῳ καὶ ἄλλην ἁμαρτάδα μηδεμίαν ἔχοι τὸ οἴδημα τῆς ἄλλης διαίτης, τούτου χρὴ τὴν κάτω κοιλίην ὑποκαθῆραι φαρμάκῳ ὅ τι χόλην ἄγει· καὶ οὕτω καταρθέντος, ὅ τε πυρετὸς ἀφίησι καὶ τὸ οἴδημα καθίσταται καὶ ὑγιὴς γίνεται. τὸ δὲ φάρμακον χρὴ διδόναι πρὸς τὴν δύναμιν τοῦ ἀνθρώπου ὁρῶν,
17 ὡς ἂν ἔχῃ ἰσχύος.

XXI. Περὶ δὲ πρίσιος, ὅταν καταλάβῃ ἀνάγκη πρίσαι ἄνθρωπον, ὧδε γινώσκειν. ἢν ἐξ ἀρχῆς λαβὼν τὸ ἴημα πρίῃς, οὐ χρὴ ἐκπρίειν τὸ ὀστέον πρὸς τὴν μήνιγγα αὐτίκα· οὐ γὰρ συμφέρει τὴν μήνιγγα ψιλὴν εἶναι τοῦ ὀστέου ἐπὶ πολὺν χρόνον κακοπαθοῦσαν, ἀλλὰ τελευτῶσά πῃ καὶ διεμύδησεν.[3] ἔστι δὲ καὶ ἕτερος κίνδυνος, ἢν αὐτίκα ἀφαιρῇς πρὸς τὴν μήνιγγα ἐκπρίσας τὸ ὀστέον, τρῶσαι ἐν τῷ ἔργῳ τῷ πρίονι τὴν
10 μήνιγγα. ἀλλὰ χρὴ πρίοντα, ἐπειδὰν ὀλίγον[4] πάνυ δέῃ διαπεπρίσθαι, καὶ ἤδη κινῆται τὸ ὀστέον, παύσασθαι πρίοντα, καὶ ἐὰν ἐπὶ τὸ αὐτόματον ἀποστῆναι τὸ ὀστέον· ἐν γὰρ τῷ διαπριωτῷ ὀστέῳ καὶ ἐπιλελειμμένῳ τῆς πρίσιος οὐκ ἐπιγένοιτο κακὸν οὐδέν, λεπτὸν γὰρ τὸ λειπόμενον ἤδη γίνεται. τὰ δὲ λοιπὰ ἰῆσθαι χρή, ὡς ἂν δοκῇ συμφέρειν τῷ ἕλκει.

[1] ἐπιλαμβάνῃ. [2] τά τε Reinhold.
[3] σαπεῖσα διεμύδησεν Scaliger; but this is surgically the wrong order. Reinhold suggests διεμύδησε καὶ τελευτῶσα ἐσάπη.

and fever also seizes him with a rigor, but the lesion itself has a healthy appearance in the part affecting the scalp and skull, and the parts about the wound look healthy except for the oedema of the face, and the oedema is not further complicated by an error in regimen, in this case you should cleanse the bowel with a cholagogue. After such purging the fever departs, the oedema subsides and the patient gets well. In giving the drug you should have an eye to the patient's vigour, what strength he has.

XXI. As to trephining when it is necessary to trephine a patient, keep the following in mind. If you operate after taking on the treatment from the beginning, you should not, in trephining, remove the bone at once down to the membrane, for it is not good for the membrane to be denuded of bone and exposed to morbid influences for a long time, or it may end by becoming macerated.[1] There is also another danger that, if you immediately remove the bone by trephining down to the membrane, you may, in operating, wound the membrane with the trephine. You should rather stop the operation when there is very little left to be sawn through, and the bone is movable; and allow it to separate of its own accord. For no harm will supervene in the trephined bone, or in the part left unsawn, since what remains is thin enough. For the rest the treatment should be such as may seem beneficial to the lesion.

[1] "Becomes macerated, and finally putrefies." R.

---

[4] ὀλίγου.

Πρίοντα δὲ χρὴ πυκινὰ ἐξαιρεῖν τὸν πρίονα τῆς θερμασίης εἵνεκα τοῦ ὀστέου, καὶ ὕδατι ψυχρῷ
20 ἐναποβάπτειν. θερμαινόμενος γὰρ ὑπὸ τῆς περιόδου ὁ πρίων καὶ τὸ ὀστέον ἐκθερμαίνων καὶ ἀναξηραίνων κατακαίει, καὶ μέζον ποιεῖ ἀφίστασθαι τὸ ὀστέον τὸ περιέχον τὴν πρῖσιν ἢ ὅσον μέλλει ἀφίστασθαι. καὶ ἢν αὐτίκα βούλῃ ἐκπρίσαι τὸ πρὸς τὴν μήνιγγα, ἔπειτα ἀφελεῖν τὸ ὀστέον, ὡσαύτως χρὴ πυκινά τε ἐξαιρεῖν τὸν πρίονα καὶ ἐναποβάπτειν τῷ ὕδατι τῷ ψυχρῷ.

Ἢν δὲ μὴ ἐξ ἀρχῆς λαμβάνῃς τὸ ἴημα, ἀλλὰ παρ' ἄλλου παραδέχῃ ὑστερίζων τῆς ἰήσιος,
30 πρίονι χρὴ χαρακτῷ[1] ἐκπρίειν μὲν αὐτίκα τὸ ὀστέον πρὸς τὴν μήνιγγα, θαμινὰ δὲ ἐξαιρεῦντα τὸν πρίονα σκοπεῖσθαι καὶ ἄλλως καὶ τῇ μήλῃ πέριξ κατὰ τὴν ὁδὸν τοῦ πρίονος· καὶ γὰρ πολὺ θᾶσσον διαπρίεται τὸ ὀστέον, ἢν ὑπόπυόν τε ἐὸν ἤδη καὶ διάπυον πρίῃς, καὶ πολλάκις τυγχάνει ἐπιπόλαιον ἐὸν τὸ ὀστέον, ἄλλως τε καὶ ἢν ταύτῃ τῆς κεφαλῆς ᾖ τὸ τρῶμα ᾗ τυγχάνει λεπτότερον ἐὸν τὸ ὀστέον ἢ παχύτερον. ἀλλὰ φυλάσσεσθαι χρὴ ὡς μὴ λάθῃς προσβαλὼν τὸν
40 πρίονα, ἀλλ' ὅπῃ δοκεῖ πάχιστον εἶναι τὸ ὀστέον, ἐς τοῦτο αἰεὶ ἐνστηρίζειν τὸν πρίονα, θαμινὰ σκοπούμενος, καὶ πειρᾶσθαι ἀνακινέων τὸ ὀστέον ἀναβάλλειν, ἀφελὼν δὲ τὰ λοιπὰ ἰητρεύειν ὡς ἂν δοκῇ συμφέρειν τῷ ἕλκει [πρὸς τὸ γινόμενον ὁρέων].[2]

Καὶ ἤν, ἐξ ἀρχῆς λαβὼν τὸ ἴημα, αὐτίκα βούλῃ ἐκπρίσας τὸ ὀστέον ἀφελεῖν ἀπὸ τῆς μήνιγγος,

[1] "Serra acutiori" Vidius. Cf. Galen's Lexicon.
[2] Pq. omits, but see Kw.'s note.

While trephining, you should frequently take out the saw and plunge it into cold water to avoid heating the bone, for the saw gets heated by rotation, and by heating and drying the bone cauterises it and makes more of the bone around the trephined part come away than was going to do. If you want to trephine down to the membrane at once, and then remove the bone, the trephine should in like manner be often taken out and plunged in cold water.[1]

If you do not take on the cure from the be ginning, but receive it from another, coming late to the treatment, trephine the bone at once down to the membrane with a sharp-toothed trephine, taking it out frequently for inspection, and also examining with a probe around the track of the saw. For the bone is much more quickly sawn through if you operate when it is already suppurating and full of pus; and the skull is often found to have no depth, especially if the wound happens to be in the part of the head where the bone inclines to be thin rather than thick. You must be careful not to be heedless in placing the trephine, but always to fix it where the bone seems thickest. Examine often, and try by to-and-fro movements to lift up the bone; and, after removing it, treat the rest as may seem beneficial to the lesion [having regard to what has happened].

If you take on the case from the beginning, and want to trephine the bone at once completely and remove it from the membrane, you should likewise

[1] As we learn from Celsus, VIII. 3, and Heliodorus in *Oribasius* XLVI. 11, the trephine was rotated by a bow and cord, not by a handle as in modern times.

ὡσαύτως χρὴ πυκινά τε σκοπεῖσθαι τῇ μήλῃ τὴν
περίοδον τοῦ πρίονος, καὶ ἐς τὸ παχύτατον αἰεὶ
50 τοῦ ὀστέου τὸν πρίονα ἐνστηρίζειν, καὶ ἀνακινέων
βούλεσθαι ἀφελεῖν τὸ ὀστέον. ἢν δὲ τρυπάνῳ
χρῇ, πρὸς τὴν μήνιγγα μὴ ἀφικνεῖσθαι, ἢν ἐξ
ἀρχῆς λαμβάνων τὸ ἴημα τρυπᾷς, ἀλλ' ἐπιλιπεῖν
τοῦ ὀστέου λεπτόν, ὥσπερ καὶ ἐν τῇ πρίσει
55 γέγραπται.

often examine the circular track of the saw with the probe, always fixing the trephine in the thickest part of the bone, and aim at getting it away by to-and-fro movements. If you use a perforating trepan, do not go down to the membrane, if you perforate on taking the case from the beginning; but leave a thin layer of bone, as was directed in trephining.

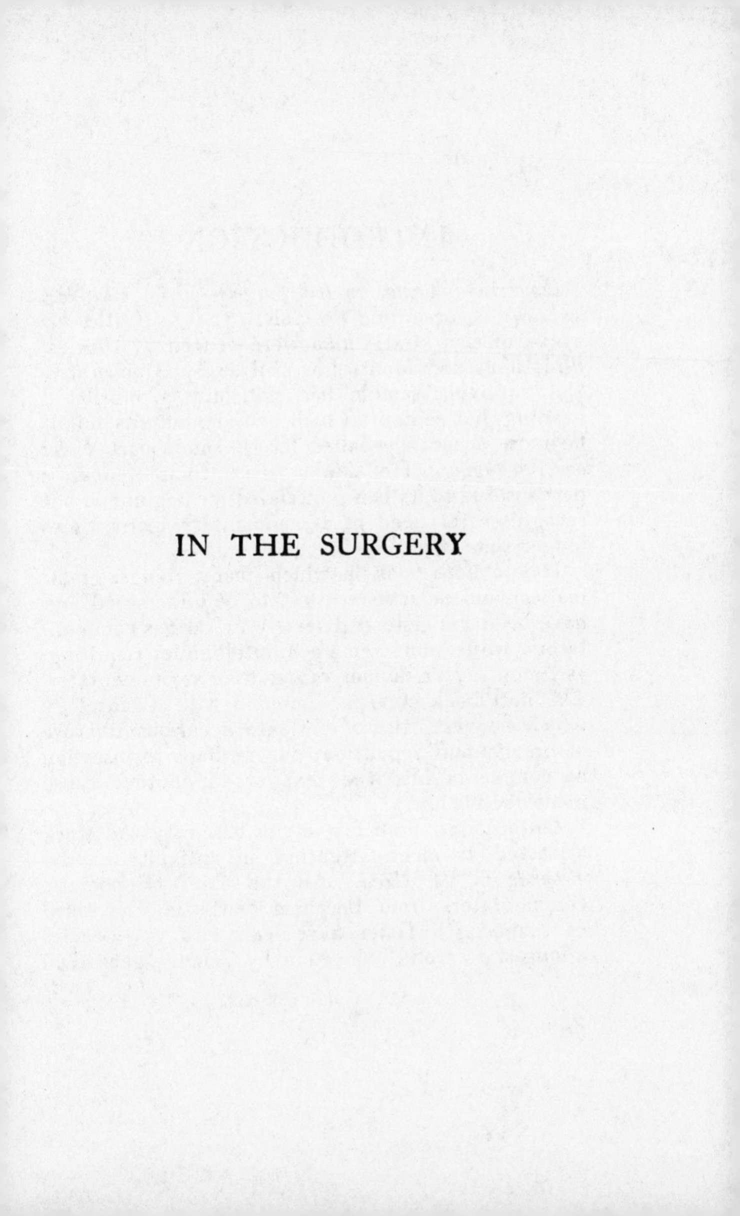

IN THE SURGERY

INTRODUCTION

Concerning Things in the Surgery—(περὶ τῶν κατ' ἰητρεῖον) is, according to Galen, the full title for works of this kind, which were written by Diocles, Philotimus and Mantias as well as by Hippocrates. Our surviving sample has not only a mutilated heading, but contents which, as Galen admits, might be more accurately called for the most part, *Notes on Bandaging.* He thinks this incompleteness is perhaps due to its being intended for beginners, but recognises its need of a commentary many times longer than itself.[1]

It is a note book in which many things, grammatical and didactic, are left to be understood and have been understood diversely by various commentators, while some remain unintelligible, requiring, as Galen says, a diviner rather than a commentator. The note book style is combined with a tautology which converts the whole into a curious mixture of brevity and repetition, due perhaps to insertion of comments into the text, or to another cause mentioned below.

On account, probably, of its obscurity the work attracted as much attention in antiquity as did *Wounds in the Head.* All the chief Hippocratic commentators from Bacchius (early in the third century B.C.) to Galen have dealt with it. Besides a long and careful exposition by Galen, a good deal

[1] XVIII(2). 629–632.

of the treatise is comprised in the preface to the *Galenic* work *On Bandages,* while the whole of the later treatise on that subject ascribed to him is taken from it and the commentary. Almost all ancient authorities considered it "genuine," though Galen suggests that it was not intended for publication and may have first been given out by Thessalus, who, according to some, was its author.

In modern times, Littré at first considered it spurious, an analysis or abridgment of some lost work, just as *Mochlicon* is certainly abridged from *Fractures–Joints,* but he afterwards changed his mind for the following reasons:—It has a peculiar connection with *Fractures:* Thus a statement in *Fractures* IV on the quantity of bandages is unintelligible unless we know their length, and this is only given in *Surgery* XII; on the other hand "ἤ" used to denote "rather than," *Surgery* XIV, seems (as Galen had observed) addressed to persons who knew *Fractures* XXII, where the context shows that it must have this sense. In *Surgery* XX, ὅτι (and still more διότι read by some) strongly suggests a note which the writer intends to enlarge upon. Littré concludes that *Surgery* is probably a "canevas" or preliminary sketch for a larger work of the kind which has perished, though part of it survives in our *Fractures,* and since *Surgery* XIX almost repeats XV, there may have been two such preliminary outlines which have been imperfectly conflated. We shall notice a similar duplication in *Mochlicon.*

Littré, however, does not entirely reject the view that *Surgery* is a later abstract or collection of memoranda from an earlier work; and the philological evidence is strongly on this side.

The verb δρᾶν is common, in fact reaches its highest frequency, in this treatise. "Depraved" infinitives with accusative participles posing as second person imperatives also occur, *e.g.* IV (where the two are combined) XII, XXIV. We naturally look for some connection with the δρᾶν (or middle) division of the books on *Epidemics*, and find that the beginning of *Epid.* IV. 45 corresponds verbally with part of *Surgery* I and II. We conclude that the work probably belongs to the second Hippocratic generation, may have been written by Thessalus son of Hippocrates, but can hardly have the same author as the great treatise *Fractures–Joints*.

Galen[1] and Palladius[2] tell us that, according to some, "In the Surgery" was the original title of the combined treatises *Fractures–Joints*, and this tradition may represent a truth. There was, perhaps, a great work on the surgery of the bones (of which we have fragments), and one or more abridgments of it, or possibly both an abridgment and a collection of memoranda in note-book style. Our *Surgery* would represent the beginning of the latter, our *Mochlicon* the end of the former, while the duplications may be due to an imperfect mixture of the two.

There are other curious resemblances between *Surgery* and *Fractures*. Thus, *Surgery* XVI seems condensed from *Fractures* IV, but while the writer of the latter says he has only seen over-extension in the case of a child, the epitomist has "over-extension is harmful except in children."

[1] XVIII(2). 323.
[2] *In. Hp. Fract.* Preface.

Surgery XVIII corresponds to *Fractures* VI, but it is only by reference to the latter that we can discover that splints are to be applied on the seventh day, and not at the seventh dressing, which is the more natural translation. The writer was, perhaps, relying upon memory, but this appears to be further evidence that *Surgery* is a later epitome, not a preliminary outline.

ΚΑΤ' ΙΗΤΡΕΙΟΝ

I. Ἢ ὅμοια ἢ ἀνόμοια, ἐξ ἀρχῆς ἀπὸ τῶν μεγίστων, ἀπὸ τῶν ῥηΐστων, ἀπὸ τῶν πάντη πάντως γινωσκομένων, ἃ καὶ ἰδεῖν καὶ θιγεῖν καὶ ἀκοῦσαι ἔστιν· ἃ καὶ τῇ ὄψει καὶ τῇ ἁφῇ καὶ τῇ ἀκοῇ καὶ τῇ ῥινὶ καὶ τῇ γλώσσῃ καὶ τῇ γνώμῃ ἔστιν αἰσθέσθαι· ἅ, οἷς γινώσκομεν, ἅπασιν 7 ἔστι γνῶναι.

II. Τὰ δὲ ἐς χειρουργίην κατ' ἰητρεῖον· ὁ ἀσθενέων, ὁ δρῶν, οἱ ὑπηρέται, τὰ ὄργανα, τὸ φῶς, ὅπου, ὅπως· ὅσα, οἷσιν, ὅπως,[1] ὁπότε· τὸ 4 σῶμα, τὰ ἄρμενα· ὁ χρόνος, ὁ τρόπος, ὁ τόπος.

III. Ὁ δρῶν, ἢ καθήμενος ἢ ἑστέως, συμμέτρως πρὸς ἑωυτόν, πρὸς τὸ χειριζόμενον, πρὸς τὴν αὐγήν.

Αὐγῆς μὲν οὖν δύο εἴδεα, τὸ μὲν κοινόν, τὸ δὲ τεχνητόν· τὸ μὲν οὖν κοινὸν οὐκ ἐφ' ἡμῖν, τὸ δὲ τεχνητὸν καὶ ἐφ' ἡμῖν. ὧν ἑκατέρου δισσαὶ[2] χρήσιες, ἢ πρὸς αὐγὴν ἢ ὑπ' αὐγήν. ὑπ' αὐγὴν μὲν οὖν ὀλίγη τε ἡ χρῆσις καταφανής τε ἡ μετριότης· τὰ δὲ πρὸς αὐγήν, ἐκ τῶν παρεουσέων, 10 ἐκ τῶν συμφερουσέων αὐγέων πρὸς τὴν λαμπροτάτην τρέπειν τὸ χειριζόμενον, πλὴν ὁπόσα λαθεῖν δεῖ ἢ ὁρᾶν αἰσχρόν, οὕτω δὲ τὸ μὲν χειριζόμενον ἐναντίον τῇ αὐγῇ, τὸν δὲ χειρίζοντα ἐναντίον τῷ χειριζομένῳ, πλὴν ὥστε μὴ ἐπισκο-

[1] οἷς· ὡς. But Galen read ὅπως twice (XVIII(2). 669).
[2] δύο αἱ.

IN THE SURGERY

I. [Examination: look for] what is like or unlike the normal, beginning with the most marked signs and those easiest to recognise, open to all kinds of investigation, which can be seen, touched and heard, which are open to all our senses, sight, touch, hearing, the nose, the tongue and the understanding, which can be known by all our sources of knowledge.

II. Operative requisites in the surgery; the patient, the operator, assistants, instruments, the light, where and how placed; their number, which he uses how and when; the (patient's?) person and the apparatus; time manner and place.[1]

III. The operator whether seated or standing should be placed conveniently to himself, to the part being operated upon and to the light.

Now, there are two kinds of light, the ordinary and the artificial, and while the ordinary is not in our power the artificial is in our power. Each may be used in two ways, as direct light and as oblique light. Oblique light is rarely used, and the suitable amount[2] is obvious. With direct light, so far as available and beneficial, turn the part operated upon towards the brightest light—except such parts as should be unexposed and are indecent to look at—thus while the part operated upon faces the light, the surgeon faces the part, but not so as to overshadow

[1] "Part affected," according to Galen: XVIII(2). 674.

[2] This is the usual meaning of μετριότης. See *Fractures* V.

τάζειν· οὕτω γὰρ ἂν ὁ μὲν δρῶν ὁρῴη, τὸ δὲ χειριζόμενον οὐχ ὁρῷτο.

Πρὸς ἑωυτὸν δέ, καθημένῳ μὲν πόδες ἐς τὴν ἄνω ἴξιν κατ' ἰθὺ γούνασι· διάστασιν δὲ ὀλίγον συμβεβῶτες. γούνατα δὲ ἀνωτέρω βουβώνων
20 σμικρόν, διάστασιν δέ, ἀγκώνων θέσει,[1] καὶ παραθέσει· ἱμάτιον εὐσταλέως, εὐκρινέως, ἴσως, ὁμοίως ἀγκῶσιν ὤμοισιν.

Πρὸς δὲ τὸ χειριζόμενον, τοῦ μὲν πρόσω καὶ ἐγγὺς [ὅριον,][2] καὶ τοῦ ἄνω καὶ τοῦ κάτω, καὶ ἔνθα ἢ ἔνθα ἢ μέσον. τοῦ μὲν πρόσω καὶ ἐγγὺς ὅριον, ἀγκῶνας ἐς μὲν τὸ πρόσθεν γούνατα μὴ ἀμείβειν, ἐς δὲ τὸ ὄπισθεν πλευράς· τοῦ δὲ ἄνω μὴ ἀνωτέρω μάζων ἄκρας χεῖρας ἔχειν· τοῦ δὲ κάτω, μὴ κατωτέρω ἢ ὡς τὸ στῆθος ἐπὶ γούνασιν ἔχοντα,
30 χεῖρας ἄκρας ἔχειν ἐγγωνίους πρὸς βραχίονας. τὰ μὲν κατὰ μέσον οὕτως· τὰ δὲ ἔνθα ἢ ἔνθα, μὴ ἔξω τῆς ἕδρης, κατὰ λόγον δὲ τῆς ἐπιστροφῆς προσβαλλόμενον τὸ σῶμα, καὶ τοῦ σώματος τὸ ἐργαζόμενον.

Ἑστεῶτα δέ, ἰδεῖν μὲν καὶ ἐπ' ἀμφοτέρων βεβαῶτα ἐξ ἴσου τῶν ποδῶν ἅλις, δρᾶν δὲ τῷ ἑτέρῳ ἐπιβεβῶτα, μὴ τῷ κατὰ τὴν δρῶσαν χεῖρα· ὕψος γουνάτων[3] πρὸς βουβῶνας ὡς ἐν ἕδρῃ· καὶ τὰ ἄλλα ὅρια τὰ αὐτά.

40 Ὁ δὲ χειριζόμενος τῷ χειρίζοντι τῷ ἄλλῳ τοῦ σώματος μέρει ὑπηρετείτω, ἢ ἑστεὼς ἢ καθήμενος ἢ κείμενος, ὅπως[4] ἂν ῥήϊστα ὃ δεῖ σχῆμα ἔχων διατελῇ, φυλάσσων ὑπόρρυσιν, ὑπόστασιν, ἔκ-

[1] ἀγκῶσιν, θέσει.
[2] Omit Pq. Litt. and codd. : except V.
[3] ὕψος· γούνατα Kw. ὕψος γούνατος Littré. [4] ὡς.

it. For the operator will in this way get a good view and the part treated not be exposed to view.

As regards himself, when seated his feet should be in a vertical line straight up as regards the knees, and be brought together with a slight interval. Knees a little higher than the groins and the interval between them such as may support and leave room for the elbows. Dress well drawn together, without creases, even and corresponding on elbows and shoulders.

As regards the part operated upon, there is limit for far and near, up and down, to either side and middle. The far and near limit is such that the elbows need not pass in front of the knees or behind the ribs, and for up and down, that the hands are not held above the breasts, or lower than that, when the chest is on the knees, the forearms are kept at right angles to the arms. Such is the rule as regards the median position but deviation to either side is made by throwing forward the body, or its active part, with a suitable twist, without moving the seat.[1]

If he stands, he should make the examination with both feet fairly level, but operate with the weight on one foot (not that on the side of the hand in use); height of knees[2] in the same relation to groins as when seated, and the other limits the same.

Let the patient assist the surgeon with the other (free) part of his body standing, sitting or lying so as to maintain most easily the proper posture, on his guard against slipping, collapse, displacement, pen-

[1] According to Galen, the anatomical "seat" or pelvis.

[2] The other foot is on some elevated support: see *Fractures* VIII. Galen XVIII(2). 700.

τρεψιν, καταντίαν, ὡς ὃ δεῖ σώζηται καὶ σχῆμα καὶ εἶδος τοῦ χειριζομένου ἐν παρέξει, ἐν χειρι-
46 σμῷ, ἐν τῇ ἔπειτα ἕξει.

IV. Ὄνυχας μήτε ὑπερέχειν μήτε ἐλλείπειν δακτύλων κορυφάς·[1] ἐς χρῆσιν ἀσκεῖν δακτύλοισι μὲν ἄκροισι, τὰ πλεῖστα λιχανῷ πρὸς μέγαν· ὅλῃ δὲ καταπρηνεῖ, ἀμφοτέρῃσι δὲ ἐναντίῃσιν. δακτύλων εὐφυΐη· μέγα τὸ ἐν μέσῳ τῶν δακτύλων, καὶ ἀπεναντίον τὸν μέγαν τῷ λιχανῷ. νοῦσος δέ, δι' ἣν καὶ βλάπτονται, τοῖσιν ἐκ γενεῆς ἢ ἐν τροφῇ εἴθισται ὁ μέγας ὑπὸ τῶν ἄλλων δακτύλων κατέχεσθαι δῆλον. τὰ ἔργα
10 πάντα ἀσκεῖν ἑκατέρῃ δρῶντα, καὶ ἀμφοτέρῃσιν ἅμα—ὅμοιαι γάρ εἰσιν ἀμφότεραι—στοχαζόμενον ἀγαθῶς, καλῶς, ταχέως, ἀπόνως, εὐρύ-
13 θμως, εὐπόρως.

V. Ὄργανα μὲν καὶ ὅτε, καὶ οἵως, εἰρήσεται. ὅπου δεῖ μὴ ἐμποδὼν τῷ ἔργῳ μηδὲ ἐκποδὼν τῇ ἀναιρέσει, παρὰ τὸ ἐργαζόμενον δὲ τοῦ σώματος ἔστω· ἄλλος δὲ ἢν διδῷ, ἕτοιμος ὀλίγῳ πρότερον
5 ἔστω, ποιείτω δέ, ὅταν κελεύῃς.

VI. Οἱ δὲ περὶ τὸν ἀσθενέοντα, τὸ μὲν χειριζόμενον παρεχόντων, ὡς ἂν δοθῇ·[2] τὸ δὲ ἄλλο σῶμα κατεχόντων, ὡς ὅλον ἀτρεμῇ, σιγῶντες,
4 ἀκούοντες τοῦ ἐφεστεῶτος.

VII. Ἐπιδέσιος δύο εἴδεα, εἰργασμένον καὶ ἐργαζόμενον. ἐργαζόμενον μὲν ταχέως, ἀπόνως, εὐπόρως, εὐρύθμως. ταχέως μὲν ἀνύειν τὰ ἔργα·

[1] κορυφῆς. [2] δοκῇ.

[1] The meaning can only be fully understood after reading *Fractures*.

dency, so that the position and form of the part treated may be properly preserved during presentation, operation, and the attitude afterwards.[1]

IV. The nails neither to exceed nor come short of the finger tips. Practise using the finger ends especially with the forefinger opposed to the thumb, with the whole hand held palm downwards, and with both hands opposed. Good formation of fingers: one with wide intervals and with the thumb opposed to the forefinger, but there is obviously a harmful disorder in those who, either congenitally or through nurture, habitually hold down the thumb under the fingers. Practise all the operations, performing them with each hand and with both together—for they are both alike—your object being to attain ability, grace, speed, painlessness, elegance and readiness.

V. As to instruments, the time and manner of their use will be discussed. Their proper position is such as neither to be in the way of the operation nor to be out of the way when wanted; their place is by the operator's hand,[2] but if an assistant gives them, let him be ready a little beforehand, and act when you bid him.

VI. Let those who look after the patient present the part for operation as you want it, and hold fast the rest of the body so as to be all steady, keeping silence and obeying their superior.

VII. Of bandaging there are two aspects, completed and in process of application. As regards application, speedily, painlessly, with resource and neatness. Speedily to bring the operation to an end,

[2] This seems to refer to the surgeon, as above, not to the part operated on (τὸ χειριζόμενον).

ἀπόνως δὲ ῥηϊδίως δρᾶν· εὐπόρως[1] δέ, ἐς πᾶν ἑτοίμως· εὐρύθμως δὲ ὁρῆσθαι ἡδέως· ἀφ' ὧν δὲ ταῦτα ἀσκημάτων εἴρηται. εἰργασμένον δὲ ἀγαθῶς, καλῶς· καλῶς μὲν ἁπλῶς εὐκρινέως· ἢ ὅμοια καὶ ἴσα, ἴσως καὶ ὁμοίως· ἢ ἄνισα καὶ ἀνόμοια ἀνίσως καὶ ἀνομοίως. τὰ μὲν εἴδεα
10 ἁπλοῦν [εὔκυκλον][2] σκέπαρνον, σιμὸν, ὀφθαλμός, καὶ ῥόμβος καὶ ἡμίτομον· ἅρμοζον τὸ εἶδος τῷ
12 εἴδει καὶ τῷ πάθει τοῦ ἐπιδεομένου.

VIII. Ἀγαθῶς δὲ δύο εἴδεα τοῦ ἐπιδεομένου· ἰσχύος μὲν ἢ πιέξει, ἢ πλήθει ὀθονίων. τὸ μὲν οὖν αὐτὴ ἡ ἐπίδεσις ἰῆται, τὸ δὲ τοῖσιν ἰωμένοισιν ὑπηρετεῖ. ἐς μὲν οὖν ταῦτα νόμος· ἐν δὲ τούτοισι μέγιστα ἐπιδέσιος· πίεξις μὲν ὥστε τὰ ἐπικείμενα μὴ ἀφεστάναι, μηδὲ ἐρηρεῖσθαι [κάρτα],[3] ἀλλ' ἡρμόσθαι μέν, προσηναγκάσθαι δὲ μή, ἧσσον μὲν τὰ ἔσχατα, ἥκιστα δὲ τὰ μέσα. ἅμμα καὶ ῥάμμα νεμόμενον μὴ κάτω, ἀλλ' ἄνω, ἐν παρέξει καὶ
10 σχέσει καὶ ἐπιδέσει καὶ πιέξει. ἀρχὰς βάλλεσθαι μὴ ἐπὶ τὸ ἕλκος, ἀλλ' ἔνθα τὸ ἅμμα. τὸ δὲ ἅμμα μήτε ἐν τρίβῳ μήτε ἐν ἔργῳ, μήτε ἐκεῖσε ὅπου ἐνεόν, ὡς μὴ ἐς τὸ ἐνεὸν κείσεται.[4] ἅμμα δὲ καὶ
14 ῥάμμα μαλθακόν, μὴ μέγα.

[1] εὐπορίη . . . εὐρυθμίη.
[2] εὔκυκλον or ἔγκυκλον was inserted as explanation of ἁπλοῦν by Artemidorus and Dioscorides. Cf. Galen, XVIII(2). 729.
[3] Added by Littré from Galen *de Fasc.*
[4] Kw.'s reading of this obscure passage.

---

[1] So Galen.
[2] As Galen remarks, there is no "second" unless we take it to include all other good qualities; some apply it to the two objects of bandaging.
[3] A puzzle to commentators as contrasted with later directions, cf. XII.

painlessly to do it with ease, with resource ready for anything, with neatness that it may be pleasant to look at. Exercises for attaining these ends have been mentioned. Completed bandaging should be well and neatly done. Neatly means smoothly, well distributed,[1] evenly and alike where the parts are even and similar, unevenly and unlike where they are unlike and uneven. As to kinds, simple (circular), oblique (adze like), very oblique (reversed?), the eye, the rhomb, the half rhomb, (use) the form suited to the shape and the affection of the part bandaged.

VIII. "Well" has two aspects when applied to the part bandaged: first[2] firmness got either by tension or by the number of bandages. Now, the bandaging may either cure by itself or assist the curative agents. There is a rule for this and it includes the most important elements of bandaging. Pressure so that the applications neither fall away nor are very tight, fitting to the part without forcible compression, less at the ends and least in the middle.[3] Knot and thread suture carried upwards and not downwards in presentation, attitude, bandaging and compression.[4] The ends (for tying) to be put, not over the wound, but where the knot is to be. The knot where there is neither friction nor motion, and not where it will be useless, lest its purpose be not served.[5] Knot and suture soft and not large.

[4] ἕξει "fixation" is what we should expect, but the whole is obscure.

[5] A much discussed passage. Perhaps means not close to the edge of the dressing lest it slip off. Heliodorus (*Orib.* XLVIII. 70) and Galen seem to ignore the last six words, but both say that ἐνεόν=κενεόν "useless." Can it be a pun, "not where there is a void lest it be void of use"? As Galen says, we should expect "not over a hollow" such as the armpit.

IX. Εὖ γε μήν ἐστι γνῶναι ὅτι ἐς τὰ κατάντη καὶ ἀπόξη φεύγει πᾶς ἐπίδεσμος, οἷον κεφαλῆς μὲν τὸ ἄνω, κνήμης δὲ τὸ κάτω. ἐπιδεῖν δεξιὰ ἐπ' ἀριστερά, ἀριστερὰ δὲ ἐπὶ δεξιά, πλὴν τῆς κεφαλῆς, ταύτην κατ' ἴξιν. τὰ δὲ ὑπεναντία[1] ἀπὸ δύο ἀρχέων· ἢν δὲ ἀπὸ μιῆς, ἐφ' [ἑκάτερα][2] ὅπερ ὅμοιον ἐς τὸ μόνιμον, οἷον τὸ μέσον τῆς κεφαλῆς, ἢ ὅ τι ἄλλο τοιοῦτον. τὰ δὲ κινεύμενα, οἷον ἄρθρα, ὅπῃ μὲν συγκάμπτεται, ὡς ἥκιστα
10 καὶ εὐσταλέστατα περιβάλλειν, οἷον ἰγνύην· ὅπῃ δὲ περιτείνεται, ἁπλᾶ τε καὶ πλατέα, οἷον μύλη· προσπεριβάλλειν δὲ καταλήψιος μὲν τῶν περὶ ταῦτα εἵνεκα, ἀναλήψιος δὲ τοῦ συμπάντος ἐπιδέσμου, ἐν τοῖσιν ἀτρεμέουσι καὶ λαπαρωτέροισι τοῦ σώματος, οἷον τὸ ἄνω καὶ τὸ κάτω τοῦ γούνατος· ὁμολογεῖ δέ, ὤμου μὲν ἡ περὶ τὴν ἑτέρην μασχάλην περιβολή, βουβῶνος δὲ ἡ περὶ τὸν ἕτερον κενεῶνα, καὶ κνήμης ἡ ὑπὲρ γαστροκνημίης. ὁπόσοισι μὲν ἄνω ἡ φυγή, κάτωθεν ἡ
20 ἀντίληψις, οἷσι δὲ κάτω, τοὐναντίον· οἷσι δὲ μὴ ἔστιν, οἷον κεφαλῇ, τούτων ἐν τῷ ὁμαλωτάτῳ τὰς καταλήψιας ποιεῖσθαι, καὶ ἥκιστα λοξῷ τῷ ἐπιδέσμῳ χρῆσθαι, ὡς τὸ μονιμώτατον ὕστατον περιβληθὲν τὰ πλανωδέστατα κατέχῃ. ὁπόσοισι δὲ τοῖσιν ὀθονίοισι μὴ εὐκαταλήπτως, μηδὲ εὐαναλήπτως ἔχει, ῥάμμασι τὰς ἀναλήψιας ποι-
27 εῖσθαι ἐκ καταβολῆς ἢ συρραφῆς.

[1] τὰ καθ' ἑκάτερον μέρος ὁμοίως διακείμενα.—Galen.
[2] Most MSS. omit.

IX. It is well to bear in mind that every bandage slips towards the pendent and conical parts, such as the top of the head and the bottom of the leg. Bandage parts on the right side towards the left, and those on the left to the right, except the head; do this vertically.[1] Parts with opposite sides alike[2] require a two-headed bandage, but if you bandage from one end, extend it each way so that it may have a similar relation to the fixed part, such as the middle of the head or the like. As to mobile parts, such as joints, where there is flexion the turns should be as few and as contracted as possible, as with the back of the knee, but where the part is extended, like the knee cap, spread out and broad. Make additional turns both to hold fast applications in these parts, and to support the dressing in the fixed and flatter parts of the body, such as those above and below the knee. In case of the shoulder, a turn round the opposite armpit is suitable, for the groin, one round the opposite flank, and for the leg, the part above the calf. In cases where the tendency is to slip up, the support is from below, when down the reverse. Where this is impossible, as on the head, make the hold-fasts on the smoothest part, and avoid obliquity as far as you can, so that the outermost and firmest turn may hold down the most mobile ones. Where it is not easy to get either good fixation or support with the bandages, make supports with threaded sutures in loops[3] or continuous suture.

[1] "From vertex to chin." Galen.

[2] Galen's paraphrase.

[3] Apparently our interrupted sutures, with long ends to tie. "Stitching with ligatures." Adams.

X. Ἐπιδέσματα καθαρά, κοῦφα, μαλθακά, λεπτά. ἑλίσσειν ἀμφοτέρῃσιν ἅμα, καὶ ἑκατέρῃ χωρὶς ἀσκεῖν. τῇ πρεπούσῃ δὲ ἐς τὰ πλάτη καὶ τὰ πάχη τῶν μορίων[1] τεκμαιρόμενον χρῆσθαι. ἑλίξιος κεφαλαὶ σκληραί,[2] ὁμαλαί, εὐκρινέες. τὰ δὲ δὴ μέλλοντα ἀποπίπτειν [καλῶς][3] ταχέως ἀποπεσόντων·[4] τὰ δὲ ὡς μήτε πιέξειν μήτε ἀπο-
8 πίπτειν τὰ εἰρημένα.

XI. Ὧν δὲ ἔχεται ἢ ἐπίδεσις ἢ ὑπόδεσις ἢ ἀμφότερα· ὑπόδεσις μὲν αἰτίη ὥστε ἢ ἀφεστεῶτα προστεῖλαι, ἢ ἐκπεπταμένα συστεῖλαι, ἢ συνεσταλμένα διαστεῖλαι, ἢ διεστραμμένα διορθῶσαι, ἢ τἀναντία. παρασκευὴ δέ· ὀθόνια κοῦφα, λεπτά, μαλθακά, καθαρά, πλατέα, μὴ ἔχοντα συρραφάς, μηδ' ἐξάστιας, καὶ ὑγιᾶ ὥστε τάνυσιν φέρειν καὶ ὀλίγῳ κρέσσω, μὴ ξηρά, ἀλλ' ἔγχυμα χυμῷ ᾧ ἕκαστα σύντροφα. ἀφεστεῶτα μὲν[5] ὥστε τὰ
10 μετέωρα τῆς ἕδρης ψαύειν μέν, πιέζειν δὲ μή· ἄρχεσθαι[6] δὲ ἐκ τοῦ ὑγιέος, τελευτᾶν δὲ πρὸς τὸ ἕλκος, ὥστε τὸ μὲν ὑπεὸν ἐξαθέλγηται, ἕτερον δὲ μὴ ἐπισυλλέγηται· ἐπιδεῖν τὰ μὲν ὀρθὰ ἐς ὀρθόν, τὰ δὲ λοξὰ λόξως, ἐν σχήματι ἀπόνῳ, ἐν ᾧ μήτε ἀπόσφιγξις μήτε ἀπόστασις ἔσται [τις][7] ἐξ οὗ ὅταν μεταλλάσσῃ, ἢ ἐς ἀνάληψιν ἢ ἐς θέσιν, μὴ μεταλλάξουσιν, ἀλλ' ὅμοια ταῦτα[8] ἕξουσι μύες, φλέβες, νεῦρα, ὀστέα [ᾗ

[1] ὀθονίων.

[2] σκληραί puzzled Galen. Ermerins inserts a negative, μή. The edges of a bandage should not be hard.

[3] κακίω Kw. codd. καλῶς Erm. Pq.

[4] A much discussed passage. G. says ἀποπεσόντων is a solecism, either as imperative or participle.

[5] Add προστεῖλαι.

X. Bandages, clean, light, soft, thin. Practise the rolling with both hands at once, and with each separately. Use one of suitable size, estimating by the thickness and breadth of the parts. Edges of the roll firm, not frayed, without creases. When things are really going to fall off, it is well that they do so quickly (?). Modes of bandaging such as neither compress nor fall off are those mentioned.

XI. What bandaging, whether upper or under or both, aims at. The function of an under bandage is to bring together what is separated, reduce everted wounds, separate what is adherent, adjust what is distorted, or the reverse.[1] Apparatus. Linen bandages light, thin, soft, clean, broad, without sutures or projections, sound so as to bear the tension required, and a little stronger; not dry, but soaked in a liquid suited to each case. Close a sinus[2] so that the upper parts touch the base without pressing on it, begin bandaging from the sound part and end at the open wound, so that while the contents are pressed out no more is accumulated. Bandage vertical ones[3] in a vertical direction and the oblique obliquely, in a position causing no pain, without either compression or laxity, so that when the change is made to a sling or fixation the muscles, vessels, ligaments and bones will retain their normal

[1] G. refers this to bad bandaging.

[2] A sinus is a superficial abscess which has opened and continues to discharge.

[3] G. refers this to the sinus, not to affected parts generally.

---

[6] ἦρχθαι Galen Kw. [7] Omit Galen Vulg. Kw.
[8] ὁμοιότατα Kw.

μάλιστα εὔθετα καὶ εὔσχετα].[1] ἀναλελάφθαι[2]
20 δὲ ἢ κεῖσθαι ἐν σχήματι ἀπόνῳ τῷ κατὰ φύσιν· ὧν δὲ ἂν [μὴ][3] ἀποστῇ, τἀναντία· ὧν δὲ ἐκπεπταμένα συστεῖλαι, τὰ μὲν ἄλλα τὰ αὐτά, ἐκ πολλοῦ δέ τινος δεῖ τὴν συναγωγήν, καὶ ἐκ προσαγωγῆς τὴν πίεξιν, τὸ πρῶτον ἥκιστα, ἔπειτα ἐπὶ μᾶλλον, ὅριον τοῦ μάλιστα τὸ συμψαύειν. ὧν δὲ συνεσταλμένα διαστεῖλαι, σὺν μὲν φλεγμονῇ, τἀναντία. ἄνευ δὲ ταύτης, παρασκευῇ μὲν τῇ αὐτῇ, ἐπιδέσει δὲ ἐναντίῃ. διεστραμμένα δὲ διορθῶσαι, τὰ μὲν ἄλλα κατὰ ταὐτά· δεῖ δὲ τὰ
30 μὲν ἀπεληλυθότα ἐπάγειν [τὰ δὲ ἐπεληλυθότα ἀπάγειν],[4] ἐπιδέσει, παρακολλήσει, ἀναλήψει,
32 [θέσει]·[4] τὰ δὲ ἐναντία, ἐναντίως.

XII. [Κατήγμασι δὲ] σπληνῶν μήκεα, πλάτεα, πάχεα, πλήθεα. μῆκος ὅση ἡ ἐπίδεσις· πλάτος, τριῶν ἢ τεσσάρων δακτύλων· πάχος, τριπτύχους ἢ τετραπτύχους·[5] πλῆθος, κυκλεῦντας μὴ ὑπερβάλλειν, μηδὲ ἐλλείπειν· οἷσι δὲ ἐς διόρθωσιν, μῆκος κυκλεῦντα· πάχος καὶ πλάτος τῇ ἐνδείῃ τεκμαίρεσθαι, μὴ ἀθρόα πληροῦντα.

Τῶν δὲ ὀθονίων ὑποδεσμίδες εἰσὶ δύο· τῇ πρώτῃ ἐκ τοῦ σίνεος ἐς τὸ ἄνω τελευτώσῃ·[6] τῇ
10 δὲ δευτέρῃ ἐκ τοῦ σίνεος ἐς τὸ κάτω, ἐκ τοῦ κάτω

[1] Read by Galen; not in the codd. [2] ἀναλελάμφθαι.
[3] μὴ Kw.; suggested by Galen's predecessors.
[4] Omit BV. [5] τρίπτυχα τετράπτυχα.
[6] ῃ . . . τελευτῶσα Erm Reinhold. Pq. suggests τελευτῶσι, as Ald.

[1] Restored from Galen's Commentary.
[2] G. gives three other interpretations, without the negative.

positions [in which they are best put up and supported].[1] Let the part be slung or put up in a natural comfortable position. Where there is no open sinus the reverse.[2] Where there is a gaping wound bring the parts together just as in other cases, but start the joining up at a good distance; and graduate the pressure, first very little, then increasing, the extreme limit being contact of the parts. In separating what is adherent, if there is inflammation the reverse holds good,[3] if not use the same apparatus, but bandage in the opposite way. To adjust what is distorted act generally on the same principles; what is turned out must be brought in [and what is turned in brought out] by bandaging, agglutination,[4] suspension, setting—the reverse reversely.

XII. In fractures, the length, breadth, thickness and number of compresses. Length to correspond with the bandaging, breadth, three or four fingers, thickness, folded thrice or four times. Number, sufficient to go round without overlapping or vacancy: when required to adjust the shape,[5] long enough to go round, estimating breadth and thickness by the deficiency, but not filling it up with one compress.

Of the linen bandages, the under ones[6] are two in number. Start with the first from the lesion and end upwards, but carry the second downwards from

[3] *i.e.* avoid bandaging as far as possible; Galen.

[4] Refers to turned in eyelashes.

[5] *i.e.* in conical or irregular parts: not "deformity" as Adams.

[6] This Hippocratic division of under and upper bandages did not survive. ὑποδεσμίδες remains a peculiar Hippocratic word for bandages below the pads or compresses. XVIII(2). 785 Galen.

ἐς τὸ ἄνω τελευτώσῃ τὰ κατὰ τὸ σίνος πιέζειν μάλιστα, ἥκιστα τὰ ἄκρα, τὰ δὲ ἄλλα κατὰ λόγον. ἡ δὲ ἐπίδεσις πολὺ τοῦ ὑγιέος προσλαμβανέτω.

Ἐπιδέσμων δὲ πλῆθος, μῆκος, πλάτος· πλῆθος μὲν μὴ ἡσσᾶσθαι τοῦ σίνεος, μηδὲ νάρθηξιν ἐνέρεισιν εἶναι, μηδὲ ἄχθος, μηδὲ περίρρεψιν, μηδὲ ἐκθήλυνσιν· μῆκος δὲ καὶ πλάτος, τριῶν ἢ τεσσάρων ἢ πέντε ἢ ἓξ πήχεων μὲν μῆκος, δακ-
20 τύλων δὲ πλάτος. καὶ παραιρήματος περιβολαὶ τοσαῦται ὥστε μὴ πιέζειν· μαλθακὰ δέ, μὴ παχέα· ταῦτα πάντα ὡς ἐπὶ μήκει καὶ πλάτει καὶ πάχει τοῦ παθόντος.

Νάρθηκες δὲ λεῖοι, ὁμαλοί, σιμοὶ κατ' ἄκρα, σμικρῷ μείους ἔνθεν καὶ ἔνθεν τῆς ἐπιδέσιος, παχύτατοι δὲ ᾗ ἐξήριπε τὸ κάτηγμα. ὁπόσα δὲ κυρτὰ καὶ ἄσαρκα φύσει, φυλασσόμενον τῶν ὑπερεχόντων, οἷον τὰ κατὰ δακτύλους ἢ σφυρά, ἢ τῇ θέσει ἢ τῇ βραχύτητι. παραιρήμασι δὲ
30 ἁρμόζειν, μὴ πιέζειν· τὸ πρῶτον κηρωτῇ μαλθακῇ
31 καὶ λείῃ καὶ καθαρῇ ἑλισσέτω.

XIII. Ὕδατος θερμότης, πλῆθος· θερμότης μὲν κατὰ τῆς ἑωυτοῦ χειρὸς καταχεῖν, πλῆθος δὲ χαλάσαι μὲν καὶ ἰσχνῆναι τὸ πλεῖστον ἄριστον, σαρκῶσαι δὲ καὶ ἁπαλῦναι τὸ μέτριον· μέτρον δὲ τῆς καταχύσιος, ἔτι μετεωριζόμενον δεῖ, πρὶν συμπίπτειν, παύεσθαι· τὸ μὲν γὰρ πρῶτον
7 ἀείρεται, ἔπειτα δὲ ἰσχναίνεται.

XIV. Θέσις δὲ μαλθακή, ὁμαλή, ἀνάρροπος τοῖσι ἐξέχουσι τοῦ σώματος, οἷον πτέρνῃ καὶ

---

[1] Or "where the fracture occurred."

the lesion, bringing it up again to end at the top. Make most pressure over the lesion and least at the ends, the rest in proportion. Let the bandaging include a good deal of the sound part.

Amount, length and breadth of the bandages. Amount sufficient to deal with the lesion, without either pressing in the splints, or being burdensome, or slipping round, or causing weakness. As to length and breadth, three, four, five or six cubits for length, fingers for breadth. The supporting bands in such a number of coils as not to compress, soft and not thick. All these suited to the length, breadth and thickness of the part affected.

Splints, smooth, even, tapering at the ends, a little shorter in each direction than the bandaging; thickest over the prominence at the fracture;[1] avoiding either by position or shortening the convexities naturally uncovered by flesh, such as on the fingers and ankles. Fit them on by supporting bands without pressure. Let the first dressing be made with bandages rolled in soft, smooth and clean cerate.[2]

XIII. Of water (one must consider) temperature, quantity. Temperature by pouring it over one's own hand. Quantity, for relaxation and attenuation the more the better, but for flesh forming and softening observe moderation, and for moderate douching one should stop while the part is still swollen up before it collapses, for first it swells and then becomes attenuated.

XIV. Permanent position: soft, smooth, sloping up for projecting parts as with the heel or hip, so

[2] So Galen, for cerate see Introduction. Pq. "before bandaging anoint the skin with."

ἰσχίῳ, ὡς μήτε ἀνακλᾶται [μήτε ἀποκλᾶται][1] μήτε ἐκτρέπηται,[2] σωλῆνα παντὶ τῷ σκέλει ἢ ἡμίσει· ἐς τὸ πάθος δὲ βλέπειν καὶ τὰ ἄλλα
6 ὁκόσα βλάπτει δῆλα.[3]

XV. Πάρεξις γάρ,[4] καὶ διάτασις, καὶ ἀνάπλασις, καὶ τὰ ἄλλα κατὰ φύσιν. φύσις δὲ ἐν μὲν ἔργοις, τοῦ ἔργου τῇ πρήξει, ὃ βούλεται τεκμαρτέον· ἐς δὲ ταῦτα, ἐκ τοῦ ἐλινύοντος, ἐκ τοῦ κοινοῦ, ἐκ τοῦ ἔθεος· ἐκ μὲν τοῦ ἐλινύοντος καὶ ἀφειμένου τὰς ἰθυωρίας σκέπτεσθαι, οἷον τὸ τῆς χειρός· ἐκ δὲ τοῦ κοινοῦ, ἔκτασιν, σύγκαμψιν, οἷον τὸ ἐγγὺς τοῦ ἐγγωνίου πήχεος πρὸς βραχίονα· ἐκ τοῦ ἔθεος, ὅτι οὐκ ἄλλα σχήματα
10 φέρειν δυνατώτερα. οἷον σκέλεα ἔκτασιν· ἀπὸ τούτων γὰρ ῥήϊστα πλεῖστον χρόνον ἔχοι ἂν μὴ μεταλλάσσοντα. ἐν δὲ τῇ μεταλλαγῇ ἐκ διατάσιος ὁμοιότατα ἔχουσιν[5] ἐς ἕξιν ἢ θέσιν μύες, φλέβες, νεῦρα, ὀστέα, ᾗ μάλιστα εὔθετα καὶ
15 εὔσχετα.

XVI. Διάτασις, μάλιστα τὰ μέγιστα καὶ πάχιστα, καὶ ὅπου ἀμφότερα· δεύτερα, ὧν τὸ ὑποτεταγμένον, ἥκιστα ὧν τὸ ἄνω· μᾶλλον δὲ τοῦ μετρίου βλάβη, πλὴν παιδίων· ἔχειν ἀνάντη σμικρόν· διορθώσιος παράδειγμα, τὸ ὁμώνυμον, τὸ
6 ὁμόζυγον, τὸ ὅμοιον, τὸ ὑγιές.

[1] Galen omits.

[2] ἐκτρέπεται vulg. Galen; ἐκτρίβηται Pq. The things to be feared are distortion or abrasion which would be ἐκτρίβηται; ἀποκλᾶται, which implies fracture, seems hardly possible.

ἡμίσει—Galen says ἢ is negative (ἀντ' ἀποφάσεως) as in *Iliad* I. 117, but we discover this only by reference to *Fractures* XXII.

[3] δηλαδή. [4] δέ.

as neither to be bent back [bent aside? broken off?] or distorted. Apply a hollow splint to the whole leg rather than to half. Consider the affection and also the obvious disadvantages (of this splint).

XV. Presentation, extension, setting, and the rest, according to nature. Now nature shows itself in actions, and one must judge what nature wants [1] by the performance of action: for the above matters (judge) from the state of rest, from what is normal, from the customary. From rest and relaxation estimate proper direction, for example as regards the arm: from what is normal judge extension and flexion, such as the nearly rectangular relation of the forearm to the arm; from habit infer the posture more easy to maintain than any other, such as extension in the case of the legs; for one would most easily keep such postures for the longest time without changing, and in the change after [surgical] extension the muscles, vessels, tendons and bones have the most similar relations as to habit and posture, and are thus most conveniently put up or slung.

XVI. Extension, most when the largest and thickest and when both bones [of the arm] are broken. Next in cases where it is the underneath one [ulna], least where it is the upper. Excessive tension does damage except in children.[2] Keep the limb a little raised. As model for adjustment take the homonymous,[3] corresponding, similar, sound limb.

[1] Littré-Adams "what we want."

[2] Because their tendons are more elastic, G.; but it may be a confused reference to the case in *Fract.* IV.

[3] G. says it should be "synonymous."

---

[5] ὁμοιότατα ἔχουσιν Kw. ὅμοια ταῦτα ἕξουσι Pq., as in XI.

XVII. Ἀνάτριψις δύναται λῦσαι, δῆσαι, σαρκῶσαι, μινυθῆσαι· ἡ σκληρὴ δῆσαι· ἡ μαλακὴ
3 λῦσαι· ἡ πολλὴ μινυθῆσαι· ἡ μετρίη παχῦναι.

XVIII. Ἐπιδεῖν δὲ τὸ πρῶτον· ὁ ἐπιδεδεμένος μάλιστα φάτω πεπιέχθαι κατὰ τὸ σίνος· ἥκιστα τὰ ἄκρα· ἡρμόσθαι[1] δέ, μὴ πεπιέχθαι· πλήθει, μὴ ἰσχύι· τὴν δὲ ἡμέρην ταύτην καὶ νύκτα, ὀλίγῳ μᾶλλον, τὴν δὲ ὑστέρην, ἧσσον· τρίτῃ, χαλαρά. εὑρεθήτω δὲ τῇ μὲν ὑστεραίῃ ἐν ἄκροισιν οἴδημα μαλθακόν. τῇ τρίτῃ δὲ τὸ ἐπιδεθὲν λυθέν, ἰσχνότερον, παρὰ πάσας τὰς ἐπιδέσιας τοῦτο. τῇ δὲ ὑστεραίῃ ἐπιδέσει, ἢν δικαίως ἐπιδεδεμένον
10 φανῇ, μαθεῖν δεῖ· ἐντεῦθεν δὲ μᾶλλον καὶ ἐπὶ πλέοσι πιεχθήτω· τῇ δὲ τρίτῃ ἐπὶ μᾶλλον καὶ ἐπὶ πλέοσιν. τῇ δὲ ἑβδόμῃ ἀπὸ τῆς πρώτης ἐπιδέσιος λυθέντα εὑρεθήτω ἰσχνά, χαλαρὰ τὰ ὀστέα. ἐς δὲ νάρθηκας δεθέντα, ἢν ἰσχνὰ καὶ ἄκνησμα καὶ ἀνέλκεα ᾖ, ἐὰν μέχρις εἴκοσιν ἡμερέων ἀπὸ τοῦ σίνεος· ἢν δέ τι ὑποπτεύηται, λῦσαι
17 ἐν τῷ μέσῳ· νάρθηκας διὰ τρίτης ἐρείδειν.

XIX. Ἡ ἀνάληψις, ἡ θέσις, ἡ ἐπίδεσις, ὡς ἐν τῷ αὐτῷ σχήματι διαφυλάσσειν. κεφάλαια σχημάτων, ἔθεα, φύσιες ἑκάστου τῶν μελέων· τὰ δὲ εἴδεα, ἐκ τοῦ τρέχειν, ὁδοιπορέειν, ἑστάναι, κατα-
5 κεῖσθαι, ἐκ τοῦ ἔργου, ἐκ τοῦ ἀφεῖσθαι.

XX. Ὅτι[2] χρῆσις κρατύνει, ἀργίη δὲ τήκει.

XXI. Ἡ πίεξις πλήθει, μὴ[3] ἰσχύι.

[1] ἡρμᾶσθαι. [2] τὸ δέ, ὅτι; Kw. [3] ἤ.

---

[1] Cf. *Fract.* VI. [2] *i.e.* on alternate days.
[3] G. considers XIX. a marginal note to XV.
[4] Cf. *Joints* LVIII.

XVII. Friction can produce relaxation, constriction, increase of flesh, attenuation. Hard friction constricts, soft relaxes: if long continued it attenuates, when moderate it increases flesh.

XVIII. As to the first bandaging: the patient should say there is pressure chiefly over the injury, least at the ends: that the dressing fits firmly but without compression: pressure should be got by amount of bandaging not by tension. During this day and night pressure should increase a little, but be less during the next day, and lax on the third. A soft swelling should be found on the second day at the extremities. On the third the part when unbandaged should be less swollen, and so with every dressing. At the second dressing one must find out whether it seems properly done, and then use more bandages and greater pressure; at the third still more with more coils of bandage. On the seventh day[1] after the first dressing the parts when set free should be found without swelling and the bones mobile. When put up in splints, if the parts are not swollen and are free from itching or wound, leave alone till twenty days after the injury; but if there is any suspicion remove in the interval. Make the splints firm every third day.[2]

XIX. In suspension, putting up, bandaging, take care that the part keeps the same attitude, the general principle being the habitual natural position of each limb. The kinds of attitude are derived from running, walking, standing, lying, work, relaxation.[3]

XX. (Remember) that use strengthens, disuse debilitates.[4]

XXI. The pressure by quantity (of bandages) not by force.

XXII. Ὁπόσα δὲ ἐκχυμώματα, ἢ φλάσματα, ἢ σπάσματα, ἢ οἰδήματα ἀφλέγμαντα, ἐξαρύεται αἷμα ἐκ τοῦ τρώματος, ἐς μὲν τὸ ἄνω τοῦ σώματος τὸ πλεῖστον, βραχὺ δέ τι καὶ ἐς τὸ κάτω· μὴ κατάντη τὴν χεῖρα ἔχοντα ἢ τὸ σκέλος· τιθέμενον τὴν ἀρχὴν κατὰ τὸ τρῶμα καὶ μάλιστα ἐρείδοντα, ἥκιστα τὰ ἄκρα, μέσως τὰ διὰ μέσου. τὸ ἔσχατον πρὸς τὸ ἄνω τοῦ σώματος νεμόμενον· ἐπιδέσει, πιέξει· ἄταρ καὶ ταῦτα πλήθει μᾶλλον 10 ἢ ἰσχύι. μάλιστα δὲ τούτοισιν ὀθόνια, λεπτά, κοῦφα, μαλθακά, καθαρά, πλατέα, ὑγιᾶ, ὡς ἂν 12 ἄνευ ναρθήκων· καὶ καταχύσει χρῆσθαι πλέονι.

XXIII. Τὰ δὲ ἐκπτώματα, ἢ στρέμματα, ἢ διαστήματα, ἢ ἀποσπάσματα, ἢ ἀποκλάσματα, ἢ διαστρέμματα, οἷα τὰ κυλλά, τὰ ἑτερόρροπα, ὅθεν[1] μὲν ἐξέστη, συνδιδόντα, ὅπῃ δέ, συντείνοντα, ὡς ἐς τἀναντία ῥέπῃ, ἐπιδεθέντα ἢ πρὶν ἐπιδεθῆναι, σμικρῷ μᾶλλον ἢ ὥστε ἐξ ἴσου εἶναι· καὶ τοῖσιν ἐπιδέσμοισι, καὶ τοῖσι σπλήνεσι, καὶ τοῖσιν ἀναλήμμασι, καὶ τοῖσι σχήμασι, κατατάσει, ἀνατρίψει, διορθώσει, [ταῦτα καὶ][2] κατα-10 χύσει πλέονι.

XXIV. Τὰ δὲ μινυθήματα, πολὺ προσλαμβάνοντα τοῦ ὑγιέος, ἐπιδεῖν ὡς ἂν ἐξ ἐπιδρομῆς τὰ συντακέντα πλέον ἢ αὐτὰ[3] ἐμινύθει, ἀλλοίῃ τῇ ἐπιδέσει παραλλάξαντα, ἐκκλίνει[4] ἐς τὴν αὔξησιν καὶ τὴν ἀνάπλασιν τῶν σαρκῶν ποιήσηται. βέλτιον δὲ καὶ τὰ ἄνωθεν, οἷον κνήμης καὶ τὸν μηρόν, καὶ τὸ ἕτερον σκέλος τῷ ὑγιεῖ[5] συνεπιδεῖν,

[1] ἔνθεν. [2] Omit Galen. Kw. [3] αὐτόματα.
[4] ἐκκλίνῃ. [5] τὸ ὑγιές.

[1] Includes club foot, knock knee, bandy leg.

XXII. In case of bruisings, crushings, ruptures of muscles or swellings without inflammation, blood is expressed from the injured part [by bandaging] mostly upwards, but some little downwards. This is done (with neither arm nor leg in a pendent position) by beginning the bandage at the wound and making most pressure there, least at the ends and moderate in between; the final turns being brought upwards. By bandaging, by compression—but here, too, pressure must be got by quantity of bandage rather than by force. In these cases especially, the linen bandages should be thin, light, soft, clean, broad and sound, as one would use without splints; use also copious douching.

XXIII. [Bandaging as regards] dislocations, sprains, separations, avulsions, fractures near joints or distortions, such as deformities to either side:[1] yielding on the side from which it deviates, bracing up on the side towards which it deviates, so that when it is put up, or before it is put up, it is not straight but has a slight inclination the opposite way. The treatment includes use of bandages, compresses, suspension, postures, extension, friction, adjustment; and in addition copious douching.

XXIV. [Bandaging as regards] atrophied parts: Apply the bandage, taking in a good deal of the sound parts in a way that the wasted tissues may gain more by afflux than they lose spontaneously; by changing to a different mode of bandaging[2] it may divert (the tissues) towards growth and bring about flesh formation. It is a rather good plan to bandage the upper parts also, such as the top of the leg and the thigh, also the sound leg that it may be

[2] From that described in XXII. A very obscure passage.

ὡς ὁμοιότερον ᾖ καὶ ὁμοίως ἐλινύῃ, καὶ ὁμοίως τῆς τροφῆς ἀποκλείηται καὶ δέχηται. ὀθονίων
10 πλήθει, μὴ πιέξει· ἀνιέντα πρῶτον τὸ μάλιστα δεόμενον, καὶ ἀνατρίψει χρώμενον σαρκούσῃ καὶ
12 καταχύσει· ἄνευ ναρθήκων.

XXV. Τὰ δὲ ἑρμάσματα καὶ ἀποστηρίγματα, οἷον στήθει, πλευρῇσι, κεφαλῇ, καὶ τοῖσιν ἄλλοισιν, ὅσα τοιαῦτα· τὰ μὲν σφυγμῶν ἕνεκεν, ὡς μὴ ἐνσείηται· τὰ δὲ καὶ τῶν διαστασίων τῶν κατὰ τὰς ἁρμονίας ἐν τοῖσι [τῶν] κατὰ τὴν κεφαλὴν ὀστέων[1] ἐρεισμάτων χάριν· ἐπί τε βηχῶν ἢ πταρμῶν, ἢ ἄλλης κινήσιος, οἷον[2] κατὰ θώρηκα καὶ κεφαλὴν ἀποστηρίγματα γίγνεται. τούτων ἁπάντων αἱ αὐταὶ συμμετρίαι τῆς ἐπιδέ-
10 σιος· ᾗ μὲν γὰρ τὰ σίνη μάλιστα πεπιέχθαι· ὑποτιθέναι οὖν [εἴριον][3] μαλθακὸν ἅρμοζον τῷ πάθει· ἐπιδεῖν δὲ μὴ μᾶλλον πιεζεῦντα ἢ ὥστε τοὺς σφυγμοὺς μὴ ἐνσείειν, μηδὲ μᾶλλον ἢ ὥστε τῶν διεστηκότων τὰ ἔσχατα τῶν ἁρμονίων συμψαύειν ἀλλήλων, μηδὲ τὰς βῆχας καὶ τοὺς πταρμοὺς ὥστε κωλύειν, ἀλλ' ὥστε ἀποστήριγμα
17 εἶναι ὡς μήτε διαναγκάζηται, μήτε ἐνσείηται.

[1] ὀστέοις omit τῶν. [2] οἷα τά.
[3] Littré and Pq. omit and add τι after μαλθακόν.

in a like state, and share alike in rest and the deprivation or reception of nutriment. Use plenty of bandages, not compression; relaxing first where it is most needed, using friction of the flesh-forming kind and douching—no splints.

XXV. Supports attached or separate,[1] such as those for chest, ribs, head and other such parts; sometimes used because of pulsations[2] that the part may not be shaken; at other times, in cases of separation of the commissures in the bones of the head, as supports: also in case of coughings, sneezings and other movements they serve as separate supports (cushions?) for the chest and head. The suitable modes of bandaging in all these cases are the same, for where the lesion is there should be the chief pressure. Put something[3] soft underneath suited to the affection. Do not make the bandaging tighter than suffices to prevent the pulsations from shaking the part, or than is necessary to bring the edges of the separated commissures into touch with one another; nor is it intended to prevent coughings and sneezings,[4] but to act as a support for the avoidance both of forcible separation and shaking.

[1] So Galen, who says the words are usually synonymous.

[2] Includes everything from twitchings to respiratory movements. G.

[3] Reading μαλθακόν τι.

[4] The text seems corrupt, but it can hardly mean "so tight as to prevent sneezing"!

FRACTURES, JOINTS, MOCHLICON

INTRODUCTION

There is no question as to the relationship of these three treatises. *Fractures* and *Joints* probably once formed a single work, and are certainly by the same author,[1] while *Mochlicon* is composed of an abbreviation of those parts of them which treat of dislocations. In antiquity no one doubted that *Fractures* and *Joints* were by the great Hippocrates, except a few who attributed them to another man of the same name, his grandfather, the son of Gnosidicus.[2] Galen, in all his lists, classes them first, or nearly first, among the γνησιώτατα[3] or "most genuine" works. Of the two things we know for certain about the teaching of Hippocrates, Plato's statement that he held it impossible to understand the body without studying nature as a whole has proved too vague to be attached to any particular treatise, but the condemnation by his kinsman Ctesias of his reduction of the hip-joint (unless it refers to verbal teaching or to some work which has vanished) must apply, as Galen says,[4] to *Joints*, where the subject is treated in detail.

[1] This seems sufficiently proved by the fact that references are made from *Joints* to *Fractures* in exactly the same terms as to the earlier parts of *Joints*: *e.g.* J LXVII, LXXII, ὡς καὶ πρόσθεν εἴρηται. εἴρηται [εἴρηκα B. Apoll.] καὶ πρόσθεν, which refer to F XXXI and XIII respectively. Reference to another treatise is put differently: *e.g.* ἐν ἑτέρῳ λόγῳ J XLV.

[2] Galen, XV. 456. [3] XVII(1). 577. [4] XVIII(1). 731.

The work was known to, and in part paraphrased by, Diocles,[1] who was probably adult before Hippocrates died, and there is no record that he doubted its authorship. We may therefore, perhaps, conclude that nothing in the *Corpus* has a better claim to be by Hippocrates himself than *Fractures–Joints,* and proceed to discuss them in some detail.

The question asked in antiquity was: Why does *Fractures* contain a good deal about dislocations (joints) while *Joints* has some sections on fractures? To which Galen replies that Hippocrates cared less for words than for things, and fractures and dislocations often come together. This answer is not quite satisfactory, for the weak point of the work is precisely the absence of any clear account of fracture-dislocations: besides, it seems probable to most careful readers that the result is mainly due to a work on fractures and dislocations having been broken up and put together again in disorder.

We may perhaps indicate this most clearly and briefly by taking *Mochlicon,* in which a natural order is preserved, as our guide, showing at the same time its relationship to the older treatise, or treatises. The order of *Mochlicon* is face, upper and lower limbs from above downwards, spine and ribs, though, like other Hippocratic works, it ends in a confused mass of rough notes.

M II–III, nose and ear, are derived from J XXXV–XL. M IV, lower jaw, from J XXX–XXXI. M V epitomizes in one chapter the remarkable account of shoulder dislocations, J I–XII. M VI is from J XIII, on dislocation of the outer end of the collar-bone considered as avulsion of the acromion.

[1] Apollonius, 13; Galen, XVIII(1). 519. Cf. Littré I. 334.

We are surprised to find that M VII–XIX are not an epitome but a verbal repetition of J XVII–XXIX. They are derived mainly (VII–XV) from F XXXVIII–XLVII, on the elbow; XVI–XVIII, on the wrist, have no extant original, and XIX, on the fingers, does not appear to be an abridgment of the long account in J LXXX.

There seems no reasonable doubt, from the nature of the case, the style of the writing and peculiarities of language, that the epitome was made by the author of *Mochlicon* and afterwards transferred to *Joints* to fill up a vacancy. A reader of the latter observes a sudden change of style, the appearance of new words (ἐξαίφνης for ἐξαπίνης) and a whole string of depraved infinitives;[1] but the section is in perfect harmony with the rest of *Mochlicon*.

M XX–XXIV abbreviate the very full account of thigh dislocations in J LI–LX, while the directions for reduction, given at length in J LXX–LXXVIII, are condensed into M XXV.

M XXVI–XXXI on knee, ankle and foot repeat the phenomenon of VII–XIX. They correspond verbally with J LXXXII–LXXXVII and are epitomized from *Fractures* X–XIV—except XXVI, on the knee, which is, in part, from F XXXVII. We shall find that J LXXXII–LXXXVII form part of an appendix to the original treatise.

M XXXII condenses the account of club foot given in J LXII.

M XXXIII–XXXV deal with compound disloca-

[1] We may note that, according to our text, M XII has the more normal nominatives which have become accusatives on transference to J XXII.

tions, loss or amputation of parts, gangrene and necrosis. They are derived from J LXIII–LXIX.

M XXXVI feebly represents the long account of spinal curvature in J XLI–XLVI, also fracture and contusion of the ribs, J XLIX.

In XXXVII M begins to go to pieces. It is based partly on J XLI, partly on J L, and the rest of the treatise is a mass of confused notes on dislocations and fractures, often hardly intelligible, but obviously all taken from *Fractures–Joints*. Imbedded in it is a paragraph (XXXIX) on disease of the palate corresponding almost verbally with passages in *Epidemics* II, IV, and VI; and interesting as showing that *Mochlicon*, like *Surgery*, has some connection with the middle division of this series.

Fractures and *Joints* may now be summarized briefly. About one-fourth of *Fractures* deals with dislocations. The first seven chapters treat fracture of the forearm in detail as a typical case. Chapter VIII fracture of the upper arm: IX–XXIII dislocations of the foot and ankle, and fractures of the lower limb. We are surprised to be told in chapter IX that dislocation of the wrist has already been mentioned. The remainder is devoted partly (XXIV–XXXVII) to compound fractures, and partly (XXXVIII–XLVIII) to dislocations of the elbow, with a few words on dislocation of the knee (XXXVIII) and fracture of the olecranon.

Joints begins similarly with a sample case, dislocation of the shoulder-joint, described in great detail (I–XII). Then comes fracture of the collar-bone and its dislocation (XIII–XVI). Next (XVII–XXIX) is the interpolation from *Mochlicon*, on elbow, wrist, and finger-joints. Injuries of the jaw, nose

and ear (XXX–XL) are given great attention, doubtless owing to the vigorous boxing methods then in use. XL–L treat of the spine and ribs in detail, and show much anatomical knowledge. LI–LXI include the celebrated account of dislocation of the hip and its results, and LXII has the excellent description of club foot. In LXIII–LXIX we are diverted to the consideration of compound dislocations, amputation, necrosis and gangrene, and finally return to the hip-joint and its reduction in LXXI–LXXVIII.

According to Galen, chapter LXXVIII is the last, and his commentary ends here. So does that of Apollonius, except for some rough notes, most of which occur at the end of our *Mochlicon*.

This view is confirmed by the nature of chapter LXXIX, which is a brief introduction to the study of dislocations, and would come more appropriately at the beginning.

Chapter LXXX looks like the original account of finger-joint dislocation; but was unknown to Apollonius, who says (on chapter XXIX) that Hippocrates made only a few remarks on the subject owing to its simplicity, and proceeds to supplement them by an extract from *Diocles*, which seems almost certainly based upon LXXX, and to form part of the "paraphrase" mentioned by Galen. We may perhaps conjecture that chapter LXXX was lost and discovered again after its place had been occupied. The rest of the appendix is an epitome of knee, foot and ankle lesions supplied from *Mochlicon*, the originals having somehow got into *Fractures*.

The answer to the question of antiquity is, then,

that the great work on Fractures and Dislocations got into disorder soon after it was written, and that parts were lost, either temporarily (as J LXXX) or permanently, as with the original account of the wrist. The excellences of its *disjecta membra* speak for themselves, and have been recognized by all surgeons ancient and modern. An editor has the less agreeable task of dealing with defects and difficulties.

Many questions which occur to a modern reader are unlikely to receive satisfactory answers. Why does Hippocrates say that the fibula is longer than the tibia and projects above it [1] (apparently because he saw and exaggerated its analogy with the ulna) and that twenty days are "very many" for consolidation of a broken collar-bone, whereas we allow three to six weeks? [2] Why does he assert with emphasis that inward dislocation of the thigh-bone is much the most frequent,[3] and all antiquity (together with Ambrose Paré) [4] agree with him, whereas all modern evidence is to the contrary? Why does he ignore injuries of the knee-cap, and the use of that ancient instrument the safety-pin? These problems and other statements which will surprise the surgeon, such as the cure of hump back by varicose veins and the frequency of dislocation of the knee, must

[1] *Fractures*, XII, XXXVII.
[2] *Joints*, XIV. [3] *Joints*, LI.
[4] So Adams (558). In his chapter on hip dislocation (XVI. 38) Paré says "le plus souvent en dehors et en dedans, en devant et en derrière rarement." He may have held the modern view (*dehors* comes first) but have been unwilling to contradict such authorities as Hippocrates, Celsus and Galen. Possibly some grip in ancient wrestling made the internal form then more frequent.

remain unsolved. Two subjects, however, require further consideration: the accounts of elbow and ankle dislocations. The former is treated by most editors at some length, and it is generally admitted that the latest and longest discussion (that of Petrequin) throws light on the subject. He points out that some difficulties are removed by supposing the Hippocratic attitude of the arm to be that with the bend of the elbow turned inwards, not forwards, and since Hippocrates speaks of dislocation of the humerus or upper arm (the convex from the concave), whereas we speak of dislocation of the forearm, a double correction is necessary, his inwards and outwards becoming our backwards and forwards respectively. Similarly, with lateral dislocation, the Hippocratic forwards and backwards become our inwards and outwards. This seems the best that can be done, though it brings the two surgical editors, Petrequin and Adams, into violent contradiction on some points.

The second puzzle is why—though Herodotus knows exactly what happened to the astragalus of Darius when he sprained his ankle—does Hippocrates never mention the bone, and give us a very obscure account of ankle dislocation? In part, doubtless, it is the layman rushing in where the specialist fears to tread; but the existence of a duplicate epitome of each of these subjects will enable us to discuss them further in the text.

Soranus tells us that the father of rhetoric, Gorgias, was one of the teachers of the father of medicine, and so long as such works as *The Art* and *Breaths* were considered genuine, they might have been adduced either as showing the result of this teach-

ing, or as possibly giving origin to such a legend. But the story may very well be correct, for Gorgias and Hippocrates were both in Thessaly about the same time, and the physician may have admired not only the fine constitution of the elder man, which was destined to prolong his life well beyond a century, but also his fine language, and have taken some lessons in composition. But if we look for traces of rhetoric in what are now considered possibly genuine works, we are surprised to find them most prominent in the great surgical treatises. *Fractures–Joints* abound, if not in purple patches, at least in purple spots, as if the writer was trying to make use of recently acquired knowledge of rhetorical forms. Attention was called to this by Diels, and it has been more fully worked out by Krömer. Some rhetorical forms show through even the worst translation, and the reader will easily discover at least twelve examples of the rhetorical query. Plays upon words are also frequent and obvious in the Greek, though difficult to repeat in English. Of special interest is the frequent occurrence of chiasmus and other forms of the evenly balanced sentence. A short sample of either may be found respectively in *Fractures*, XLVII : πολλῶν μὲν γὰρ ἂν κώλυμα εἴη, ὠφελίη δὲ ὀλίγων, and *Joints*, XLVI : ἀλλὰ καὶ οὕτως ἂν ἀποθάνοι, παραχρῆμα δὲ οὐκ ἀποθάνοι.

The latter, with the allied form of anaphora, or needless but ornate repetition of the same word (*e.g.* of ἄλλο in *Fractures*, II ; ἧσσον, *Joints*, XI) may remind readers of the less artistic repetitions common in *Wounds in the Head*, and suggest that in spite of diversity of style it may be by the same author. We notice also a similarity of doctrine,

especially the statement that contusions of bones are usually more serious than fractures, applied respectively to skull and ribs.

Too much weight may, perhaps, be given to this. Thus Littré (IV. 566) notes a resemblance between *Fractures*, XXXI, and *Diet in Acute Diseases*, VII. In both there is a disapproval, expressed in very similar language, of any marked interference, operative or dietetic respectively, during the third, fourth, or fifth days. He considers that the identity in sense and form of criticism, together with "the identity of the epoch," is enough to prove identity of authorship. He might have added that there is a number of curious terms common to *Diet in Acute Diseases* and *Fractures–Joints*: *e.g.* ἄγχιστα, in the sense of μάλιστα, and ἠδελφισμένος, ἄπαρτι, τὸ ἐπίπαν.[1] But there are differences which raise doubts. Thus the favourite drink of the author of *Fractures–Joints* is oxyglyphy (hydromel, prepared by boiling squeezed-out honey-combs).[2] *Diet in Acute Diseases* never mentions this, though it has much to say about the closely allied oxymel and melicrate, which are ignored in *Fractures—Joints*.

The most formidable opponent of the Hippocratic authorship was H. Diels, whose main contention is that ancient writers did not refute one another by name, nor mention those whom they copied. Therefore, probably, neither Ctesias nor Diocles named Hippocrates. That they refer to him is only Galen's assumption. Reasons to the contrary are adduced by Krömer, and seem equally potent.[3] The "paraphrase" of Diocles at least shows that the work was

[1] See Kühlewein *op. cit.* p. 6. [2] Galen, XVIII(2). 466
[3] *Op. cit.* p. 7.

well known early in the fourth century, which is sufficient to refute the second argument usually brought against its Hippocratic origin, that the writer knows too much anatomy, and in particular distinguishes clearly between arteries and veins. If we may trust Caelius Aurelianus, their distinction was known to Euryphon,[1] who was older than Hippocrates, while the writer's ability to give a good account of the shoulder-joint and spine, and promise of further details, is only what we should expect from what Galen says about the anatomical studies of the old Asclepiadae.[2]

Still, we must agree with Diels that this last attempt to demonstrate at least one genuine work of Hippocrates may be met by the ancient warning, *δοκὸς δ' ἐπὶ πᾶσι τέτυκται*, or rather that the whole sentence of Xenophanes may appropriately be applied to the Hippocratic problem, "Even if one hit upon the truth, he would not be sure he had done so, for guess-work is spread over all things."

[1] *T. P.* 2. 10. [2] *Anat. Adm.* 2. 1

ΠΕΡΙ ΑΓΜΩΝ

I. Ἐχρῆν τὸν ἰητρὸν τῶν ἐκπτωσίων τε καὶ καταγμάτων ὡς ἰθύτατα τὰς κατατάσιας ποιεῖσθαι· αὕτη γὰρ ἡ δικαιοτάτη φύσις. ἢν δέ τι ἐγκλίνῃ ἢ τῇ ἢ τῇ, ἐπὶ τὸ πρηνὲς ῥέπειν· ἐλάσσων γὰρ ἡ ἁμαρτὰς ἢ ἐπὶ τὸ ὕπτιον. οἱ μὲν οὖν μηδὲν προβουλεύονται οὐδὲν ἐξαμαρτάνουσιν ὡς ἐπὶ τὸ πολύ· αὐτὸς γὰρ ἐπιδησόμενος[1] τὴν χεῖρα ἀπορέγει οὕτως ὑπὸ τῆς δικαίης φύσιος ἀναγκαζόμενος· οἱ δὲ ἰητροὶ σοφιζόμενοι δῆθεν
10 ἔστιν ἄρα ἐφ' οἷς[2] ἁμαρτάνουσι. σπουδὴ μὲν οὖν οὐ πολλὴ χεῖρα κατεηγυῖαν χειρίσαι, καὶ παντὸς δὲ ἰητροῦ, ὡς ἔπος εἰπεῖν· ἀναγκάζομαι δὲ ἐγὼ πλείω γράφειν περὶ αὐτοῦ[3] ὅτι οἶδα ἰητροὺς σοφοὺς δόξαντας εἶναι ἀπὸ σχημάτων χειρὸς ἐν ἐπιδέσει, ἀφ' ὧν ἀμαθέας αὐτοὺς ἐχρῆν δοκεῖν εἶναι. ἄλλα[4] γὰρ πολλὰ οὕτω ταύτης τῆς τέχνης κρίνεται· τὸ γὰρ ξενοπρεπὲς οὔπω συνιέντες, εἰ χρηστόν, μᾶλλον ἐπαινέουσιν ἢ τὸ σύνηθες, ὃ ἤδη οἴδασιν ὅτι χρηστόν, καὶ τὸ ἀλλόκοτον ἢ τὸ
20 εὔδηλον. ῥητέον οὖν ὁπόσας ἂν ἐθέλω τῶν ἁμαρτάδων τῶν ἰητρῶν, τὰς μὲν ἀποδιδάξαι, τὰς δὲ διδάξαι[· ἄρξομαι δὲ][5] περὶ τῆς φύσιος τῆς

[1] ὁ ἐπιδεόμενος. [2] ἔστιν οἷ.
[3] αὐτῆς.

ON FRACTURES

I. In dislocations and fractures, the practitioner should make extensions in as straight a line as possible, for this is most conformable with nature;[1] but if it inclines at all to either side, it should turn towards pronation (palm down) rather than supination (palm up), for the error is less. Indeed, those who have no preconceived idea make no mistake as a rule, for the patient himself holds out the arm for bandaging in the position impressed on it by conformity with nature. The theorizing practitioners are just the ones who go wrong. In fact the treatment of a fractured arm is not difficult, and is almost any practitioner's job, but I have to write a good deal about it because I know practitioners who have got credit for wisdom by putting up arms in positions which ought rather to have given them a name for ignorance. And many other parts of this art are judged thus: for they praise what seems outlandish before they know whether it is good, rather than the customary which they already know to be good; the bizarre rather than the obvious. One must mention then those errors of practitioners as to the nature of the arm on which I want to give positive

[1] Galen makes this a general statement; but the writer is apparently speaking of the forearm, which he had already mentioned in a lost introduction.

---

[4] ἀλλά. [5] Omit Kw. BMV.

χειρός· καὶ γὰρ ἄλλων ὀστέων τῶν κατὰ τὸ
24 σῶμα δίδαγμα ὅδε ὁ λόγος ἐστίν.

II. Τὴν μὲν οὖν χεῖρα, περὶ οὗ[1] ὁ λόγος, ἔδωκέ τις καταδῆσαι πρηνέα[2] ποιήσας· ὁ δὲ ἠνάγκαζεν οὕτως ἔχειν ὥσπερ οἱ τοξεύοντες, ἐπὴν τὸν ὦμον ἐμβάλλωσι, καὶ οὕτως ἔχουσαν ἐπέδει, νομίζων ἑωυτῷ εἶναι τοῦτο αὐτῇ τὸ κατὰ φύσιν· καὶ μαρτύριον ἐπήγετο τά τε ὀστέα ἅπαντα τὰ ἐν τῷ πήχει, ὅτι ἰθυωρίην κατάλληλα εἶχε,[3] τήν τε ὁμοχροίην, ὅτι αὐτὴ καθ' ἑωυτὴν τὴν ἰθυωρίην ἔχει οὕτω καὶ ἐκ τοῦ ἔξωθεν μέρεος καὶ ἐκ τοῦ
10 ἔσωθεν· οὕτω δὲ ἔφη καὶ τὰς σάρκας καὶ τὰ νεῦρα πεφυκέναι, καὶ τὴν τοξικὴν ἐπήγετο μαρτύριον. ταῦτα λέγων καὶ ταῦτα ποιέων σοφὸς ἐδόκει εἶναι· τῶν δὲ ἄλλων τεχνέων ἐπελελήθει καὶ ὁπόσα ἰσχύι ἐργάζονται καὶ ὁπόσα τεχνήμασιν, οὐκ εἰδὼς ὅτι ἄλλο ἐν ἄλλῳ τὸ κατὰ φύσιν σχῆμά ἐστιν, καὶ ἐν τῷ αὐτῷ ἔργῳ ἕτερα τῆς δεξιῆς χειρὸς σχήματα κατὰ φύσιν ἐστί, καὶ ἕτερα τῆς ἀριστερῆς, ἢν οὕτω τύχῃ. ἄλλο μὲν γὰρ σχῆμα ἐν ἀκοντισμῷ κατὰ φύσιν, ἄλλο δὲ ἐν
20 σφενδονήσει, ἄλλο δὲ ἐν λιθοβολίῃσι, ἄλλο ἐν πυγμῇ, ἄλλο ἐν τῷ ἐλινύειν. ὁπόσας δ' ἄν τις τέχνας εὕροι ἐν ᾗσιν οὐ τὸ αὐτὸ σχῆμα τῶν χειρῶν κατὰ φύσιν ἐστὶν καὶ[4] ἐν ἑκάστῃ τῶν τεχνων, ἄλλα[4] πρὸς τὸ ἅρμενον ὃ ἔχῃ ἕκαστος, καὶ πρὸς

[1] οὗ because it is an idiom or phrase not referring specially to ἡ χείρ.
[2] ἐπιδῆσαι καταπρηνέα.
[3] ἔχει κατάλληλα.
[4] ἀλλὰ (omitting καί).

---

[1] Commentators, from Galen downwards, point out the absurdity of teaching "errors." Ermerins got rid of it in

and negative instruction,[1] for this discourse is an instruction on other bones of the body also.

II. To come to our subject, a patient presented his arm to be dressed in the attitude of pronation, but the practitioner made him hold it as the archers do when they bring forward the shoulder,[2] and he put it up in this posture, persuading himself that this was its natural position. He adduced as evidence the parallelism of the forearm bones, and the surface also, how that it has its outer and inner parts in a direct line, declaring this to be the natural disposition of the flesh and tendons, and he brought in the art of the archer as evidence. This gave an appearance of wisdom to his discourse and practice, but he had forgotten the other arts and all those things which are executed by strength or artifice, not knowing that the natural position varies in one and another, and that in doing the same work it may be that the right arm has one natural position and the left another. For there is one natural position in throwing the javelin, another in using the sling, another in casting a stone, another in boxing, another in repose. How many arts might one find in which the natural position of the arms is not the same, but they assume postures in accordance with the apparatus

his usual bold manner by reading τὰ for τάς. Diels considered it a glaring hysteron-proteron which can be simply remedied by reversal, and this is practically done in the translation. It seems a play upon words at which the writer is more successful elsewhere. See chap. XXX end.

[2] Galen says the archer held his left arm back downwards or nearly so; but this is contrary to ancient representations. What the writer chiefly objects to is putting up a broken forearm with the elbow extended.

τὸ ἔργον ὃ ἂν ἐπιτελέσασθαι θέλῃ, σχηματίζονται αἱ χεῖρες· τοξικὴν δὲ ἀσκέοντι εἰκὸς τοῦτο τὸ σχῆμα κράτιστον εἶναι τῆς ἑτέρης χειρός· τοῦ γὰρ βραχίονος τὸ γιγγλυμοειδές, ἐν τῇ τοῦ πήχεος βαθμίδι ἐν τούτῳ τῷ σχήματι ἐρεῖδον ἰθυωρίην
30 ποιεῖ τοῖσιν ὀστέοισιν τοῦ πήχεος καὶ τοῦ βραχίονος, ὡς ἂν ἓν εἴη τὸ πᾶν· καὶ ἡ ἀνάκλασις τοῦ ἄρθρου κέκλασται[1] ἐν τούτῳ τῷ σχήματι. εἰκὸς μὲν οὖν οὕτως ἀκαμπτότατόν τε καὶ τετανώτατον εἶναι τὸ χωρίον, καὶ μὴ ἡσσᾶσθαι, μηδὲ συνδιδόναι, ἑλκομένης τῆς νευρῆς ὑπὸ τῆς δεξιῆς χειρός· καὶ οὕτως ἐπὶ πλεῖστον μὲν τὴν νευρὴν ἑλκύσει, ἀφήσει δὲ ἀπὸ στερεωτάτου καὶ ἀθροωτάτου· ἀπὸ τῶν τοιούτων γὰρ ἀφεσίων τῶν τοξευμάτων, ταχεῖαι καὶ αἱ ἰσχύες καὶ τὰ μήκεα γίνονται.
40 ἐπιδέσει δὲ καὶ τοξικῇ οὐδὲν κοινόν. τοῦτο μὲν γάρ, εἰ ἐπιδήσας ἔχειν τὴν χεῖρα οὕτως ἔμελλε,[2] πόνους ἂν ἄλλους πολλοὺς προσετίθει μείζονας τοῦ τρώματος· τοῦτο δ', εἰ συγκάμψαι ἐκέλευεν, οὔτε τὰ ὀστέα οὔτε τὰ νεῦρα οὔτε αἱ σάρκες ἔτι ἐν τῷ αὐτῷ ἐγίνοντο, ἀλλὰ ἄλλῃ μετεκοσμεῖτο κρατέοντα τὴν ἐπίδεσιν· καὶ τί ὄφελός ἐστι τοξικοῦ σχήματος; καὶ ταῦτα ἴσως οὐκ ἂν ἐξημάρτανε σοφιζόμενος, εἰ εἴα τὸν τετρωμένον
49 αὐτὸν τὴν χεῖρα παρασχέσθαι.

III. Ἄλλος δ' αὖ τις τῶν ἰητρῶν ὑπτίην τὴν χεῖρα δούς, οὕτω κατατείνειν ἐκέλευε,[3] καὶ οὕτως ἔχουσαν ἐπέδει, τοῦτο νομίζων τὸ κατὰ φύσιν εἶναι, τῷ τε χροῒ σημαινόμενος καὶ τὰ ὀστέα νομίζων κατὰ φύσιν εἶναι οὕτως, ὅτι φαίνεται τὸ ἐξέχον ὀστέον τὸ παρὰ τὸν καρπὸν ᾗ ὁ σμικρὸς

[1] τέταται Kw. (τετασθαι B').

each man uses and the work he wants to accomplish! As to the practiser of archery, he naturally finds the above posture strongest for one arm: for the hinge-like end of the humerus in this position being pressed into the cavity of the ulna makes a straight line of the bones of the upper arm and forearm, as if the whole were one, and the flexure of the joint is extended (abolished) in this attitude. Naturally then the part is thus most inflexible and tense, so as neither to be overcome or give way when the cord is drawn by the right hand. And thus he will make the longest pull, and shoot with the greatest force and frequency, for shafts launched in this way fly strongly, swiftly and far. But there is nothing in common between putting up fractures and archery. For, first, if the operator, after putting up an arm, kept it in this position, he would inflict much additional pain, greater than that of the injury, and again, if he bade him bend the elbow, neither bones, tendons, nor flesh would keep in the same position, but would rearrange themselves in spite of the dressings. Where, then, is the advantage of the archer position? And perhaps our theorizer would not have committed this error had he let the patient himself present the arm.

III. Again, another practitioner handing over the arm back downwards had it extended thus and then put it up in this position, supposing it to be the natural one from surface indications: presuming also that the bones are in their natural position because the prominent bone at the wrist on the little finger

[2] ἐκελεύεν. [3] ἐκέλευσε.

δάκτυλος, κατ' ἰθυωρίην εἶναι τοῦ ὀστέου, ἀφ' ὀτέου[1] τὸν πῆχυν οἱ ἄνθρωποι μετρέουσιν ταῦτα τὰ μαρτύρια ἐπήγετο ὅτι κατὰ φύσιν οὕτως ἔχει,
10 καὶ ἐδόκει εὖ λέγειν.

Ἀλλὰ τοῦτο μέν, εἰ ὑπτίη ἡ χεὶρ κατατείνοιτο, ἰσχυρῶς πονοίη ἄν· γνοίη δ' ἄν τις τὴν ἑωυτοῦ χεῖρα κατατείνας ὡς ἐπώδυνον τὸ σχῆμα. ἐπεὶ καὶ ἀνὴρ ἥσσων κρέσσονα διαλαβὼν οὕτως εὖ[2] τῇσιν ἑωυτοῦ χερσίν, ὡς κλᾶται ὁ ἀγκὼν ὕπτιος, ἄγοι ἂν ὅπη ἐθέλοι· οὔτε γὰρ εἰ ξίφος ἐν ταύτῃ τῇ χειρὶ ἔχοι, ἔχοι ἂν ὅ τι χρήσαιτο τῷ ξίφει· οὕτω βίαιον τοῦτο τὸ σχῆμά ἐστιν. τοῦτο δέ, εἰ ἐπιδήσας τις ἐν τούτῳ τῷ σχήματι ἐῴη, μέζων μὲν
20 πόνος, εἰ περιΐοι, μέγας δὲ καὶ εἰ κατακέοιτο. τοῦτο δέ, εἰ συγκάμψει τὴν χεῖρα, ἀνάγκη πᾶσα[3] τούς τε μύας καὶ τὰ ὀστέα ἄλλο σχῆμα ἔχειν. ἠγνόει δὲ καὶ τάδε τὰ ἐν τῷ σχήματι χωρὶς τῆς ἄλλης λύμης· τὸ γὰρ ὀστέον τὸ παρὰ τὸν καρπὸν ἐξέχον, τὸ κατὰ τὸν σμικρὸν δάκτυλον, τοῦτο μὲν τοῦ πήχεός ἐστιν· τὸ δὲ ἐν τῇ συγκάμψει ἐὸν ἀπό τευ[4] τὸν πῆχυν οἱ ἄνθρωποι μετρέουσι, τοῦτο δὲ τοῦ βραχίονος ἡ κεφαλή ἐστιν. ὁ δὲ ᾤετο τωὐτὸ ὀστέον εἶναι τοῦτό τε κἀκεῖνο, πολλοὶ δὲ καὶ
30 ἄλλοι· ἔστι δὲ ἐκείνῳ τῷ ὀστέῳ τωὐτὸ ὁ ἀγκὼν καλούμενος, ᾧ ποτὶ[5] στηριζόμεθα. οὕτως οὖν ὑπτίην ἔχοντι τὴν χεῖρα, τοῦτο μὲν τὸ ὀστέον διεστραμμένον φαίνεται, τοῦτο δὲ τὰ νεῦρα τὰ ἀπὸ τοῦ καρποῦ τείνοντα ἐκ τοῦ ἔσω μέρεος καὶ ἀπὸ τῶν δακτύλων, ταῦτα ὑπτίην ἔχοντι τὴν χεῖρα διεστραμμένα γίνεται· τείνεται[6] γὰρ ταῦτα τὰ νεῦρα

[1] ἀπ' ὅτευ. [2] ἐν. [3] Kw. omits.

side appears to be in line with the bone from which men measure the forearm (cubit). He adduced this as evidence for the naturalness of the position, and seemed to speak well.

But, to begin with, if the arm were kept extended in supination it would be very painful; anyone who held his arm extended in this position would find how painful it is. In fact, a weaker person grasping a stronger one firmly so as to get his elbow extended in supination might lead him whither he chose, for if he had a sword in this hand he would be unable to use it, so constrained is this attitude. Further, if one put up a patient's arm in this position and left him so, the pain, though greater when he walked about, would also be great when he was recumbent. Again, if he shall bend the arm, it is absolutely necessary for both the muscles and bones to have another position. Besides the harm done, the practitioner was ignorant of the following facts as to the position. The projecting bone at the wrist on the side of the little finger belongs indeed to the ulna, but that at the bend of the elbow from which men measure the cubit is the head of the humerus, whereas he thought the one and the other belonged to the same bone, and so do many besides. It is the so-called elbow on which we lean that belongs to this bone.[1] In a patient with the forearm thus supinated, first, the bone is obviously distorted, and secondly, the cords stretching from the wrist on its inner side and from the fingers also undergo distortion in this supine position, for

[1] *i.e.* the olecranon process is part of the ulna.

[4] ἀπ' ὅτευ. [5] ὅν ποτί. [6] τείνει.

πρὸς τὸ τοῦ βραχίονος ὀστέον, ὅθεν ὁ πῆχυς μετρεῖται. αὗται τοσαῦται καὶ τοιαῦται αἱ ἁμαρτάδες καὶ ἀγνοίαι τῆς φύσιος τῆς χειρός. εἰ
40 δέ, ὡς ἐγὼ κελεύω, χεῖρα κατεηγυῖαν κατατείνοι τις, ἐπιστρέψει μὲν τὸ ὀστέον ἐς ἰθύ, τὸ κατὰ τὸν σμικρὸν δάκτυλον, τὸ ἐς τὸν ἀγκῶνα τεῖνον, ἰθυωρίην δὲ ἕξει τὰ νεῦρα τὰ ἀπὸ τοῦ καρποῦ πρὸς τοῦ βραχίονος τὰ ἄκρα τείνοντα· ἀναλαμβανομένη δὲ ἡ χεὶρ ἐν παραπλησίῳ σχήματι ἔσται, ἐν ᾧ περ καὶ ἐπιδεομένη, ἄπονος μὲν ὁδοιπορέοντι, ἄπονος δὲ κατακειμένῳ καὶ ἀκάματος. καθίννυσθαι δὲ χρὴ τὸν ἄνθρωπον οὕτως, ὅπως ᾖ τὸ ἐξέχον τοῦ ὀστέου πρὸς τὴν λαμπροτά-
50 την τῶν παρεουσέων αὐγέων, ὡς μὴ λάθῃ τὸν χειρίζοντα ἐν τῇ κατατάσει, εἰ ἱκανῶς ἐξίθυνται. τοῦ γε μὴν ἐμπείρου οὐδ' ἂν τὴν χεῖρα λάθοι ἐπαγομένην τὸ ἐξέχον· ἀτὰρ καὶ ἀλγεῖ μάλιστα κατὰ
54 τὸ ἐξέχον ψαυόμενον.

IV. Τῶν δὲ ὀστέων τοῦ πήχεος, ὧν μὴ ἀμφότερα κατέηγε,[1] ῥάων ἡ ἴησις, ἢν τὸ ἄνω ὀστέον τετρωμένον ᾖ καί περ παχύτερον ἐόν· ἅμα μὲν ὅτι τὸ ὑγιὲς ὑποτεταμένον γίνεται ἀντὶ θεμελίου, ἅμα δὲ ὅτι εὐκρυπτότερον γίνεται, πλὴν εἰ[2] τὸ ἐγγὺς τοῦ καρποῦ· παχείη γὰρ ἡ τῆς σαρκὸς ἐπίφυσις ἡ ἐπὶ τὸ ἄνω. τὸ δὲ κάτω ὀστέον ἄσαρκον καὶ οὐκ εὐσύγκρυπτον, καὶ κατατάσιος ἰσχυροτέρης δεῖται. ἢν δὲ μὴ τοῦτο συντριβῇ, ἀλλὰ τὸ ἕτερον,
10 φαυλοτέρη[3] ἡ κατάτασις ἀρκεῖ. ἢν δὲ ἀμφότερα κατεήγῃ, ἰσχυροτάτης κατατάσιος δεῖται· παιδίου μὲν γὰρ ἤδη εἶδον καταταθέντα μᾶλλον ἢ ὡς

[1] κατέηγεν, . . . εἰ . . . τέτρωται. [2] ᾖ.
[3] ἐλαφροτέρη.

these cords extend to the bone of the upper arm from which the cubit is measured. Such and so great are these errors and ignorances concerning the nature of the arm. But if one does extension of a fractured arm as I direct, he will both turn the bone stretching from the region of the little finger to the elbow so as to be straight[1] and will have the cords stretching from the wrist to the (lower) end of the humerus in a direct line; further, the arm when slung will keep about the same position as it was in when put up, and it will give the patient no pain when he walks, no pain when he lies down and no sense of weariness. The patient should be so seated that the projecting part of the bone is turned towards the brightest light available, that the operator may not overlook the proper degree of extension and straightening. Of course the hand of an experienced practitioner would not fail to recognise the prominence (at the fracture) by touch; also there is a special tenderness at the prominence when palpated.

IV. When the bones of the forearm are not both fractured the cure is easier if the upper bone (radius) is injured, though it is the thicker, both because the sound bone lying underneath acts as a support and because it is better covered, except at the part near the wrist, for the fleshy growth on the upper bone is thick; but the lower bone (ulna) is fleshless, not well covered, and requires stronger extension. If it is not this bone but the other that is broken, rather slight extension suffices: if both are broken very strong extension is requisite In the case of a child I have seen the bones ex-

[1] *i.e.* the styloid process in line with the olecranon.

ἔδει, οἱ δὲ πλεῖστοι ἧσσον τείνονται ἢ ὡς δεῖ. χρὴ δ' ἐπὴν τείνωσι, τὰ θέναρα προσβάλλοντα διορθοῦν· ἔπειτα χρίσαντα κηρωτῇ μὴ πάνυ πολλῇ, ὡς μὴ περιπλέῃ τὰ ἐπιδέσματα, οὕτως ἐπιδεῖν ὅπως μὴ κατωτέρω ἄκρην τὴν χεῖρα ἕξει τοῦ ἀγκῶνος, ἀλλὰ σμικρῷ τινὶ ἀνωτέρω, ὡς μὴ τὸ αἷμα ἐς ἄκρον ἐπιρρέῃ, ἀλλὰ ἀπολαμβάνηται·
20 ἔπειτα ἐπιδεῖν τῷ ὀθονίῳ, τὴν ἀρχὴν βαλλόμενος κατὰ τὸ κάτηγμα· ἐρείδων μὲν οὖν,[1] μὴ πιέζων δὲ κάρτα. ἐπὴν δὲ περιβάλῃ κατὰ τωὐτὸ δὶς ἢ τρίς, ἐπὶ τὸ ἄνω νεμέσθω ἐπιδέων, ἵνα αἱ ἐπιρροαὶ τοῦ αἵματος ἀπολαμβάνωνται, καὶ τελευτησάτω κεῖθι. χρὴ δὲ μὴ μακρὰ εἶναι τὰ πρῶτα ὀθόνια. τῶν δὲ δευτέρων ὀθονίων τὴν μὲν ἀρχὴν βάλλεσθαι ἐπὶ τὸ κάτηγμα· περιβαλών τε[2] ἅπαξ ἐς τωὐτό, ἔπειτα νεμέσθω ἐς τὸ κάτω καὶ ἐπὶ ἧσσον πιέζων, καὶ ἐπὶ μέζον διαβιβάσκων, ὡς ἂν αὐτὸ[3] ἱκανὸν
30 γένηται τὸ ὀθόνιον ἀναπαλινδρομῆσαι κεῖθι ἵνα περ τὸ ἕτερον ἐτελεύτησεν. ἐνταῦθα μὲν οὖν τὰ ὀθόνια ἐπ' ἀριστερὰ ἢ ἐπὶ δεξιὰ ἐπιδεδέσθω, ἢ ἐπὶ ὁπότερα ἂν συμφέρῃ πρὸς τὸ σχῆμα τοῦ κατεαγότος,[4] καὶ ἐφ' ὁπότερα ἂν περιρρέπειν συμφέρῃ. μετὰ δὲ ταῦτα, σπλῆνας κατατείνειν χρὴ κεχρισμένους κηρωτῇ ὀλίγῃ· καὶ γὰρ προσηνέστερον καὶ εὐθετώτερον. ἔπειτα οὕτως ἐπιδεῖν τοῖσιν ὀθονίοισιν ὡς[5] ἐναλλάξ, ὅτε μὲν ἐπὶ δεξιά, ὅτε δὲ ἐπ' ἀριστερά· καὶ τὰ μὲν πλείω κάτωθεν
40 ἀρχόμενος ἐς τὸ ἄνω ἄγειν, ἔστι δ' ὅτε καὶ ἄνωθεν ἐς τὸ κάτω. τὰ δὲ ὑπόξηρα ἀκεῖσθαι τοῖσι σπλήνεσι κυκλεῦντα· τῷ δὲ πλήθει τῶν περι-

[1] Omit οὖν. δέ. [3] αὐτῷ.

tended more than was necessary, but most patients get less than the proper amount. During extension one should use the palms of the hands to press the parts into position, then after anointing with cerate (in no great quantity lest the dressings should slip), proceed to put it up in such a way that the patient shall have his hand not lower than the elbow but a little higher; so that the blood may not flow to the extremity but be kept back. Then apply the linen bandage, putting the head of it at the fracture so as to give support, but without much pressure. After two or three turns are made at the same spot, let the bandage be carried upwards that afflux of blood may be kept back, and let it end off there. The first bandages should not be lengthy. Put the head of the second bandage on the fracture, making one turn there; then let it be carried downwards, with decreasing pressure and at wider intervals, till enough of the bandage is left for it to run back again to the place where the other ended. Let the bandages in this part of the dressing be applied either to left or right, whichever suits the form of the fracture and the direction towards which the limb ought to turn. After this, compresses should be laid along after being anointed with a little cerate; for the application is more supple and more easily made. Then put on bandages crosswise to right and left alternately, beginning in most cases from below upwards but sometimes from above downwards. Treat conical parts by surrounding them with compresses, bringing them to a level not all

[4] κατήγματος. [5] Omit ὡς.

βολέων μὴ πᾶν ἀθροὸν συνδιορθοῦντα, ἀλλὰ κατὰ μέρος. περιβάλλειν δὲ χρὴ χαλαρὰ καὶ περὶ τὸν καρπὸν τῆς χειρὸς ἄλλοτε καὶ ἄλλοτε. πλῆθος δὲ τῶν ὀθονίων ἱκανὸν τὸ πρῶτον αἱ δύο
47 μοῖραι.

V. Σημεῖα δὲ τοῦ καλῶς ἰητρευμένου ταῦτα, καὶ ὀρθῶς ἐπιδεομένου, εἰ ἐρωτῴης αὐτὸν εἰ πεπίεκται, καὶ εἰ φαίη μὲν πεπιέχθαι, ἡσύχως δέ, καὶ μάλιστα εἰ κατὰ τὸ κάτηγμα φαίη· τοιαῦτα τοίνυν φάναι χρὴ πεπρηγμένα διὰ τέλεος τὸν ὀρθῶς ἐπιδεόμενον. σημεῖα δὲ ταῦτα τῆς μετριότητος, τὴν μὲν ἡμέρην, ἣν ἂν ἐπιδεθῇ, καὶ τὴν νύκτα δοκείτω αὐτὸς ἑωυτῷ μὴ ἐπὶ ἧσσον πεπιέχθαι, ἀλλ' ἐπὶ μᾶλλον· τῇ δὲ ὑστεραίῃ
10 οἰδημάτιον ἐλθεῖν ἐς χεῖρα ἄκρην μαλθακόν· μετριότητος γὰρ σημεῖον τῆς πιέξιός σου· τελευτώσης δὲ τῆς ἡμέρης, ἐπὶ ἧσσον δοκείτω πεπιέχθαι· τῇ δὲ τρίτῃ χαλαρά σοι δοκείτω εἶναι τὰ ἐπιδέσματα. κἢν μέν τι τούτων τῶν εἰρημένων ἐλλείπῃ, γινώσκειν χρὴ ὅτι χαλαρωτέρη ἐστὶν ἡ ἐπίδεσις τοῦ μετρίου· ἢν δέ τι τῶν εἰρημένων πλεονάζῃ, χρὴ γινώσκειν ὅτι μᾶλλον ἐπιέχθη τοῦ μετρίου· καὶ τούτοισι σημαινόμενος τὸ ὕστερον ἐπιδέων ἢ χαλᾶν μᾶλλον, ἢ πιέζειν. ἀπολύσαντα
20 δὲ χρὴ τριταῖον ἐόντα κατατεινάμενον καὶ διορθωσάμενον· καὶ ἢν μετρίως τὸ πρῶτον τετυχήκῃς ἐπιδήσας, ταύτην τὴν ἐπίδεσιν χρὴ ὀλίγῳ μᾶλλον

[1] Littré inserts αὖθις ἐπιδῆσαι—and renders (as followed by Adams), "Having removed the bandages on the third day, you must make extension and adjust the fracture and bind it up again." As Petrequin remarks, this seems contrary to common sense, surgery and the express directions

at once but gradually by the number of circumvolutions. You should put additional loose turns now and then at the wrist. The two sets of bandages are a sufficient number for the first dressing.

V. These are the indications of good treatment and correct bandaging:—If you ask the patient whether the part is compressed and he says it is, but moderately and that chiefly at the fracture. A properly bandaged patient should give a similar report of the operation throughout. The following are the indications of a due moderation. During the day of the dressing and the following night the pressure should appear to the patient not to diminish but rather to increase, and on the following day a slight and soft swelling should appear in the hand; you should take this as a sign of the due mean as to pressure. At the end of the day the pressure should seem less, and on the third day you should find the bandages loose. If, then, any of the said conditions are lacking you may conclude that the bandaging was slacker than the mean, but if any of them be excessive you may conclude that the pressure was greater than the mean, and taking this as a guide make the next dressing looser or tighter. You should remove the dressing on the third day after the extension and adjustment,[1] and if your first bandaging hit the

of the author (XXXI). The limb is supposed to be set, any further adjustment being made on the seventh day. Celsus (VIII. 10. 1), Galen (*Meth. Med.* VI. 5) and Paulus (VI. 99) all follow Hippocrates, but make no mention of a second setting on the third day. Still, in the case of the leg he seems to recommend interference at every dressing; and grammar is on the side of Littré.

ἢ ἐκείνην πιέσαι. βάλλεσθαι δὲ χρὴ τὰς ἀρχὰς κατὰ τὸ κάτηγμα, ὥσπερ καὶ τὸ πρότερον· ἢν μὲν γὰρ τοῦτο πρότερον ἐπιδέῃς, ἐξειρύαται[1] ἐκ τούτου οἱ ἰχῶρες ἐς τὰς ἐσχατιὰς ἔνθα καὶ ἔνθα· ἢν δέ τι ἄλλο πρότερον πιέξῃς, ἐς τοῦτο ἐξειρύαται[1] ἐκ τοῦ πιεχθέντος· ἐς πολλὰ δὲ εὔχρηστον τὸ[2] συνιέναι. οὕτως οὖν ἄρχεσθαι μὲν αἰεὶ χρὴ τὴν
30 ἐπίδεσιν καὶ τὴν πίεξιν ἐκ τούτου τοῦ χωρίου, τὰ δὲ ἄλλα κατὰ λόγον, ὡς προσωτέρω ἀπὸ τοῦ κατήγματος ἀγάγῃς, ἐπὶ ἧσσον τὴν πίεξιν ποιεῖσθαι. χαλαρὰ δὲ παντάπασι μηδέποτε περιβάλλειν, ἀλλὰ προσπεπτωκυῖα. ἔπειτα δὲ πλείοσιν ὀθονίοισι χρὴ ἐπιδεῖν ἑκάστην τῶν ἐπιδεσίων. ἐρωτώμενος δὲ φάτω ὀλίγῳ μᾶλλόν οἱ πεπιέχθαι, ἢ τὸ πρότερον, καὶ μάλιστα φάτω κατὰ τὸ κάτηγμα καὶ τὰ ἄλλα δὲ κατὰ λόγον· καὶ ἀμφὶ τῷ οἰδήματι, καὶ ἀμφὶ τῷ πονέειν, καὶ
40 ἀμφὶ τῷ ῥηΐζειν, κατὰ λόγον τῆς προτέρης ἐπιδέσιος γινέσθω. ἐπὴν δὲ τριταῖος ᾖ, χαλαρώτερά οἱ δοκείτω εἶναι τὰ ἐπιδέσματα· ἔπειτα ἀπολύσαντα χρὴ αὖθις ἐπιδῆσαι, ὀλίγῳ μᾶλλον πιέζοντα, καὶ ἐν πᾶσι τοῖσιν ὀθονίοισιν οἷσί περ ἤμελλεν ἐπιδεῖσθαι· ἔπειτα δὲ πάντα αὐτὸν ταῦτα καταλαβέτω, ἅπερ καὶ ἐν τῇσι πρώτῃσι
47 περιόδοισι τῶν ἐπιδεσίων.

VI. Ἐπὴν δὲ τριταῖος γένηται, ἑβδομαῖος δὲ ἀπὸ τῆς πρώτης ἐπιδέσιος, ἢν ὀρθῶς ἐπιδέηται, τὸ μὲν οἴδημα ἐν ἄκρῃ τῇ χειρὶ ἔσται, οὐδὲ τοῦτο λίην μέγα· τὸ δ' ἐπιδεόμενον χωρίον ἐν πάσῃσι τῇσιν ἐπιδέσεσιν ἐπὶ τὸ λεπτότερον καὶ ἰσχνότερον εὑρεθήσεται, ἐν δὲ τῇ ἑβδόμῃ καὶ πάνυ λεπτόν,

[1] ἐξείργαται bis. See note, p. 158.

proper mean this one should be a little tighter. The heads of the bandages should be applied over the fracture as before, for if you did this before, the serous effusions were driven thence into the outer parts on both sides, but if you formerly made the pressure anywhere else, they were driven into this place (the fracture) from the part compressed. It is useful for many things to understand this. It shows that one should always begin the bandaging and compression at this point, and, for the rest, in proportion as you get further from the point of fracture make the pressure less. Never make the turns altogether slack, but closely adherent. Further, one should use more bandages at each dressing, and the patient when asked should say he felt a little more pressure than before, especially at the point of fracture, and the rest in proportion. And as regards the swelling, feeling of pain and relief, things should be in accord with the previous dressing. When the third day comes, he should find the dressings rather loose. Then after undoing them he should bandage again with a little more pressure and with all the bandages that he is going to use, and afterwards the patient should experience all those symptoms which he had in the first periods of bandaging.

VI. When the third day is reached (the seventh from the first dressing), if he is being properly bandaged, there will be the swelling on the hand, but it will not be very marked. As to the part bandaged, it will be found to be thinner and more shrunken at each dressing, and on the seventh day

---

[a] τοῦτο.

καὶ τὰ ὀστέα τὰ κατεηγότα ἐπὶ μᾶλλον κινεύμενα καὶ εὐπαράγωγα ἐς κατόρθωσιν. καὶ ἢν ᾖ ταῦτα τοιαῦτα, κατορθωσάμενον χρὴ ἐπιδῆσαι ὡς ’ς νάρ-
10 θηκας, ὀλίγῳ μᾶλλον πιέσαντα ἢ τὸ πρότερον, ἢν μὴ πόνος τις πλείων ᾖ ἀπὸ τοῦ οἰδήματος τοῦ ἐν ἄκρῃ τῇ χειρί. ἐπὴν δ' ἐπιδήσῃς τοῖσιν ὀθονίοισι, τοὺς νάρθηκας περιθεῖναι χρὴ καὶ περιλαβεῖν ἐν τοῖσι δεσμοῖσι ὡς χαλαρωτάτοισιν, ὁπόσον ἠρεμεῖν, ὥστε μηδὲν συμβάλλεσθαι ἐς τὴν πίεξιν τῆς χειρὸς τὴν τῶν ναρθήκων πρόσθεσιν. μετὰ δὲ ταῦτα, ὅ τε πόνος, αἵ τε ῥαστῶναι αἱ αὐταὶ γινέσθωσαν αἵ περ καὶ ἐν τῇσι πρώτῃσι[1] περιόδοισι τῶν ἐπιδεσίων. ἐπὴν δὲ τριταῖος ἐὼν φῇ
20 χαλαρὸν εἶναι, τότ' ἔπειτα χρὴ τοὺς νάρθηκας ἐρείσασθαι, μάλιστα μὲν κατὰ τὸ κάτηγμα, ἀτὰρ καὶ τἄλλα κατὰ λόγον, ᾗπερ καὶ ἡ ἐπίδεσις ἐχάλα ἄρα[2] μᾶλλον ἢ ἐπίεζεν. παχύτατον δὲ χρὴ εἶναι τὸν νάρθηκα ᾗ ἐξέστη τὸ κάτηγμα, μὴ μὴν πολλῷ. ἐπιτηδεύειν δὲ χρὴ μάλιστα μὲν κατ' ἰθυωρίην τοῦ μεγάλου δακτύλου, ὡς μὴ κείσεται ὁ νάρθηξ, ἀλλὰ τῇ ἢ τῇ, μηδὲ κατὰ τὴν τοῦ σμικροῦ ἰθυωρίην, ᾗ τὸ ὀστέον ὑπερέχει ἐν τῷ καρπῷ, ἀλλὰ τῇ ἢ τῇ· ἢν δὲ ἄρα πρὸς τὸ κάτηγμα
30 συμφέρῃ κεῖσθαι κατὰ ταῦτά τινας τῶν ναρθήκων, βραχυτέρους αὐτοὺς χρὴ τῶν ἄλλων ποιεῖν, ὡς μὴ ἐξικνέωνται πρὸς τὰ ὀστέα τὰ ὑπερέχοντα παρὰ τὸν καρπόν· κίνδυνος γὰρ ἑλκώσιος καὶ νεύρων ψιλώσιος. χρὴ δὲ διὰ τρίτης ἐρείδειν τοῖσι νάρθηξι πάνυ ἡσυχῇ, οὕτω τῇ γνώμῃ ἔχοντα, ὡς οἱ νάρθηκες φυλακῆς εἵνεκα τῆς

[1] προτέρῃσι.

it will be quite thin, while the fractured bones will be more mobile and ready for adjustment. If this is so, after seeing to the adjustment you should bandage as for splints, making a little more pressure than before, unless there is any increase of pain from the swelling on the hand. When you dress with the bandages you should apply the splints round the limb and include them in ligatures as loose as possible consistently with firmness, so that the addition of the splints may contribute nothing to the compression of the arm. After this the pain and the relief following it should be the same as in the previous periods of bandaging. When, on the third day, he says it is loose, then indeed you should tighten up the splints, especially at the fracture, and the rest in proportion where the dressing also was loose rather than tight. The splint should be thicker where the fracture projects, but not much so, and you should take special care that it does not lie in the line of the thumb, but on one side or the other, nor in the line of the little finger where the bone projects at the wrist, but on one side or the other. If, indeed, it is for the benefit of the fracture that some of the splints should be placed thus, you should make them shorter than the rest, so that they do not reach as far as the bones which project at the wrist, for there is risk of ulceration and denuding of tendons. You should tighten the splints every third day[1] very slightly, bearing in mind that they are put there to maintain

[1] *i.e.* every other day.

---

[2] Pq. ἐχαλάρα codd.; but this is not Greek. Kw. omits ἄρα.

ἐπιδέσιος προσκέονται[1] ἀλλ' οὐ τῆς πιέξιος
38 εἵνεκεν ἐπιδέδενται.[2]

VII. Ἢν μὲν οὖν εὖ εἰδῇς ὅτι ἱκανῶς τὰ ὀστέα ἀπίθυνται ἐν τῇσι προτέρῃσι ἐπιδέσεσι, καὶ μητε κνησμοί τινες λυπέωσι, μήτε τις ἕλκωσις μηδεμία ὑποπτεύηται εἶναι, ἐᾶν χρὴ ἐπιδεδέσθαι ἐν τοῖσι νάρθηξι, ἔστ' ἂν ὑπὲρ εἴκοσιν ἡμέρας γένηται. ἐν τριήκοντα δὲ μάλιστα τῇσι συμπάσῃσι κρατύνεται ὀστέα τὰ ἐν τῷ πήχει τὸ ἐπίπαν· ἀτρεκὲς δὲ οὐδέν· μάλα γὰρ καὶ φύσις φύσεος καὶ ἡλικίη ἡλικίης διαφέρει. ἐπὴν δὲ λύσῃς, ὕδωρ θερμὸν
10 καταχέαι χρὴ καὶ μετεπιδῆσαι, ἧσσον μὲν ὀλίγῳ πιέσαντα ἢ τὸ πρόσθεν, ἐλάσσοσι δὲ τοῖσιν ὀθονίοισιν ἢ τὸ πρότερον· καὶ ἔπειτα διὰ τρίτης ἡμέρης λύσαντα ἐπιδεῖν, ἐπὶ μὲν ἧσσον πιέζοντα, ἐπὶ δὲ ἐλάσσοσι τοῖσιν ὀθονίοισιν. ἐπὴν δέ, ὅταν τοῖσι νάρθηξι δεθῇ, ὑποπτεύῃς τὰ ὀστέα μὴ ὀρθῶς κεῖσθαι, ἢ ἄλλο τι ὀχλέῃ τὸν τετρωμένον, λῦσαι ἐν τῷ ἡμίσει[3] τοῦ χρόνου ἢ ὀλίγῳ πρόσθεν, καὶ αὖθις μετεπιδῆσαι. δίαιτα δὲ τούτοισιν οἷσιν ἂν μὴ ἕλκεα ἐξ ἀρχῆς γένηται ἢ ὀστέα ἔξω
20 ἐξίσχῃ, ἀρκεῖ ὑποφαύλη. [σμικρόν τι καὶ γὰρ][4] ἐνδεέστερον[5] χρὴ διαιτᾶν ἄχρις ἡμερέων δέκα, ἅτε δὴ καὶ ἐλινύοντας· καὶ ὄψοισιν ἁπαλοῖσι χρῆσθαι ὁπόσα τῇ διεξόδῳ μετριότητα παρασχήσει, οἴνου δὲ καὶ κρεηφαγίης ἀπέχεσθαι· ἔπειτα μέντοι ἐκ προσαγωγῆς ἀνακομίζεσθαι. οὗτος ὁ λόγος ὥσπερ νόμος κεῖται δίκαιος περὶ κατηγμάτων ἰήσιος, ὥς τε χειρίζειν χρή, ὥς τε ἀποβαίνειν ἀπὸ τῆς δικαίης χειρίξιος· ὅ τι δ' ἂν μὴ οὕτως ἀποβαίνῃ, εἰδέναι χρὴ ὅτι ἐν τῇ

[1] προσκέωνται Vulg.: προσκέαται Kw.

the dressing, but not bound in for the sake of pressure.

VII. If you are convinced that the bones are sufficiently adjusted in the former dressings, and there is no painful irritation nor any suspicion of a sore, you should leave the part put up in splints till over the twentieth day. It takes about thirty days altogether as a rule for the bone of the forearm to unite. But there is nothing exact about it, for both constitutions and ages differ greatly. When you remove the dressing, douche with warm water and replace it, using a little less pressure and fewer bandages than before; and after this, remove and re-apply every other day with less pressure and fewer bandages. If, in any case where splints are used, you suspect that the bones are not properly adjusted, or that something else is troubling the patient, remove the dressing and replace it in the middle of the interval or a little sooner. Light diet suffices in those cases where there is no open wound at the first, or protrusion of the bone, for it should be slightly restricted for the first ten days, seeing that the patients are resting; and soft foods should be taken such as favour a due amount of evacuation. Avoid wine and meat, but afterwards gradually feed him up. This discourse gives a sort of normal rule for the treatment of fractures, how one should handle them surgically, and the results of correct handling. If any of the results are not as described, you may

---

[2] ἐπιδέωνται Vulg. : ἐπιδεδέαται Kw.
[3] μεσηγύ.
[4] So Galen and some MSS. Omit Littré, Erm. Kw.
[5] ἐνδεέστερον δέ.

30 χειρίξει τι ἐνδεὲς πεποίηται ἢ πεπλεόνασται. ἔτι δὲ τάδε χρὴ προσσυνιέναι ἐν τούτῳ τῷ ἁπλῷ τρόπῳ, ἃ οὐ κάρτα ἐπιμελέονται οἱ ἰητροί, καίτοι πᾶσαν μελέτην καὶ πᾶσαν ἐπίδεσιν οἷά τε διαφθείρειν ἐστί, μὴ ὀρθῶς ποιεύμενα· ἢν γὰρ τὰ μὲν ὀστέα ἄμφω κατηγῇ, ἢ τὸ κάτω μοῦνον, ὁ δὲ ἐπιδεδεμένος ἐν ταινίῃ τινὶ τὴν χεῖρα ἔχῃ ἀναλελαμμένην,[1] τυγχάνῃ δὲ ἡ ταινίη κατὰ τὸ κάτηγμα πλείστη ἐοῦσα, ἔνθεν δὲ καὶ ἔνθεν ἡ χεὶρ ἀπαιωρῆται, τοῦτον ἀνάγκη τὸ 40 ὀστέον εὑρεθῆναι διεστραμμένον ἔχοντα πρὸς τὸ ἄνω μέρος· ἢν δέ, κατεηγότων τῶν ὀστέων οὕτως, ἄκρην τε τὴν χεῖρα ἐν τῇ ταινίῃ ἔχῃ καὶ παρὰ τὸν ἀγκῶνα, ὁ δὲ ἄλλος πῆχυς [μὴ][2] μετέωρος ᾖ, οὕτως[3] εὑρεθήσεται τὸ ὀστέον ἐς τὸ κάτω μέρος διεστραμμένως ἔχον. χρὴ οὖν, ἐν ταινίῃ πλάτος ἐχούσῃ, μαλθακῇ, τὸ πλεῖστον τοῦ πήχεος καὶ 47 τὸν καρπὸν τῆς χειρὸς ὁμαλῶς αἰωρεῖσθαι.

VIII. Ἢν δὲ ὁ βραχίων καταγῇ, ἢν μέν τις ἀποτανύσας τὴν χεῖρα ἐν τούτῳ τῷ σχήματι διατείνῃ, ὁ μῦς τοῦ βραχίονος κατατεταμένος ἐπιδεθήσεται· ἐπὴν δ' ἐπιδεθεὶς συγκάμψῃ τὸν ἀγκῶνα, ὁ μῦς τοῦ βραχίονος ἄλλο σχῆμα σχήσει. δικαιοτάτη οὖν βραχίονος κατάτασις ἥδε· ξύλον πηχυαῖον ἢ ὀλίγῳ βραχύτερον, ὁποῖοι οἱ στειλαιοί εἰσι τῶν σκαφίων, κρεμάσαι χρὴ ἔνθεν καὶ ἔνθεν, σειρῇ δήσαντα· καθίσαντα δὲ τὸν 10 ἄνθρωπον ἐπὶ ὑψηλοῦ τινός, τὴν χεῖρα ὑπερκεῖσθαι, ὡς ὑπὸ τῇ μασχάλῃ γένηται ὁ στειλαιὸς ἔχων συμμέτρως, ὥστε μόλις δύνασθαι καθίν-

[1] ἀναλελαμμένος.

be sure there has been some defect or excess in the surgical treatment. You should acquaint yourself further with the following points in this simple method, points with which practitioners do not trouble themselves very much, though they are such as (if not properly seen to) can bring to naught all your carefulness in bandaging. If both bones are broken, or the lower (ulna) only, and the patient, after bandaging, has his arm slung in a sort of scarf, this scarf being chiefly at the point of fracture, while the arm on either side is unsupported, he will necessarily be found to have the bone distorted towards the upper side; while if, when the bones are thus broken, he has the hand and part near the elbow in the scarf, while the rest of the arm is unsupported, this patient will be found to have the bone distorted towards the lower side. It follows that as much as possible of the arm and wrist should be supported evenly in a soft broad scarf.

VIII. When the humerus is fractured, if one extends the whole arm and keeps it in this posture, the muscle of the arm [1] will be bandaged in a state of extension, but when the bandaged patient bends his arm the muscle will assume another posture. It follows that the most correct mode of extension of the arm is this:—One should hang up a rod, in shape like a spade handle and of a cubit in length or rather shorter, by a cord at each end. Seat the patient on a high stool and pass his arm over the rod so that it comes evenly under the armpit in such a position that the

[1] Biceps.

[2] Omit; but Galen defends both readings (xviii(2). 415).
[3] οὗτος . . . διεστραμμένον ἔχων.

νυσθαι τὸν ἄνθρωπον, σμικροῦ δέοντα μετέωρον εἶναι· ἔπειτα θέντα τι ἄλλο ἔφεδρον, καὶ ὑποθέντα σκύτινον ὑποκεφάλαιον, ἢ ἓν ἢ πλείω, ὅπως συμμέτρως σχήσει ὕψεος τοῦ πήχεος πλαγίου πρὸς ὀρθὴν γωνίην, ἄριστον μὲν σκῦτος πλατὺ καὶ μαλθακὸν ἢ ταινίην πλατέην ἀμφιβάλλοντα, τῶν μεγάλων τι σταθμίων ἐξαρτῆσαι, 20 ὅ τι μετρίως ἕξει κατατείνειν· εἰ δὲ μή, τῶν ἀνδρων ὅστις ἐρρωμένος, ἐν τούτῳ τῷ σχήματι τοῦ πήχεος ἐόντος παρὰ τὸν ἀγκῶνα καταναγκαζέτω ἐς τὸ κάτω. ὁ δὲ ἰητρὸς ὀρθὸς μὲν ἐὼν χειριζέτω, τὸν ἕτερον πόδα ἐπὶ ὑψηλοτέρου τινὸς ἔχων, κατορθώσας δὲ τοῖσι θέναρσι τὸ ὀστέον· ῥηϊδίως δὲ κατορθώσεται· ἀγαθὴ γὰρ ἡ κατάστασις,[1] ἤν τις καλῶς παρασκευάσηται. ἔπειτα ἐπιδείτω, τάς τε ἀρχὰς βαλλόμενος ἐπὶ τὸ κάτηγμα, καὶ τἄλλα πάντα ὥσπερ πρότερον 30 παρῃνέθη, χειριζέτω· καὶ ἐρωτήματα ταὐτὰ ἐρωτάτω· καὶ σημείοισι χρήσθω τοῖσιν αὐτοῖσι, εἰ μετρίως ἔχει, ἢ οὔ· καὶ διὰ τρίτης ἐπιδείτω, καὶ ἐπὶ μᾶλλον πιεζέτω. καὶ ἑβδομαῖον ἢ ἐναταῖον ἐν νάρθηξι δησάτω· καὶ ἢν ὑποπτεύσῃ μὴ καλῶς κεῖσθαι τὸ ὀστέον μεσηγὺ τούτου τοῦ χρόνου, λυσάτω, καὶ εὐθετισάμενος μετεπιδησάτω.

Κρατύνεται δὲ μάλιστα βραχίονος ὀστέον ἐν τεσσαράκοντα ἡμέρῃσιν. ἐπὴν δὲ ταύτας 40 ὑπερβάλῃ, λύειν χρή, καὶ ἐπὶ ἧσσον πιέζειν τοῖσιν ὀθονίοισι καὶ ἐπὶ ἐλάσσοσιν ἐπιδεῖν. δίαιταν δὲ ἀκριβεστέρην τινὰ ἢ τὸ πρότερον διαιτᾶν, καὶ πλείω χρόνον· τεκμαίρεσθαι δὲ πρὸς τοῦ οἰδήματος τοῦ ἐν ἄκρῃ τῇ χειρί, τὴν ῥώμην

man can hardly sit and is almost suspended. Then placing another stool, put one or more leather cushions under the forearm as may suit its elevation when flexed at a right angle. The best plan is to pass some broad soft leather or a broad scarf round the arm and suspend from it heavy weights sufficient for due extension; failing this, let a strong man grasp the arm in this position at the elbow and force it downwards. As to the surgeon, he should operate standing with one foot on some elevated support, adjusting the bone with the palms of his hands. The adjustment will be easy, for there is good extension [1] if it is properly managed. Then let him do the bandaging, putting the heads of the bandages on the fracture and performing all the rest of the operation as previously directed. Let him ask the same questions, and use the same indications to judge whether things are right or not. He should bandage every third day and use greater pressure, and on the seventh or ninth day put it up in splints. If he suspects the bone is not in good position, let him loosen the dressings towards the middle of this period,[2] and after putting it right, re-apply them.

The bone of the upper arm usually consolidates in forty days. When these are passed one should undo the dressings and diminish the pressure and the number of bandages. A somewhat stricter diet and more prolonged (is required here) than in the former case. Make your estimate from the swelling in the hand, having an eye to the patient's strength.

[1] Reading *κατάτασις*.
[2] *i.e.* the period in splints.

---

[1] *κατάτασις* Galen Kw.

όρεων. προσσυνίεναι δὲ χρὴ καὶ τάδε, ὅτι ὁ βραχίων κυρτὸς πέφυκεν ἐς τὸ ἔξω μέρος· ἐς τοῦτο τοίνυν τὸ μέρος φιλεῖ διαστρέφεσθαι, ἐπὴν μὴ καλῶς ἰητρεύηται· ἀτὰρ καὶ τἄλλα πάντα
50 ὀστέα ἐς ὅπερ πέφυκε διεστραμμένα, ἐς τοῦτο καὶ ἰητρευόμενα φιλεῖ διαστρέφεσθαι, ἐπὴν κατεαγῇ. χρὴ τοίνυν, ἐπὴν τοιοῦτόν τι ὑποπτεύηται, ταινίῃ πλατέῃ προσεπιλαμβάνειν τὸν βραχίονα κύκλῳ περὶ τὸ στῆθος περιδέοντα· καὶ ἐπὴν ἀναπαύεσθαι μέλλῃ, μεσηγὺ τοῦ ἀγκῶνος καὶ τῶν πλευρέων σπλῆνά τινα πολύπτυχον πτύξαντα ὑποτιθέναι, ἢ ἄλλο τι ὃ τούτῳ ἔοικεν· οὕτω γὰρ ἂν ἰθὺ[1] τὸ κύρτωμα τοῦ ὀστέου γένοιτο· φυλάσσεσθαι δὲ μέντοι χρή, ὅπως μὴ ᾖ ἄγαν ἐς τὸ
60 ἔσω μέρος.

IX. Ποὺς δὲ ἀνθρώπου ἐκ πολλῶν καὶ σμικρῶν ὀστέων συγκεῖται, ὥσπερ καὶ χεὶρ ἄκρη· κατάγνυται μὲν οὐ πάνυ τι ταῦτα τὰ ὀστέα, ἢν μὴ σὺν τῷ χρωτὶ[2] τιτρωσκομένῳ ὑπὸ ὀξέος τινὸς ἢ βαρέος· τὰ μὲν οὖν τιτρωσκόμενα, ἐν ἑλκωσίων μέρει εἰρήσεται ὡς χρὴ ἰητρεύειν. ἢν δέ τι κινηθῇ ἐκ τῆς χώρης, ἢ τῶν δακτύλων ἄρθρον ἢ ἄλλο τι τῶν ὀστέων τοῦ ταρσοῦ καλουμένου, ἀναγκάζειν μὲν χρὴ ἐς τὴν ἑωυτοῦ χώρην
10 ἕκαστον, ὥσπερ καὶ τὰ ἐν τῇ χειρὶ εἴρηται·[3] ἰητρεύειν δὲ κηρωτῇ καὶ σπλήνεσι καὶ ὀθονίοισι ὥσπερ καὶ τὰ κατήγματα, πλὴν τῶν ναρθήκων, τὸν μὲν αὐτὸν τρόπον πιεζεῦντα, διὰ τρίτης δὲ ἐπιδέοντα· ὑποκρινέσθω δὲ ὁ ἐπιδεόμενος παραπλήσια, οἷά περ καὶ ἐν τοῖσι κατήγμασι, καὶ περὶ τοῦ πεπιέχθαι καὶ περὶ τοῦ χαλαρὸν εἶναι.[4]

[1] ἀλορδότατον B. Kw. ἰθὺ MV Pq. Littré.

One must also bear in mind that the humerus is naturally convex outwards, and is therefore apt to get distorted in this direction when improperly treated. In fact, all bones when fractured tend to become distorted during the cure towards the side to which they are naturally bent. So, if you suspect anything of this kind, you should pass round it an additional broad band, binding it to the chest, and when the patient goes to bed, put a many-folded compress, or something of the kind, between the elbow and the ribs, thus the curvature of the bone will be rectified. You must take care, however, that it is not bent too much inwards.

IX. The human foot, like the hand, is composed of many small bones. These bones are not often broken, unless the tissues are also wounded by something sharp or heavy. The proper treatment of the wounded parts will be discussed in the section on lesions of soft parts.[1] But if any of the bones be displaced, whether a joint of the toes or some bone of what is called the tarsus, you should press each back into its proper place just in the way described as regards the bones of the hand. Treat as in cases of fracture with cerate, compresses and bandages, but without splints, using pressure in the same way and changing the dressings every other day. The patient's answers both as to pressure and relaxation should be similar to those in cases of fracture. All

[1] Rather "compound fractures," cf. XXIV, XXV. Galen defines ἕλκος as a lesion of a soft part.

---

[2] χρώς = τὸ σαρκῶδες (Galen).
[3] A lost chapter, condensed in *Moch.* XVI, *Joints* XXVI.
[4] χαλᾶν.

ὑγιέα δὲ γίνεται ἐν εἴκοσιν ἡμέρῃσι τελέως ἅπαντα, πλὴν ὁπόσα κοινωνεῖ τοῖσι τῆς κνήμης ὀστέοισι καὶ αὐτῇ τῇ ἴξει.[1] συμφέρει δὲ κατα-
20 κεῖσθαι τοῦτον τὸν χρόνον. ἀλλὰ γὰρ οὐ τολμέουσιν ὑπερορῶντες τὸ νόσημα, ἀλλὰ περιέρχονται πρὶν ὑγιέες γενέσθαι. διὰ τοῦτο καὶ οἱ πλεῖστοι οὐκ ἐξυγιαίνουσι τελέως. ἀλλὰ πολλάκις αὐτοὺς ὁ πόνος ὑπομιμνήσκει· εἰκότως, ὅλον γὰρ τὸ ἄχθος τοῦ σώματος οἱ πόδες ὀχέουσι. ὁπόταν οὖν μήπω ὑγιέες ἐόντες ὁδοιπορέωσι, φλαύρως συναλθάσσεται[2] τὰ ἄρθρα τὰ κινηθέντα· διὰ τοῦτο ἄλλοτε καὶ ἄλλοτε ὁδοι-
29 πορέοντες ὀδυνῶνται τὰ πρὸς τῇ κνήμῃ.

X. Τὰ δὲ κοινωνέοντα τοῖσι τῆς κνήμης ὀστέοισι μείζω τε τῶν ἑτέρων ἐστί, καὶ κινηθέντων τούτων πολυχρονιωτέρη ἡ ἄλθεξις. ἴησις μὲν οὖν ἡ αὐτή· ὀθονίοισι δὲ πλείοσι χρῆσθαι καὶ σπλήνεσι, καὶ ἐπὶ πᾶν ἔνθεν καὶ ἔνθεν ἐπιδεῖν· πιέζειν δὲ ὥσπερ καὶ τἄλλα πάντα, ταύτῃ μάλιστα ᾗ ἐκινήθη, καὶ τὰς πρώτας περιβολὰς τῶν ὀθονίων κατὰ ταῦτα ποιεῖσθαι· ἐν δὲ ἑκάστῃ τῶν ἀπολυσίων ὕδατι πολλῷ θερμῷ χρῆσθαι· ἐν πᾶσι δὲ
10 πολλὸν ὕδωρ καταχεῖν τοῖσι κατ' ἄρθρα σίνεσιν. αἱ δὲ πιέξιες καὶ αἱ χαλάσιες ἐν τοῖσιν αὐτοῖσι χρόνοισι τὰ αὐτὰ σημεῖα δεικνυόντων ἅπερ ἐπὶ τοῖσι πρόσθεν καὶ τὰς μετεπιδέσιας ὡσαύτως χρὴ ποιεῖσθαι. ὑγιέες δὲ τελέως οὗτοι γίνονται ἐν τεσσεράκοντα ἡμέρῃσι μάλιστα, ἢν τολμέωσι κατακεῖσθαι· ἢν δὲ μή, πάσχουσι ταῦτα ἃ καὶ
17 πρότερον, καὶ ἐπὶ μᾶλλον.

XI. Ὅσοι δὲ πηδήσαντες ἀφ' ὑψηλοῦ τινὸς

[1] κατ' αὐτὴν τὴν ἴξιν.

these bones are completely healed in twenty days, except those which are connected with the leg-bones in a vertical line. It is good to lie up during this period, but patients, despising the injury, do not bring themselves to this, but go about before they are well. This is the reason why most of them do not make a complete recovery, and the pain often returns; naturally so, for the feet carry the whole weight. It follows that when they walk about before they are well, the displaced joints heal up badly; on which account they have occasional pains in the parts near the leg.

X.[1] The bones which are in connection with those of the leg are larger than the others,[2] and when they are displaced healing takes much longer. Treatment, indeed, is the same, but more bandages and pads should be used, also extend the dressings completely in both directions. Use pressure, as in all cases so here especially, at the point of displacement, and make the first turns of the bandage there. At each change of dressing use plenty of warm water; indeed, douche copiously with warm water in all injuries of joints. There should be the same signs as to pressure and slackness in the same periods as in the former cases, and the change of dressings should be made in the same way. These patients recover completely in about forty days, if they bring themselves to lie up; failing this, they suffer the same as the former cases, and to a greater degree.

XI. Those who, in leaping from a height, come

[1] Displacement of the astragalus?
[2] "Those of the wrist." Adams.

[2] συναλθεῖται.

ἐστηρίξαντο τῇ πτέρνῃ ἰσχυρῶς, τούτοις διΐστανται μὲν τὰ ὀστέα, φλέβια δὲ ἐκχυμοῦνται ἀμφιφλασθείσης τῆς σαρκὸς ἀμφὶ τὸ ὀστέον, οἴδημα δὲ ἐπιγίνεται καὶ πόνος πολύς. τὸ γὰρ ὀστέον τοῦτο οὐ σμικρόν ἐστι, καὶ ὑπερέχει μὲν ὑπὸ τὴν ἰθυωρίην τῆς κνήμης, κοινωνεῖ δὲ φλεψὶ καὶ νεύροισι ἐπικαίροισι· ὁ τένων δὲ ὀπίσθιος τούτῳ προσήρτηται τῷ ὀστέῳ. τούτους χρὴ
10 ἰητρεύειν μὲν κηρωτῇ καὶ σπλήνεσι καὶ ὀθονίοισιν· ὕδατι δὲ θερμῷ πλείστῳ ἐπὶ τούτοισι χρῆσθαι καὶ ὀθονίων πλειόνων ἐπὶ τούτοισι δεῖ καὶ ἄλλως ὡς βελτίστων καὶ προσηνεστάτων. καὶ ἢν μὲν τύχῃ ἁπαλὸν τὸ δέρμα φύσει ἔχον τὸ ἀμφὶ τῇ πτέρνῃ,[1] ἐᾶν οὕτως· ἢν δὲ παχὺ καὶ σκληρόν, οἷα μετεξέτεροι ἴσχουσιν, κατατάμνειν χρὴ ὁμαλῶς καὶ διαλεπτύνειν, μή διατιτρώσκοντα. ἐπιδεῖν δὲ ἀγαθῶς οὐ πάντος ἀνδρός ἐστι τὰ τοιαῦτα· ἢν γὰρ τις ἐπιδέῃ, ὥσπερ καὶ τὰ ἄλλα
20 τὰ κατὰ τὰ σφυρὰ ἐπιδεῖται, ὅτε μὲν περὶ τὸν πόδα περιβαλλόμενος, ὅτε δὲ περὶ τὸν τένοντα, αἱ ἀποσφίγξιες αὗται χωρίζουσι τὴν πτέρνην ᾗ τὸ φλάσμα ἐγένετο· καὶ οὕτω κίνδυνος σφακελίσαι τὸ ὀστέον τὸ τῆς πτέρνης· καίτοι ἢν σφακελίσῃ, τὸν αἰῶνα πάντα ἱκανὸν ἀντίσχειν τὸ νόσημα. καὶ γὰρ τἄλλα ὅσα μὴ ἐκ τοιούτου τρόπου σφακελίζει, ἀλλ' ἐν κατακλίσει μελανθείσης τῆς πτέρνης ὑπὸ ἀμελείης τοῦ σχήματος ἢ ἐν κνήμῃ τρώματος γενομένου ἐπικαίρου καὶ
30 χρονίου καὶ κοινοῦ τῇ πτέρνῃ, ἢ ἐν μηρῷ ἢ ἐπ' ἄλλῳ νοσήματι ὑπτιασμοῦ χρονίου γενομένου, ὁμῶς καὶ τοῖσι τοιούτοισι χρόνια, καὶ ὀχλώδεα καὶ πολλάκις ἀναρρηγνύμενα, ἢν μὴ χρηστῇ μὲν

down violently on the heel, get the bones separated, while there is extravasation from the blood-vessels since the flesh is contused about the bone. Swelling supervenes and severe pain, for this bone is not small, it extends beyond the line of the leg, and is connected with important vessels and cords. The back tendon[1] is inserted into this bone. You should treat these patients with cerate, pads and bandages, using an abundance of hot water, and they require plenty of bandages, the best and softest you can get. If the skin about the heel is naturally smooth, leave it alone, but if thick and hard as it is in some persons, you should pare it evenly and thin it down without going through to the flesh. It is not every man's job to bandage such cases properly, for if one applies the bandage, as is done in other lesions at the ankle, taking one turn round the foot and the next round the back tendon, the bandage compresses the part and excludes the heel where the contusion is, so that there is risk of necrosis of the heel-bone; and if there is necrosis the malady may last the patient's whole life. In fact, necrosis from other causes, as when the heel blackens while the patient is in bed owing to carelessness as to its position, or when there is a serious and chronic wound in the leg connected with the heel, or in the thigh, or another malady involving prolonged rest on his back—all these necroses are equally[2] chronic and troublesome, and often break out afresh if not treated with most

[1] Tendo Achillis.

[2] ὁμῶς, Littré's emendation for ὅμως, "nevertheless" (Kw. and codd.).

---

[1] τὴν πτέρνην.

μελέτῃ θεραπευθῇ, πολλῇ δὲ ἡσυχίῃ, ὥς τά γε σφακελίζοντα· ἐκ τοῦ τοιούτου δὲ τρόπου σφακελίζοντα καὶ κινδύνους μεγάλους τῷ σώματι παρέχει πρὸς τῇ ἄλλῃ λύμῃ. καὶ γὰρ πυρετοὶ ὑπεροξέες, συνεχέες, τρομώδεες, λυγγώδεες, γνώμης ἁπτόμενοι, καὶ ὀλιγήμεροι κτείνοντές τε·
40 γένοιντο δ' ἂν καὶ φλεβῶν αἱμορρόων πελιώσιες ναρκώσιες[1] καὶ γαγγραινώσιες ὑπὸ τῆς πιέξιος· γένοιτο δ' ἂν ταῦτα ἔξω τοῦ ἄλλου σφακελισμοῦ. ταῦτα μὲν οὖν εἴρηται, οἷα τὰ ἰσχυρότατα φλάσματα γίνεται· τὰ μέντοι πλεῖστα ἡσυχαίως ἀμφιφλᾶται καὶ οὐδεμίη πολλὴ σπουδὴ τῆς μελέτης, ἀλλ' ὅμως ὀρθῶς γε δεῖ χειρίζειν. ἐπὴν μέντοι ἰσχυρὸν δόξῃ εἶναι τὸ ἔρεισμα, τά τε εἰρημένα ποιεῖν χρή, καὶ τὴν ἐπίδεσιν τὴν πλείστην ποιεῖσθαι ἀμφὶ τὴν πτέρνην περι-
50 βάλλοντα, ἄλλοτε πρὸς τὰ ἄκρα τοῦ ποδὸς ἀντιπεριβάλλοντα, ἄλλοτε πρὸς τὰ μέσα, ἄλλοτε πρὸς τὰ περὶ τὴν κνήμην· προσεπιδεῖν δὲ καὶ τὰ πλησίον πάντα ἔνθεν καὶ ἔνθεν, ὥσπερ καὶ πρόσθεν εἴρηται· καὶ ἰσχυρὴν μὲν μὴ ποιεῖσθαι τὴν πίεξιν, ἐν πολλοῖσι δὲ τοῖσιν ὀθονίοισιν. ἄμεινον δὲ καὶ ἐλλέβορον πιπίσκειν[2] αὐθημερὸν ἢ τῇ ὑστεραίῃ· ἀπολῦσαι δὲ τριταῖον καὶ αὖθις μετεπιδῆσαι. σημεῖα δὲ τάδε, εἰ παλιγκοταίνει ἢ οὔ· ἐπὴν μὲν τὰ ἐκχυμώματα
60 τῶν φλεβῶν καὶ τὰ μελάσματα καὶ τὰ ἐγγὺς ἐκείνων ὑπέρυθρα γίνηται καὶ ὑπόσκληρα, κίνδυνος παλιγκοτῆσαι· ἀλλ' ἢν μὲν ἀπύρετος ᾖ, φαρμακεύειν ἄνω χρή, ὥσπερ εἴρηται, καὶ ὅσα ἂν μὴ συνεχῆ[3] πυρεταίνηται·[4] ἢν δὲ συνεχῆ πυρεταίνηται, μὴ φαρμακεύειν, ἀπέχειν δὲ σιτίων καὶ

skilful attention and long rest. Necroses of this sort, indeed, besides other harm, bring great dangers to the body, for there may be very acute fevers, continuous and attended by tremblings, hiccoughs and affections of the mind, fatal in a few days. There may also be lividity and congestion of the large blood-vessels, loss of sensation and gangrene due to compression, and these may occur without necrosis of the bone. The above remarks apply to very severe contusions, but the parts are often moderately contused and require no very great care, though, all the same, they must be treated properly. When, however, the crushing seems violent the above directions should be observed, the greater part of the bandaging being about the heel, taking turns sometimes round the end of the foot, sometimes about the middle part, and sometimes carrying it up the leg. All the neighbouring parts in both directions should be included in the bandage, as explained above; and do not make strong pressure, but use many bandages. It is also good to give a dose of hellebore on the first and second days. Remove the bandage and re-apply it on the third day. The following are signs of the presence and absence of aggravations. When there are extravasations from the blood-vessels, and blackenings, and the neighbouring parts become reddish and rather hard, there is danger of aggravation. Still, if there is no fever you should give an emetic as was directed; also in cases where the fever is not continuous; but if there is continued fever, do not give an evacuant, but avoid food, solid

[1] ναυσιώσιες (regurgitations), Galen and most MSS., but hard to accept.
[2] πῖσαι. [3] συνεχεῖ. [4] πυρεταίνῃ bis.

ῥοφημάτων, ποτῷ δὲ χρῆσθαι ὕδατι καὶ μὴ οἴνῳ, ἀλλὰ τῷ ὀξυγλυκεῖ. ἢν δὲ μὴ μέλλῃ παλιγκοταίνειν τὰ ἐκχυμώματα καὶ τὰ μελάσματα καὶ τὰ περιέχοντα, ὑπόχλωρα γίνεται καὶ οὐ σκληρά·
70 ἀγαθὸν τοῦτο τὸ μαρτύριον ἐν πᾶσι τοῖσιν ἐκχυμώμασι, τοῖσι μὴ μέλλουσι παλιγκοταίνειν· ὅσα δὲ σὺν σκληρύσμασι πελιοῦται, κίνδυνος μὲν μελανθῆναι. τὸν δὲ πόδα ἐπιτηδεύειν χρὴ ὅκως ἀνωτέρω τοῦ ἄλλου σώματος ἔσται τὰ πλεῖστα ὀλίγον. ὑγιὴς δ' ἂν γένοιτο ἐν ἑξήκοντα
76 ἡμέρῃσιν, εἰ ἀτρεμεῖ.[1]

XII. Ἡ δὲ κνήμη δύο ὀστέα ἔχει,[2] τῇ μὲν συχνῷ λεπτότερον τὸ ἕτερον τοῦ ἑτέρου, τῇ δὲ οὐ πολλῷ λεπτότερον· συνέχεται δὲ ἀλλήλοισι τὰ πρὸς τοῦ ποδός, καὶ ἐπίφυσιν κοινὴν ἔχει, ἐν ἰθυωρίῃ δὲ τῆς κνήμης οὐ συνέχεται· τὰ δὲ πρὸς τοῦ μηροῦ συνέχεται, καὶ ἐπίφυσιν ἔχει, καὶ ἡ ἐπίφυσις διάφυσιν· μακρότερον δὲ τὸ [ἕτερον] ὀστέον σμικρῷ τῷ [3] κατὰ τὸν σμικρὸν δάκτυλον· καὶ ἡ μὲν φύσις τοιαύτη τῶν ὀστέων τῶν ἐν τῇ
10 κνήμῃ.

XIII. Ὀλισθάνει δὲ ἔστιν ὅτε τὰ μὲν πρὸς τοῦ ποδός, ὅτε μὲν σὺν τῇ ἐπιφύσει ἀμφότερα τὰ ὀστέα, ὅτε δὲ ἡ ἐπίφυσις ἐκινήθη, ὅτε δὲ τὸ ἕτερον ὀστέον. ταῦτα δὲ ὀχλώδεα μὲν ἧσσον ἢ τὰ ἐν τῷ καρπῷ τῶν χειρῶν, εἰ τολμῷεν ἀτρεμεῖν οἱ ἄνθρωποι. ἴησις δὲ παραπλησίη, οἵη περ ἐκείνων· τήν τε γὰρ ἐμβολὴν χρὴ ποιεῖσθαι ἐκ κατατάσιος, ὥσπερ ἐκείνων, ἰσχυροτέρης δὲ δεῖται τῆς κατατάσιος, ὅσῳ καὶ ἰσχυρότερον τὸ σῶμα
10 ταύτῃ. ἐς δὲ τὰ πλεῖστα μὲν ἀρκέουσιν ἄνδρες

[1] ἀτρεμέοι. [2] ἐστίν.

or fluid, and for drink use water and not wine, but hydromel may be taken.[1] If there is not going to be aggravation, the effusions and blackenings and the parts around become yellowish and not hard. This is good evidence in all extravasations that they are not going to get worse, but in those which turn livid and hard there is danger of gangrene. One must see that the foot is, as a rule, a little higher than the rest of the body. The patient will recover in sixty days if he keeps at rest.

XII. The leg has two bones, one much more slender than the other at one end, but not so much at the other end. The parts near the foot are joined together and have a common epiphysis. In the length of the leg they are not united, but the parts near the thigh-bone are united and have an epiphysis, and the epiphysis has a diaphysis.[2] The bone on the side of the little toe is slightly the longer. This is the disposition of the leg-bones.

XIII. The bones are occasionally dislocated at the foot end, sometimes both bones with the epiphysis, sometimes the epiphysis is displaced, sometimes one of the bones. These dislocations give less trouble than those of the wrist, if the patients can bring themselves to lie up. The treatment is similar to that of the latter, for reduction is to be made by extension as in those cases, but stronger extension is requisite since the body is stronger in this part. As a rule two men suffice, one pulling one way and one

[1] A decoction of honeycomb in water, cf. Galen xviii(2). 466.
[2] Spinous process or medial projection.

---

[3] Pq. τῷ for το codd.: omitting ἕτερον cf. XVIII, XXXVII.

δύο, ὁ μὲν ἔνθεν, ὁ δὲ ἔνθεν τείνοντες. ἢν δὲ μὴ ἰσχύωσιν, ἰσχυροτέρην ῥηΐδιόν ἐστι ποιεῖν τὴν κατάτασιν· ἢ γὰρ πλήμνην κατορύξαντα χρή, ἢ ἄλλο τι ὅ τι τούτῳ ἔοικεν, μαλθακόν τι περὶ τὸν πόδα περιβάλλειν· ἔπειτα πλατέσι βοείοισιν ἱμᾶσιν περιδήσαντα τὸν πόδα τὰς ἀρχὰς τῶν ἱμάντων ἢ πρὸς ὕπερον ἢ πρὸς ἕτερον ξύλον προσδήσαντα, τὸ ξύλον πρὸς τὴν πλήμνην ἄκρον ἐντιθέντα ἐπανακλᾶν,[1] τοὺς δὲ ἀντιτείνειν 20 ἄνωθεν, τῶν τε ὤμων ἐχομένους καὶ τῆς ἰγνύης. ἔστι δὲ καὶ τὸ ἄνω τοῦ σώματος ἀνάγκῃ προσλαβεῖν τοῦτο μὲν ἢν βουλῇ, ξύλον στρογγύλον, λεῖον, κατορύξας βαθέως, μέρος τι αὐτοῦ ὑπερέχον τοῦ ξύλου μεσηγὺ τῶν σκελέων ποιήσασθαι παρὰ τὸν περίναιον, ὡς κωλύῃ ἀκολουθεῖν τὸ σῶμα τοῖσι πρὸς ποδῶν τείνουσιν· ἔπειτα πρὸς τὸ τεινόμενον σκέλος μὴ ῥέπειν, τὸν δέ τινα πλάγιον παρακαθήμενον ἀπωθεῖν τὸν γλουτόν, ὡς μὴ περιέλκηται τὸ σῶμα. 30 τοῦτο δὲ καὶ ἢν βούλῃ, περὶ τὰς μασχάλας ἔνθεν καὶ ἔνθεν τὰ ξύλα παραπέπηγεν,[2] αἱ δὲ χεῖρες παρατεταμέναι φυλάσσονται,[3] προσεπιλαμβανέτω[4] δέ τις κατὰ τὸ γόνυ, καὶ οὕτως ἀντιτείνοιτο. τοῦτο δ' ἢν παρὰ τὸ γόνυ βούληται,[5] ἄλλους ἱμάντας περιδήσας καὶ περὶ τὸν μηρόν, πλήμνην ἄλλην ὑπὲρ κεφαλῆς κατορύξας, ἐξαρτήσας τοὺς ἱμάντας ἔκ τινος ξύλου, τὸ ξύλον στηρίζων ἐς τὴν πλήμνην τἀναντία τῶν πρὸς ποδῶν ἕλκειν. τοῦτο δ' ἢν βούλῃ, ἀντὶ τῶν 40 πλημνέων δοκίδα ὑποτείνας ὑπὸ τὴν κλίνην μετρίην, ἔπειτα πρὸς τῆς δοκίδος ἔνθεν καὶ ἔνθεν τὴν κεφαλὴν στηρίζων καὶ ἀνακλῶν τὰ ξύλα,

the other, but if they cannot do it, it is easy to make the extension more powerful. Thus, one should fix a wheel-nave or something similar in the ground, put a soft wrapping round the foot, and then binding broad straps of ox-hide about it attach the ends of the straps to a pestle or some other rod. Put the end of the rod into the wheel-nave and pull back, while assistants hold the patient on the upper side grasping both at the shoulders and hollow of the knee. The upper part of the body can also be fixed by an apparatus. First, then you may fix a smooth, round rod deeply in the ground with its upper part projecting between the legs at the fork, so as to prevent the body from giving way when they make extension at the foot. Also it should not incline towards the leg which is being extended, but an assistant seated at the side should press back the hip so that the body is not drawn sideways. Again, if you like, the pegs may be fixed at either armpit, and the arms kept extended along the sides. Let someone also take hold at the knee, and so counter-extension may be made. Again, if one thinks fit, one may likewise fasten straps about the knee and thigh, and fixing another wheel-nave in the ground above the head, attach the straps to a rod; use the nave as a fulcrum for the rod and make extension counter to that at the feet. Further, if you like, instead of the wheel-naves, stretch a plank of suitable length under the bed, then, using the head of the plank at each end as fulcrum, draw back the rods and make exten-

---

[1] ἐνθέντα ἀνακλᾶν. [2] παραπεπήγῃ. [3] φυλάσσωνται.
[4] παρεπιλαμβάνηται. [5] βούλῃ.

κατατείνειν τοὺς ἱμάντας· ἢν δὲ θέλῃς, ὀνίσκους καταστήσας ἔνθεν καὶ ἔνθεν, ἐπ' ἐκείνων τὴν κατάτασιν ποιεῖσθαι. πολλοὶ δὲ καὶ ἄλλοι τρόποι κατατασίων· ἄριστον δέ, ὅστις ἐν πόλει μεγάλῃ ἰητρεύει, κεκτῆσθαι ἐσκευασμένον ξύλον, ἐν ᾧ πᾶσαι αἱ ἄναγκαι ἔσονται πάντων μὲν κατηγμάτων, πάντων δὲ ἄρθρων ἐμβολῆς ἐκ 50 κατατάσιος καὶ μοχλεύσιος· ἀρκεῖ δὲ τὸ ξύλον, ἢν ᾖ τοιοῦτον οἷον οἱ τετράγωνοι στῦλοι οἷοι δρύϊνοι γίνονται, μῆκος καὶ πλάτος καὶ πάχος.

Ἐπὴν δὲ ἱκανῶς κατατανύσῃς, ῥηΐδιον ἤδη τὸ ἄρθρον ἐμβαλεῖν· ὑπεραιωρεῖται γὰρ ἐς ἰθυωρίην ὑπὲρ τῆς ἀρχαίης ἕδρης. κατορθοῦσθαι οὖν χρὴ τοῖσι θέναρσι τῶν χειρῶν, τοῖσι μὲν ἐς τὸ ἐξεστηκὸς ἐρείδοντα, τοῖσι δὲ ἐπὶ θάτερα κατώτε- 58 ρον τοῦ σφυροῦ ἀντερείδοντα.

XIV. Ἐπὴν δ' ἐμβάλῃς, ἢν μὲν οἷόν τε ᾖ, κατατεταμένον ἐπιδεῖν χρή· ἢν δὲ κωλύηται ὑπὸ τῶν ἱμάντων, ἐκείνους λύσαντα ἀντικατατείνειν, ἔστ' ἂν ἐπιδήσῃς. ἐπιδεῖν δὲ τὸν αὐτὸν τρόπον καὶ τὰς ἀρχὰς ὡσαύτως βαλλόμενον κατὰ τὸ ἐξεστηκός, καὶ τὰς περιβολὰς τὰς πρώτας πλείστας κατὰ τοῦτο ποιεῖσθαι, καὶ τοὺς σπλῆνας πλείστους κατὰ τοῦτο, καὶ τὴν πίεξιν μάλιστα κατὰ τωὐτό· προσεπιδεῖν δὲ καὶ ἔνθεν 10 καὶ ἔνθεν ἐπὶ συχνόν· μᾶλλον δέ τι τοῦτο τὸ ἄρθρον πεπιέχθαι χρὴ ἐν τῇ πρώτῃ ἐπιδέσει ἢ τὸ ἐν τῇ χειρί. ἐπὴν δὲ ἐπιδήσῃς, ἀνωτέρω μὲν τοῦ ἄλλου σώματος ἐχέτω τὸ ἐπιδεθέν, τὴν δὲ θέσιν δεῖ ποιεῖσθαι οὕτως, ὅπως ἥκιστα ἀπαιω-

sion on the straps. And if you choose, set up windlasses at either end and make the extension by them. There are also many other methods for extensions. The best thing for anyone who practises in a large city is to get a wooden apparatus comprising all the mechanical methods for all fractures and for reduction of all joints by extension and leverage. This wooden apparatus will suffice if it be like the quadrangular supports such as are made of oak[1] in length, breadth and thickness.

When you make sufficient extension it is then easy to reduce the joint for it is elevated in a direct line above its old position. It should therefore be adjusted with the palms of the hands, pressing upon the projecting part with one palm and with the other making counter pressure below the ankle on the opposite side.[2]

XIV. After reduction, you should, if possible, apply a bandage, while the limb is kept extended. If the straps get in the way, remove them and keep up counter extension while bandaging. Bandage in the same way (as for fractures) putting the heads of the bandages on the projecting part and making the first and most turns there, also most of the compresses should be there and the pressure should come especially on this part. Also extend the dressing considerably to either side. This joint requires somewhat greater pressure at the first bandaging than does the wrist. After dressing let the bandaged part be higher than the rest of the body, and put it up in a position in which the foot is as little as

[1] Adams' "threshing boards"—Littré's τρίβολοι, a rash suggestion which he afterwards withdrew.

[2] The nature of these dislocations is discussed on pp. 425 ff.

ῥηθήσεται ὁ πούς. τὸν δὲ ἰσχνασμὸν τοῦ σώματος οὕτως ποιεῖσθαι, ὁποίην τινὰ δύναμιν ἔχει καὶ τὸ ὀλίσθημα· τὰ μὲν γὰρ σμικρόν, τὰ δὲ μέγα ὀλισθάνει. τὸ ἐπίπαν δὲ ἰσχναίνειν μᾶλλον καὶ ἐπὶ πλείω χρόνον χρὴ ἐν τοῖσι κατὰ
20 τὰ σκέλεα τρώμασι ἢ ἐν τοῖσι κατὰ τὰς χεῖρας·[1] καὶ γὰρ μέζω καὶ παχύτερα ταῦτα ἐκείνων· καὶ δὴ καὶ ἀναγκαῖον ἐλινύειν τὸ σῶμα καὶ κατακεῖσθαι. μετεπιδῆσαι δὲ τὸ ἄρθρον οὔτε τι κωλύει τριταῖον οὔτε κατεπείγει· καὶ τὰ ἄλλα πάντα παραπλησίως χρὴ ἰητρεύειν, ὥσπερ καὶ τὰ παροιχόμενα. καὶ ἢν μὲν τολμᾷ ἀτρέμα κατακεῖσθαι, ἱκαναὶ τεσσαράκοντα ἡμέραι, ἢν μοῦνον ἐς τὴν ἑωυτῶν χώρην τὰ ὀστέα αὖθις καθίζηται· ἢν δὲ μὴ θέλῃ ἀτρεμεῖν, χρῷτο μὲν
30 ἂν οὐ ῥᾳδίως[2] τῷ σκέλει, ἐπιδεῖσθαι δὲ ἀναγκάζοιτ᾽ ἂν πολὺν χρόνον. ὁπόσα μέντοι τῶν ὀστέων μὴ τελέως ἵζει ἐς τὴν ἑωυτῶν χώρην, ἀλλά τι ἐπιλείπει, τῷ χρόνῳ λεπτύνεται ἰσχίον καὶ μηρὸς καὶ κνήμη· καὶ ἢν μὲν ἔσω ὀλίσθῃ, τὸ ἔξω μέρος λεπτύνεται, ἢν δὲ ἔξω, τὸ ἔσω· τὰ
36 πλεῖστα δὲ ἐς τὸ ἔσω ὀλισθάνει.

XV. Ἐπὴν δὲ κνήμης ὀστέα ἀμφότερα καταγῇ ἄνευ ἑλκώσιος, κατατάσιος ἰσχυροτέρης δεῖται. τείνειν[3] τούτων τῶν τροπων ἐνίοισι τῶν προειρημένων τισί, ἢν μεγάλαι αἱ παραλλάξιες ἔωσιν. ἱκαναὶ δὲ καὶ αἱ ἀπὸ τῶν ἀνδρῶν κατατάσιες· τὰ πλεῖστα γὰρ ἀρκέοιεν ἂν δύο ἄνδρες ἐρρωμένοι, ὁ μὲν ἔνθεν, ὁ δὲ ἔνθεν ἀντιτείνοντες. τείνειν δὲ ἐς τὸ ἰθὺ χρὴ κατὰ φύσιν καὶ κατὰ τὴν

[1] κατὰ χεῖρα. [2] βραδέως, omit οὐ. [3] κατατείνειν.

possible unsupported.[1] The patient should undergo a reducing process corresponding to his strength and to the displacement, for the displacement may be small or great. As a rule the reducing treatment should be stricter and more prolonged in injuries about the leg region than in those about the arm region, for the former parts are larger and stouter than the latter. And it is especially needful for the body to be at rest and lie up. As to rebandaging the joint on the third day, there is neither hindrance nor urgency, and one should conduct all the other treatment as in the previous cases. If the patient brings himself to keep at rest and lie up, forty days are sufficient, provided only that the bones are back again in their places. If he will not keep at rest, he will not easily recover the use of the leg and will have to use bandages for a long time. Whenever the bones are not completely replaced but there is something wanting, the hip, thigh and leg gradually become atrophied. If the dislocation is inwards the outer part is atrophied, if outwards, the inner: now most dislocations are inwards.[2]

XV. When both leg-bones are broken without an external wound, stronger extension is required. If there is much overlapping make extension by some of those methods which have been described. But extensions made by man-power are also sufficient, for in most cases two strong men are enough, one pulling at each end. The traction should be in a straight line in accordance with the natural direction

[1] Not merely prevented from hanging down, but kept at right angles to the leg (cf. Galen).

[2] *i.e.* of the foot outwards and the leg inwards.

ἰθυωρίην τῆς κνήμης καὶ τοῦ μηροῦ, καὶ ἢν κνήμης
10 ὀστέα κατεηγυίης κατατείνῃς, καὶ ἢν μηροῦ.
καὶ ἐπιδεῖν δὲ οὕτως ἐκτεταμένων ἀμφοτέρων,
ὁπότερον ἂν τούτων ἐπιδέῃς· οὐ γὰρ ταὐτὰ
συμφέρει σκέλεΐ τε καὶ χειρί· πήχεος μὲν γὰρ
καὶ βραχίονος ἐπὴν ἐπιδεθῶσιν ὀστέα κατεηγότα,
ἀναλαμβάνεται ἡ χείρ, καὶ ἢν ἐκτεταμένα
ἐπιδέῃς, τὰ σχήματα τῶν σαρκῶν ἑτεροιοῦται
ἐν τῇ συγκάμψει τοῦ ἀγκῶνος· ἀδύνατος γὰρ
ὁ ἀγκὼν ἐκτετάσθαι πολὺν χρόνον· οὐ γὰρ
πολλάκις ἐν τοιούτῳ εἴθισται ἐσχηματίσθαι,
20 ἀλλ' ἐν τῷ συγκεκάμφθαι· καὶ δὴ καὶ ἅτε
δυνάμενοι οἱ ἄνθρωποι περιϊέναι συγκεκάμφθαι
κατὰ τὸν ἀγκῶνα δέονται. σκέλος δὲ ἔν τε τῇσιν
ὁδοιπορίῃσιν καὶ ἐν τῷ ἑστάναι εἴθισται ὅτε μὲν
ἐκτετάσθαι, ὅτε δὲ σμικροῦ δεῖν ἐκτέτασθαι· καὶ
εἴθισται καθεῖσθαι ἐς τὸ κάτω κατὰ τὴν φύσιν,
καὶ δὴ καὶ πρὸς τὸ ὀχέειν τὸ ἄλλο σῶμα· διὰ τοῦτο
εὔφορον αὐτῷ ἐστὶ τὸ ἐκτετάσθαι, ὅταν ἀνάγκην[1]
ἔχῃ· καὶ δὴ καὶ ἐν τῇσι κοιτῇσι πολλάκις ἐν τῷ
σχήματι τούτῳ ἐστὶν [ἐν τῷ ἐκτετάσθαι].[2] ἐπὴν
30 δὲ δὴ τρωθῇ, ἀνάγκη[3] καταδουλοῦται τὴν
γνώμην, ὅτι ἀδύνατοι μετεωρίζεσθαι γίνονται,
ὥστε οὐδὲ μέμνηνται περὶ τοῦ συγκαμφθῆναι
καὶ ἀναστῆναι, ἀλλ' ἀτρεμέουσι[4] ἐν τούτῳ τῷ
σχήματι κείμενοι. διὰ οὖν ταύτας τὰς προφάσιας χειρὸς καὶ σκέλεος οὔτε ἡ κατάτασις οὔτε
ἡ ἐπίδεσις τοῦ σχήματος συμφέρει ἡ αὐτή. ἢν
μὲν οὖν ἱκανὴ ἡ κατάτασις ἡ ἀπὸ τῶν ἀνδρῶν ᾖ,
οὐ δεῖ μάτην πονεῖσθαι—καὶ γὰρ σολοικότερον
μηχανοποιεῖν μηδὲν δέον—ἢν δὲ μὴ ἱκανὴ ἡ κατά-
40 τασις ἀπὸ τῶν ἀνδρῶν, καὶ τῶν ἄλλων τινὰ τῶν

of the leg and thigh, both when it is being made for fractures of the leg bones and of the thigh. Apply the bandage while both[1] are extended, whichever of the two you are dressing, for the same treatment does not suit both leg and arm. For when fractures of the forearm and upper arm are bandaged, the arm is slung, and if you bandage it when extended the positions of the fleshy parts are altered by bending the elbow. Further, the elbow cannot be kept extended a long time, since it is not used to that posture, but to that of flexion. And besides, since patients are able to go about after injuries of the arm, they want it flexed at the elbow. But the leg both in walking and standing is accustomed to be sometimes extended and sometimes nearly so, and it is naturally directed downwards and, what is more, its function is to support the body. Extension therefore is easily borne when necessary and indeed it frequently has this position in bed. If then it is injured, necessity brings the mind into subjection, because patients are unable to rise, so that they do not even think of bending their legs and getting up, but keep lying at rest in this posture. For these reasons, then, the same position either in making extension or bandaging is unsuitable for both arm and leg. If, then, extension by man-power is enough, one should not take useless trouble, for to have recourse to machines when not required is rather absurd. But if extension by man-power is not enough,

[1] *i.e.* thigh and leg.

---

[1] ἀνάγκῃ.
[2] Seems an obvious gloss. Most editors omit.
[3] καὶ ἡ ἀνάγκη.
[4] τολμῶσιν.

ἀναγκέων προσφέρειν, ἥντινά γε προσχωρέῃ.[1] ὅταν δὲ δὴ ἱκανῶς καταταθῇ, ῥηΐδιον ἤδη κατορθώσασθαι τὰ ὀστέα καὶ ἐς τὴν φύσιν ἀγαγεῖν, τοῖσι θέναρσι τῶν χειρῶν ἀπευθύνοντα καὶ
45 ἐξευκρινέοντα.

XVI. Ἐπὴν δὲ κατορθώσῃς, ἐπιδεῖν τοῖσιν ὀθονίοισι κατατεταμένον, ἤν τ᾽ ἐπὶ δεξιὰ ἤν τ᾽ ἐπ᾽ ἀριστερὰ περιφέρειν συμφέρῃ αὐτοῖσι τὰ πρῶτα ὀθόνια· βάλλεσθαι δὲ τὴν ἀρχὴν τοῦ ὀθονίου κατὰ τὸ κάτηγμα, καὶ περιβάλλεσθαι κατὰ τοῦτο τὰς πρώτας περιβολάς· κἄπειτα νέμεσθαι ἐπὶ τὴν ἄνω κνήμην ἐπιδέων, ὥσπερ ἐπὶ τοῖσιν ἄλλοισι κατήγμασι εἴρηται. τὰ δὲ ὀθόνια πλατύτερα χρὴ εἶναι καὶ μακρότερα καὶ πλέω
10 πολὺ αὖ τὰ[2] κατὰ τὸ σκέλος τῶν ἐν τῇ χειρί. ἐπὴν δ᾽ ἐπιδήσῃς, καταθεῖναι ἐφ᾽ ὁμαλοῦ τινὸς καὶ μαλθακοῦ, ὥστε μὴ διεστράφθαι ᾗ τῇ ᾗ τῇ, μήτε λορδὸν μήτε κυφὸν εἶναι· μάλιστα δὲ συμφέρει προσκεφάλαιον, ἢ λίνεον ἢ ἐρίνεον, μὴ σκληρόν, λαπαρὸν μέσον κατὰ μῆκος ποιήσαντα, ἢ ἄλλο τι ὃ τούτῳ ἔοικεν.

Περὶ γὰρ τῶν σωλήνων τῶν ὑποτιθεμένων ὑπὸ τὰ σκέλεα τὰ κατεηγότα, ἀπορέω ὅ τι συμβουλεύσω· ἢ ὑποτιθέναι χρὴ ἢ οὔ; ὠφελέουσι μὲν γάρ,
20 οὐχ ὅσον δὲ οἱ ὑποτιθέντες οἴονται· οὐ γὰρ ἀναγκάζουσι οἱ σωλῆνες ἀτρεμεῖν, ὡς οἴονται· οὔτε γὰρ τῷ ἄλλῳ σώματι στρεφομένῳ ἢ ἔνθα ἢ ἔνθα ἐπαναγκάζει ὁ σωλὴν μὴ ἐπακολουθεῖν τὸ σκέλος, ἢν μὴ ἐπιμελῆται αὐτὸς ὤνθρωπος· οὔτε αὖ τὸ[3] σκέλος ἄνευ τοῦ σώματος κωλύει ὁ σωλὴν κινηθῆναι ᾗ τῇ ᾗ τῇ· ἀλλὰ μὴν ἀστερ-

[1] προχωρῇ.

bring in some of the mechanical aids, whichever may be useful.[1] When once sufficient extension is made, it becomes fairly easy to adjust the bones to their natural position by straightening them and making coaptation with the palms of the hands.

XVI. After adjustment, apply the bandages while the limb is extended, making the turns with the first bandage, either to right or left as may be suitable. Put the head of the bandage at the fracture and make the first turns there, and then carry the bandaging to the upper part of the leg as was directed for the other fractures. The bandages should be broader and longer and much more numerous for the leg parts than those of the arm. On completing the dressing, put up the limb on something smooth and soft so that it does not get distorted to either side or become concave or convex. The most suitable thing to put under is a pillow of linen or wool, not hard, making a median longitudinal depression in it, or something that resembles this.

As for the hollow splints which are put under fractured legs I am at a loss what to advise as regards their use. For the good they do is not so great as those who use them suppose. The hollow splints do not compel immobility as they think, for neither does the hollow splint forcibly prevent the limb from following the body when turned to either side, unless the patient himself sees to it, nor does it hinder the leg itself apart from the body from moving this way or that. Besides, it is, of course,

[1] ἥντινα Littré; ἥν vulg.: "if any is of use."

---

[2] For αὐτὰ (codd.); cf. below, line 25. τά Kw.
[3] αὐτό.

γέστερον ξύλον ὑποτετάσθαι, ἢν μὴ ὁμῶς ἄν[1] τις μαλθακόν τι ἐς αὐτὸ ἐντεθῇ· εὐχρηστότατον δέ ἐστιν ἐν τῇσι μεθυποστρώσεσι καὶ ἐν τῇσιν ἐς
30 ἄφοδον προχωρήσεσιν. ἔστιν οὖν σὺν σωλῆνι καὶ ἄνευ σωλῆνος, καὶ καλῶς καὶ αἰσχρῶς κατασκευάσασθαι. πιθανώτερον δὲ τοῖσι δημότῃσίν ἐστι καὶ τὸν ἰητρὸν ἀναμαρτητότερον εἶναι, ἢν σωλὴν ὑποκέηται· καίτοι ἀτεχνέστερόν γέ ἐστιν. δεῖ μὲν γὰρ ἐφ᾽ ὁμαλοῦ καὶ μαλθακοῦ κεῖσθαι πάντῃ πάντως ἐς ἰθύ· ἐπεί τοί γε ἀνάγκη κρατηθῆναι τὴν ἐπίδεσιν ὑπὸ τῆς διαστροφῆς τῆς ἐν τῇ διαθέσει, ὅποι ἂν ῥέπῃ καὶ ὁπόσα ἂν ῥέπῃ. ὑποκρινέσθω δὲ ὁ ἐπιδεδεμένος
40 ταὐτα, ἅπερ καὶ πρότερον εἴρηται· καὶ γὰρ τὴν ἐπίδεσιν χρὴ τοιαύτην εἶναι καὶ τὸ οἴδημα οὕτως ἐξαείρεσθαι ἐς τὰ ἄκρεα καὶ τὰς χαλάσιας οὕτω, καὶ τὰς μετεπιδέσιας διὰ τρίτης· καὶ εὑρισκέσθω ἰσχνότερον τὸ ἐπιδεόμενον, καὶ τὰς ἐπιδέσιας ἐπὶ μᾶλλον ποιεῖσθαι καὶ πλέοσι τοῖσιν ὀθονίοισιν· περιλαμβάνειν τε τὸν πόδα χαλαρῶς, ἢν μὴ ἄγαν ἐγγὺς ᾖ τοῦ γούνατος τὸ τρῶμα. κατατείνειν δὲ μετρίως καὶ ἐπικατορθοῦν ἐφ᾽ ἑκάστῃ ἐπιδέσει χρὴ τὰ ὀστέα· ἢν γὰρ ὀρθῶς μὲν ἰητρεύηται, κατὰ λόγον δὲ τὸ οἴδημα χωρῇ, ἔτι[2] μὲν λεπτότερον καὶ
50 ἰσχνότερον τὸ ἐπιδεόμενον χωρίον ἔσται, ἔτι δὲ αὖ παραγωγότερα τὰ ὀστέα, ἀνακούοντα τῆς κατατάσιος μᾶλλον. ἐπὴν δὲ ἑβδομαῖος ἢ ἐνναταῖος ἢ ἑνδεκαταῖος γένηται, τοὺς νάρθηκας προστιθέναι,[3] ὥσπερ καὶ ἐπὶ τοῖσιν ἄλλοισι κατήγμασι εἴρηται. τῶν δὲ ναρθήκων τὰς ἐνέδρας χρὴ φυλάσσεσθαι κατά τε τῶν σφυρῶν τὴν ἴξιν καὶ κατὰ τὸν τένοντα τὸν ἐν τῇ κνήμῃ τοῦ ποδός.

rather unpleasant to have wood under the limb unless at the same time one inserts something soft. But it is very useful in changing the bed clothes, and in getting up to go to stool. It is thus possible either with or without the hollow splint to arrange the matter well or clumsily. Still the vulgar have greater faith in it, and the practitioner will be more free from blame if a hollow splint is applied, though it is rather bad practice. Anyhow, the limb should be on something smooth and soft and be absolutely straight, since it necessarily follows that the bandaging is overcome by any deviation in posture, whatever the direction or extent of it may be. The patient should give the same answers as those above mentioned, for the bandaging should be similar, and there should be the like swelling on the extremities, and so with the looseness and the changes of dressing every third day. So, too, the bandaged part should be found more slender and greater pressure be used in the dressings and more bandages. You should also make some slack turns round the foot if the injury is not very near the knee. One should make moderate extension and adjustment of the bones at each dressing; for if the treatment be correct and the oedema subsides regularly, the bandaged part will be more slender and attenuated while the bones on their side will be more mobile and lend themselves more readily to extension. On the seventh, ninth, or eleventh day splints should be applied as was directed in the case of other fractures, and one must be careful as to the position of the splints, both in the line of the ankles, and about the back tendon

[1] ὅμαλον Kw. in Hermes XXVII. αὖτις in text.
[2] ἐπὶ bis. [3] χρὴ προστιθέναι.

ὀστέα δὲ κνήμης κρατύνεται ἐν τεσσαράκοντα ἡμέρῃσιν, ἢν ὀρθῶς ἰητρεύηται. ἢν δὲ ὑποπτεύῃς
60 τῶν ὀστέων τι δεῖσθαί τινος διορθώσιος ἢ τινα ἕλκωσιν ὀρρωδῇς, ἐν τῷ μεσηγὺ χρόνῳ χρὴ
62 λύσαντα καὶ εὐθετισάμενον μετεπιδῆσαι.

XVII. Ἢν δὲ τὸ ἕτερον ὀστέον κατεηγῇ ἐν κνήμῃ, κατατάσιος μὲν ἀσθενεστέρης δεῖται. οὐ μὴν ἐπιλείπειν χρή, οὐδὲ βλακεύειν ἐν τῇ κατατάσει, μάλιστα μὲν τῇ πρώτῃ ἐπιδέσει κατατείνεσθαι ὅσον ἐφικνεῖται αἰεί ποτε πάντα τὰ κατήγματα, εἰ δὲ μή, ὡς τάχιστα· ὅ τι γὰρ ἂν μὴ κατὰ τρόπον ηὐθετισμένων[1] τῶν ὀστέων ἐπιδέων τις πιέζῃ, ὀδυναίτερον τὸ χωρίον γίνεται.
9 ἡ δὲ ἄλλη ἰητρείη ἡ αὐτή.

XVIII. Τῶν δὲ ὀστέων, τὸ μὲν ἔσω τοῦ ἀντικνημίου καλεομένου ὀχλωδέστερον ἐν τῇ ἰητρείῃ ἐστί, καὶ κατατάσιος μᾶλλον δεόμενον, καὶ ἢν μὴ ὀρθῶς τὰ ὀστέα τεθῇ, ἀδύνατον κρύψαι· φανερὸν γὰρ καὶ ἄσαρκον πᾶν ἐστίν· καὶ ἐπιβαίνειν ἐπὶ τὸ σκέλος πολλῷ βραδύτερον δύναιντ᾽ ἄν, τούτου κατεηγότος. ἢν δὲ τὸ ἔξω ὀστέον κατεηγῇ,[2] πολὺ μὲν εὐφορώτερον φέρουσι, πολὺ δὲ εὐκρυπτότερον, καὶ ἢν μὴ καλῶς συντεθῇ
10 (ἐπίσαρκον γάρ ἐστιν), ἐπὶ πόδας τε ταχέως ἵστανται, τὸ πλεῖστον γὰρ τοῦ ἄχθεος ὀχεῖ τὸ ἔσωθεν τοῦ ἀντικνημίου ὀστέον. ἅμα μὲν γὰρ αὐτῷ τῷ σκέλει καὶ τῇ ἰθυωρίῃ τοῦ ἄχθεος τοῦ κατὰ τὸ σκέλος, τὸ πλεῖον ἔχει τοῦ πόνου τὸ ἔσω ὀστέον· τοῦ γὰρ μηροῦ ἡ κεφαλὴ ὑπεροχεῖ τὸ ὕπερθεν τοῦ σώματος, αὕτη δὲ ἔσωθεν πέφυκε τοῦ σκέλεος καὶ οὐκ ἔξωθεν, ἀλλὰ κατὰ τὴν τοῦ

[1] εὐθετισμένων. [2] καταγῇ.

from leg to foot. The bones of the leg solidify in forty days if properly treated. If you suspect that one of the bones requires some adjustment, or are afraid of ulceration, you should unbandage the part in the interval and reapply after putting it right.

XVII. If one[1] of the leg-bones be broken, the extension required is weaker: there should, however, be no shortcoming or feebleness about it. Especially at the first dressing sufficient extension should be made in all fractures so as to bring the bones together, or, failing this, as soon as possible, for when one in bandaging uses pressure, if the bones have not been properly set, the part becomes more painful. The rest of the treatment is the same.

XVIII. Of the bones, the inner of the so-called shin is the more troublesome to treat, requiring greater extension, and if the fragments are not properly set, it cannot be hid, for it is visible and entirely without flesh. When this bone is broken, patients take longer before they can use the leg, while if the outer bone be fractured they have much less inconvenience to bear, and, even if not well set, it is much more readily concealed; for it is well covered: and they can soon stand. For the inner shin bone carries the greatest part of the weight, since both by the disposition of the leg itself and by the direct line of the weight upon the leg the inner bone has most of the work. Further, the head of the thigh-bone sustains the body from below and has its natural direction towards the inner side of the leg and not the outer, but is in the line of the shin

[1] Littré and others apply this to the fibula, but the limitation seems uncalled for.

ἀντικνημίου ἴξιν· ἅμα δὲ τὸ ἄλλο ἥμισυ τοῦ σώματος γειτονεύεται μᾶλλον ταύτῃ τῇ ἴξει,
20 ἀλλ' οὐχὶ τῇ ἔξωθεν· ἅμα δέ, ὅτε παχύτερον τὸ ἔσω τοῦ ἔξωθεν, ὥσπερ καὶ ἐν τῷ πήχει τὸ κατὰ τὴν τοῦ μικροῦ δακτύλου ἴξιν λεπτότερον καὶ μακρότερον. ἐν μέντοι τῷ ἄρθρῳ τῷ κάτω[1] οὐχ ὁμοίη ἡ ὑπότασις τοῦ ὀστέου τοῦ μακροτέρου· ἀνομοίως γὰρ ὁ ἀγκὼν καὶ ἡ ἰγνύη κάμπτεται. διὰ οὖν ταύτας τὰς προφάσιας τοῦ μὲν ἔξωθεν ὀστέου κατεηγότος,[2] ταχεῖαι αἱ ἐπιβάσιες, τοῦ δὲ
28 ἔσωθεν κατεηγότος, βραδεῖαι αἱ ἐπιβάσιες.

XIX. Ἢν δὲ τὸ τοῦ μηροῦ ὀστέον καταγῇ, τὴν κατάτασιν χρὴ ποιεῖσθαι περὶ παντός, ὅπως μὴ ἐνδεεστέρως σχήσει· πλεονασθεῖσα μὲν γὰρ οὐδὲν ἂν σίνοιτο· οὐδὲ γὰρ εἰ διεστεῶτα τὰ ὀστέα ὑπὸ τῆς ἰσχύος τῆς κατατάσιος ἐπιδέοι τις, οὐκ ἂν δύναιτο κρατεῖν ἡ ἐπίδεσις ὥστε διεστάναι, ἀλλὰ συνέλθοι ἂν πρὸς ἄλληλα τὰ ὀστέα ὅτι τάχιστα [ἂν][3] ἀφείησαν οἱ τείνοντες· παχεῖαι γὰρ καὶ ἰσχυραὶ αἱ σάρκες ἐοῦσαι,
10 κρατήσουσι τῆς ἐπιδέσιος, ἀλλ' οὐ κρατηθήσονται. περὶ οὗ οὖν ὁ λόγος, διατείνειν εὖ μάλα καὶ ἀδιαστρέπτως χρή, μηδὲν ἐπιλείποντα· μεγάλη γὰρ ἡ αἰσχύνη καὶ βλάβη βραχύτερον τὸν μηρὸν ἀποδεῖξαι. χεὶρ μὲν γάρ, βραχυτέρη γενομένη, καὶ συγκρυφθείη ἂν καὶ οὐ μέγα τὸ σφάλμα· σκέλος δὲ βραχύτερον γενόμενον χωλὸν ἀποδείξειε[4] τὸν ἄνθρωπον· τὸ γὰρ ὑγιὲς ἐλέγχει παρατιθέμενον μακρότερον ἐόν, ὥστε λυσιτελεῖ τὸν μέλλοντα κακῶς ἰητρεύεσθαι, ἀμφότερα
20 καταγῆναι τὰ σκέλεα μᾶλλον ἢ τὸ ἕτερον· ἰσόρροπος γοῦν ἂν εἴη αὐτὸς ἑωυτῷ. ἐπὴν μέντοι

bone. So, too, the corresponding half of the body is nearer the line of this bone than that of the outer one, and besides, the inner is thicker than the outer, just as in the forearm the bone on the side of the little finger is longer and more slender; but in this lower articulation the longer bone does not lie underneath in the same way, for flexion at the elbow and knee are dissimilar. For these reasons, when the outer bone is fractured patients soon get about; but when the inner one is broken they do so slowly.

XIX. If the thigh-bone is fractured, it is most important that there should be no deficiency in the extension that is made, while any excess will do no harm. In fact, even if one should bandage while the bones were separated by the force of the extension, the dressing would have no power to keep them apart, but they would come together immediately when the assistants relaxed their tension. For the fleshy part being thick and powerful will prevail over the bandaging, and not be overcome by it. To come to our subject, one should extend very strongly and without deviation leaving no deficiency, for the disgrace and harm are great if the result is a shortened thigh. The arm, indeed, when shortened may be concealed and the fault is not great, but the leg when shortened will leave the patient lame, and the sound leg being longer (by comparison) exposes the defect; so that if a patient is going to have unskilful treatment, it is better that both his legs should be broken than one of them, for then at least he will be in equilibrium. When, therefore, you have made suffi-

[1] *τῷ κάτω ἄρθῳ τούτῳ.* [2] *καταγέντος* bis.
[3] Omit B M V Kw. [4] *ἀποδείξει.*

ἱκανῶς κατατανύσῃς, κατορθωσάμενον χρὴ τοῖσι θέναρσι τῶν χειρῶν ἐπιδεῖν τὸν αὐτὸν τρόπον, ὥσπερ καὶ πρόσθεν γέγραπται, καὶ τὰς ἀρχὰς βαλλόμενον, ὥσπερ εἴρηται, καὶ νεμόμενον ἐς τὸ ἄνω τῇ ἐπιδέσει. καὶ ὑποκρινέσθω ταὐτὰ ὥσπερ καὶ πρόσθεν, καὶ πονείτω κατὰ ταὐτὰ καὶ ῥηΐζέτω· καὶ μετεπιδείσθω ὡσαύτως, καὶ ναρθήκων πρόσθεσις ἡ αὐτή. κρατύνεται δὲ ὁ μηρὸς
30 ἐν πεντήκοντα ἡμέρῃσιν.

XX. Προσσυνιέναι δὲ χρὴ καὶ τόδε, ὅτι ὁ μηρὸς γαῦσός ἐστιν ἐς τὸ ἔξω μέρος μᾶλλον ἢ ἐς τὸ ἔσω, καὶ ἐς τὸ ἔμπροσθεν μᾶλλον ἢ ἐς τοὔπισθεν· ἐς ταῦτα τοίνυν τὰ μέρεα καὶ διαστρέφεται, ἐπὴν μὴ καλῶς ἰητρεύηται· καὶ δὴ καὶ κατὰ ταῦτα ἀσαρκότερος αὐτὸς ἑωυτοῦ ἐστίν, ὥστε οὐδὲ συγκρύπτειν δύνανται, ἐν τῇ διαστροφῇ. ἢν οὖν τι τοιοῦτον ὑποπτεύῃς, μηχανοποιεῖσθαι χρὴ οἷά περ ἐν τῷ βραχίονι τῷ διεστραμμένῳ
10 παρῄνηται.[1] προσπεριβάλλειν δὲ χρὴ ὀλίγα τῶν ὀθονίων κύκλῳ ἀμφὶ τὸ ἰσχίον καὶ τὰς ἰξύας, ὅπως ἂν οἱ βουβῶνές τε καὶ τὸ ἄρθρον τὸ κατὰ τὴν πλιχάδα καλουμένην προσεπιδέηται· καὶ γὰρ ἄλλως συμφέρει, καὶ ὅπως μὴ τὰ ἄκρεα τῶν ναρθήκων σίνηται πρὸς τὰ ἀνεπίδετα προσβαλλόμενα. ἀπολείπειν δὲ χρὴ ἀπὸ τοῦ γυμνοῦ αἰεὶ τοὺς νάρθηκας καὶ ἔνθεν καὶ ἔνθεν ἱκανῶς·[2] καὶ τὴν θέσιν αἰεὶ τῶν ναρθήκων προμηθεῖσθαι χρή, ὅκως μήτε κατὰ τὸ ὀστέον τῶν ἐξεχόντων παρὰ
20 τὰ ἄρθρα φύσει πεφυκότων μήτε κατὰ τὸ
21 [ἄρθρου][3] νεῦρον ἔσται.

XXI. Τὰ δὲ οἰδήματα τὰ κατ' ἰγνύην, ἢ κατὰ πόδα, ἢ κατά τι ἄλλο ἐξαειρεύμενα[4] ὑπὸ τῆς

cient extension, you should adjust the parts with the palms of the hands and bandage in the same way as was described before, placing the head of the bandage as directed and carrying it upwards. And he should give the same answers as before, and experience the same trouble and relief. Let the change of dressing be made in the same way, and the same application of splints. The thigh-bone gets firm in forty days.

XX. One should also bear the following in mind, that the thigh-bone is curved outwards rather than inwards, and to the front rather than to the back, so it gets distorted in these directions if not skilfully treated. Futhermore it is less covered with flesh on these parts so that distortions cannot be hidden. If, then, you suspect anything of this kind, you should have recourse to the mechanical methods recommended for distortion of the upper arm. Some additional turns of bandage should be made round the hip and loins so that the groins and the joint at the so-called fork may be included, for besides other benefits, it prevents the ends of the splints from doing damage by contact with the uncovered parts. The splints should always come considerably short of the bare part at either end, and care should always be taken as to their position so that it is neither on the bone where there are natural projections about the joint, nor on the tendon.

XXI. As to the swellings which arise owing to pressure behind the knee or at the foot or elsewhere,

---

[1] Cf VIII. [2] ἱκανόν.
[3] ἄρθρον codd., except B, which omits. Kw. omits.
[4] ἐξαειρόμενα.

πιέξιος, εἰρίοισι πολλοῖσι ῥυπαροῖσιν, εὖ κατειργασμένοισιν, οἴνῳ καὶ ἐλαίῳ ῥήνας, κηρωτῇ ὑποχρίων, καταδεῖν, καὶ ἢν πιέζωσιν οἱ νάρθηκες, χαλᾶν θᾶσσον· ἰσχναίνοις δ' ἄν, εἰ ἐπάνω ἐς[1] τοὺς νάρθηκας ὀθονίοισι ἰσχνοῖσιν ἐπιδέοις τὰ οἰδήματα, ἀρξάμενος ἀπὸ τοῦ κατωτάτω ἐπὶ τὸ ἄνω νεμόμενος· οὕτω γὰρ ἂν τάχιστα ἰσχνὸν τὸ οἴδημα 10 γένοιτο, καὶ ὑπερθοίη[2] ἂν ὑπὲρ τὰ ἀρχαῖα ἐπιδέσματα· ἀλλ' οὐ χρὴ τούτῳ τῷ τρόπῳ χρῆσθαι τῆς ἐπιδέσιος, ἢν μὴ κίνδυνος ᾖ ἐν τῷ οἰδήματι φλυκταινώσιος ἢ μελασμοῦ· γίνεται δὲ οὐδὲν τοιοῦτον, ἢν μὴ ἄγαν τις πιέζῃ τὸ κάτηγμα, ἢ κατακρεμάμενον ἔχῃ, ἢ κνῆται τῇ χειρί, ἢ ἄλλο 16 τι προσπίπτῃ ἐρεθιστικὸν ἐς[3] τὸν χρῶτα.

XXII. Σωλῆνα δὲ ἢν μέν τις ὑπ' αὐτὸν τὸν μηρὸν ὑποθείη μὴ ὑπερβάλλοντα τὴν ἰγνύην, βλάπτοι ἂν μᾶλλον ἢ ὠφελέοι· οὔτε γὰρ ἂν τὸ σῶμα κωλύοι οὔτε τὴν κνήμην, ἄνευ τοῦ μηροῦ κινεῖσθαι· ἀσηρὸν γὰρ ἂν εἴη πρὸς τὴν ἰγνύην προσβαλλόμενον· καὶ ὃ ἥκιστα δεῖ, τοῦτ' ἂν ἐποτρύνοι ποιεῖν, [ἥκιστα γὰρ δεῖ][4] κατὰ τὸ γόνυ κάμπτειν· πᾶσαν γὰρ ἂν τύρβην παρέχοι τῇσιν ἐπιδέσεσιν, καὶ μηροῦ ἐπιδεδεμένου καὶ κνήμης, 10 ὅστις κατὰ τὸ γόνυ κάμπτοι. ἀνάγκη γὰρ ἂν εἴη τούτῳ τοὺς μύας ἄλλοτε καὶ ἄλλοτε ἄλλο σχῆμα ἴσχειν· ἀνάγκη δ' ἂν εἴη καὶ τὰ ὀστέα τὰ κατεηγότα κίνησιν ἔχειν. περὶ παντὸς οὖν ποιητέον τὴν ἰγνύην ἐντετάσθαι. δοκέοι ἂν [ὁμοίως][5] ὁ σωλὴν ὁ περιέχων[6] πρὸς τὸν πόδα ἀπὸ

[1] ἐπανεὶς Kw. suggested by Erm., confirmed by B.
[2] ὑπερθείη codd. ὑπερθοίη Littré. ὑπέλθοι . . . ὑπὸ B Kw.
[3] πρὸς Kw.

dress them with plenty of crude wool, well pulled out, sprinkling it with oil and wine, after anointing with cerate, and if the splints cause pressure relax them at once. You will reduce the swellings by applying slender bandages after removing[1] the splints, beginning from the lowest part and passing upwards, for so the swelling would be most rapidly reduced and flow back above the original dressing. But you should not use this method of bandaging unless there is danger of blisters forming or mortification at the swelling. Now, nothing of this kind happens unless one puts great pressure on the fracture, or the part is kept hanging down or is scratched with the hand, or some other irritant affects the skin.

XXII. As to a hollow splint, if one should pass it under the thigh itself and it does not go below the bend of the knee it would do more harm than good; for it would prevent neither the body nor the leg from moving apart from the thigh, would cause discomfort by pressing against the flexure of the knee, and incite the patient to bend the knee, which is the last thing he should do. For when the thigh and leg are bandaged, he who bends the knee causes all sorts of disturbance to the dressings, since the muscles will necessarily change their relative positions and there will also necessarily be movement of the fractured bones. Special care, then, should be taken to keep the knee extended. I should think that a hollow splint reaching [evenly?] from hip to

[1] Reading ἐπανείς.

---

[4] Kw. omits.
[5] ὁμοίως seems out of place. μοι B Kw.
[6] ὑπερέχων.

τοῦ ἰσχίου, ὠφελεῖν ὑποτιθέμενος· καὶ ἄλλως κατ' ἰγνύην ταινίην χαλαρῶς περιβάλλειν σὺν τῷ σωλῆνι, ὥσπερ τὰ παιδία ἐν τῇσι κοίτῃσι σπαργανοῦται· εἶτα ἐπὴν ὁ μηρὸς ἐς τὸ ἄνω
20 διαστρέφοιτο[1] ἢ ἐς τὸ πλάγιον, εὐκατασχετώτερον εἴη ἂν σὺν τῷ σωλῆνι οὕτως. ἢν οὖν
22 διαμπερὲς ἴῃ,[2] ποιητέος ὁ σωλὴν, ἢ οὐ ποιητέος.

XXIII. Πτέρνης δὲ ἄκρης κάρτα χρὴ ἐπιμελεῖσθαι ὡς εὐθέτως ἔχῃ, καὶ ἐν τοῖσι κατὰ κνήμην καὶ ἐν τοῖσι κατὰ μηρὸν κατήγμασιν. ἢν μὲν γὰρ ἀπαιωρῆται ὁ πους τῆς ἄλλης κνήμης ἡρματισμένης, ἀνάγκη κατὰ τὸ ἀντικνήμιον τὰ ὀστέα κυρτὰ φαίνεσθαι· ἢν δὲ ἡ μὲν πτέρνη ὑψηλοτέρη [ᾖ] τοῦ μετρίου ἠρτισμένη,[3] ἡ δὲ ἄλλη κνήμη ὑπομετέωρος ᾖ, ἀνάγκη τὸ ὀστέον τοῦτο κατὰ τὸ ἀντικνήμιον τοῦτο κοιλότερον φανῆναι
10 τοῦ μετρίου, προσέτι καὶ ἢν ἡ πτέρνη τυγχάνῃ ἐοῦσα τοῦ ἀνθρώπου φύσει μεγάλη. ἀτὰρ καὶ κρατύνεται πάντα τὰ ὀστέα βραδύτερον, ἢν μὴ κατὰ φύσιν κείμενα [ᾖ, καὶ τὰ μὴ][4] ἀτρεμέοντα ἐν τῷ αὐτῷ σχήματι καὶ αἱ πωρώσιες
15 ἀσθενέστεραι.

XXIV. Ταῦτα μὲν δή, ὅσοισι τὰ μὲν ὀστέα κατέηγεν, ἐξέχει δὲ μή, μηδὲ ἄλλως ἕλκος ἐγένετο. οἷσι δὲ καὶ τὰ ὀστέα κατέηγεν ἁπλῷ τῷ τρόπῳ καὶ μὴ πολυσχιδεῖ, αὐθήμερα ἐμβληθέντα ἢ τῇ ὑστεραίῃ, καὶ κατὰ χώρην ἰζόμενα, καὶ μὴ ἐπίδοξος ἡ ἀπόστασις παρασχίδων ὀστέων ἀπιέναι, ἢ καὶ οἷσιν ἕλκος μὲν ἐγένετο, τὰ δὲ ὀστέα τὰ κατεηγότα οὐκ ἐξίσχει, οὐδ' ὁ τρόπος τῆς κατήξιος τοιοῦτος οἷος παρασχίδας ὀστέων ἐούσας

[1] διαστρέφηται. [2] διαμπερὴς σοι.

foot would be useful, especially with a band passed loosely round at the knee to include the splint, as babies are swaddled in their cots. Then if the thigh-bone is distorted upwards (*i.e.* forwards) or sideways it will thus be more easily controlled by the hollow splint. You should, then, use the hollow splint for the whole limb or not at all.

XXIII. In fractures both of the leg and of the thigh great care should be taken that the point of the heel is in good position. For if the foot is in the air while the leg is supported, the bones at the shin necessarily present a convexity, while if the foot is propped up higher than it should be, and the leg imperfectly supported,[1] this bone in the shin part has a more hollow appearance than the normal, especially if the heel happens to be large compared with the average in man. So, too, all bones solidify more slowly if not placed in their natural position and kept at rest in the same posture, and the callus is weaker.

XXIV. The above remarks apply to those whose bones are fractured without protrusion or wound of other kind. In fractures with protrusion, where they are single and not splintered, if reduced on the same or following day, the bones keeping in place, and if there is no reason to expect elimination of splinters, or even cases in which, though there is an external wound, the broken bones do not stick out, nor is the nature of the fracture such that any

[1] ὑπομετέωρος, "rather low." Adams.

---

[3] ἠρματισμένη ᾖ.
[4] καταμένῃ Kw.'s conjecture. B M V omit ᾖ. B has καὶ τὰ μὲν μή.

10 ἐπιδόξους εἶναι ἀναπλῶσαι· τοὺς τοιούτους οἱ μὲν μήτε μέγα ἀγαθὸν μήτε μέγα κακὸν ποιοῦντες, ἰητρεύουσι τὰ μὲν ἕλκεα καθαρτικῷ τινί, ἢ πισσηρὴν ἐπιθέντες, ἢ ἔναιμον ἢ ἄλλο τι ὧν εἰώθασι ποιεῖν· ἐπάνω δὲ τοὺς οἰνηροὺς σπλῆνας ἢ εἴρια ῥυπαρὰ ἐπιδέουσιν ἢ ἄλλο τι τοιοῦτον. ἐπὴν δὲ τὰ ἕλκεα καθαρὰ γένηται καὶ ἤδη συμφύηται, τότε τοῖσιν ὀθονίοισι συχνοῖσι πειρῶνται ἐπιδεῖν καὶ νάρθηξι κατορθοῦν. αὕτη μὲν ἡ ἴησις ἀγαθόν τι ποιεῖ, κακὸν δὲ οὐ μέγα. 20 τὰ μέντοι ὀστέα οὐχ ὁμοίως δύναται ἱδρύεσθαι ἐς τὴν ἑωυτῶν χώρην, ἀλλά τινι[1] ὀγκηρότερα σώματα τοῦ καιροῦ ταύτῃ γίνεται· γένοιτο δ' ἂν βραχύτερα, ὧν ἀμφότερα τὰ ὀστέα κατέηγεν ἢ 24 πήχεος ἢ κνήμης.

XXV. Ἄλλοι δ' αὖ τινές εἰσι οἳ ὀθονίοισι τὰ τοιαῦτα ἰητρεύουσι εὐθέως καὶ ἔνθεν μὲν καὶ ἔνθεν ἐπιδέουσι τοῖσιν ὀθονίοισι, κατὰ δὲ τὸ ἕλκος αὐτὸ διαλείπουσι, καὶ ἐῶσιν ἀνεψύχθαι· ἔπειτα ἐπιτιθέασι ἐπὶ τὸ ἕλκος τῶν καθαρτικῶν τι, καὶ σπλήνεσιν οἰνηροῖσι ἢ εἰρίοισι ῥυπαροῖσι θεραπεύουσιν. αὕτη ἡ ἴησις κακή, καὶ εἰκὸς τοὺς οὕτως ἰητρεύοντας τὰ μέγιστα ἀσυνετεῖν, καὶ ἐν τοῖσιν ἄλλοισι κατήγμασι καὶ ἐν τοῖσι 10 τοιούτοισιν. μέγιστον γάρ ἐστι τὸ γινώσκειν καθ' ὁποῖον τρόπον χρὴ τὴν ἀρχὴν μὲν βάλλεσθαι τοῦ ὀθονίου, καὶ καθ' ὁποῖον μάλιστα πεπιέχθαι, καὶ οἷά τε ὠφελέονται ἢν ὀρθῶς τις βάλληται τὴν ἀρχὴν καὶ πιέζῃ ᾗ μάλιστα χρή, καὶ οἷα βλάπτονται ἢν μὴ ὀρθῶς τις βάλληται μηδὲ πιέζῃ ᾗ μάλιστα χρή, ἀλλὰ ἔνθεν καὶ ἔνθεν. εἴρηται μὲν οὖν καὶ ἐν τοῖς πρόσθεν γεγραμ-

splinters are likely to come to the surface:—in such cases they do neither much good nor much harm who treat the wound with a cleansing plaster, either pitch cerate, or an application for fresh wounds, or whatever else they commonly use, and bind over it compresses soaked in wine, or uncleansed wool or something of the kind. And after the wounds are cleansed and already united, they attempt to make adjustment with splints and use a number of bandages. This treatment does some good and no great harm. The bones, however, cannot be so well settled in their proper place, but become somewhat unduly swollen at the point of fracture.[1] If both bones are broken, either of forearm or leg, there will also be shortening.

XXV. Then there are others who treat such cases at once with bandages, applying them on either side, while they leave a vacancy at the wound itself and let it be exposed. Afterwards, they put one of the cleansing applications on the wound, and treat it with pads steeped in wine, or with crude wool. This treatment is bad, and those who use it probably show the greatest folly in their treatment of other fractures as well as these. For the most important thing is to know the proper way of applying the head of the bandage, and how the chief pressure should be made, also what are the benefits of proper application and of getting the chief pressure in the proper place, and what is the harm of not placing the bandage rightly, and of not making pressure where it should chiefly be, but at one side or the other. Now, the results of each were ex-

[1] ὀστέα for σώματα; callus develops.

---

[1] τινὶ καὶ τὰ ὀστέα.

μένοισιν, ὁποῖα ἀφ᾽ ἑκατέρων[1] ἀποβαίνει· μαρτυρεῖ δὲ καὶ αὐτὴ ἡ ἰητρείη· ἀνάγκη γὰρ τῷ οὕτως
20 ἐπιδεομένῳ τὸ οἶδος ἐξαείρεσθαι ἐς αὐτὸ τὸ ἕλκος. καὶ γὰρ εἰ ὑγιὴς χρὼς ἔνθεν καὶ ἔνθεν ἐπιδεθείη, ἐν μέσῳ δὲ διαλειφθείη, μάλιστα κατὰ τὴν διάλειψιν οἰδήσειεν ἂν καὶ ἀχροιήσειεν· πῶς οὖν οὐχὶ ἕλκος γε ταῦτα ἂν πάθοι; ἀναγκαίως οὖν ἔχει ἄχροον μὲν καὶ ἐκπεπλιγμένον τὸ ἕλκος εἶναι, δακρυῶδές τε καὶ ἀνεκπύητον, ὀστέα δέ, καὶ μὴ μέλλοντα ἀποστῆναι, ἀποστατικὰ γενέσθαι· σφυγμῶδές τε καὶ πυρῶδες τὸ ἕλκος ἂν εἴη. ἀναγκάζονται δὲ διὰ τὸ οἶδος ἐπικατα-
30 πλάσσειν· ἀσύμφορον δὲ καὶ τοῦτο τοῖσιν ἔνθεν καὶ ἔνθεν ἐπιδεομένοισιν· ἄχθος γὰρ ἀνωφελὲς πρὸς τῷ ἄλλῳ σφυγμῷ ἐπιγίνεται. τελευτῶντες δὲ ἀπολύουσι τὰ ἐπιδέσματα, ὁπόταν σφιν παλιγκοτῇ, καὶ ἰητρεύουσι τὸ λοιπὸν ἄνευ ἐπιδέσιος· οὐδὲν δὲ ἧσσον, καὶ ἤν τι ἄλλο τρῶμα τοιοῦτον λάβωσι, τῷ αὐτῷ τρόπῳ ἰητρεύουσιν· οὐ γὰρ οἴονται τὴν ἐπίδεσιν τὴν ἔνθεν καὶ ἔνθεν, καὶ τὴν ἀνάψυξιν τοῦ ἕλκεος αἰτίην εἶναι, ἀλλὰ ἄλλην τινὰ ἀτυχίην· οὐ μέντοι γε ἂν ἔγραφον
40 περὶ τούτου τοσαῦτα, εἰ μὴ εὖ μὲν ᾔδειν ἀσύμφορον ἐοῦσαν τὴν ἐπίδεσιν, συχνοὺς δὲ οὕτως ἰητρεύοντας, ἐπίκαιρον δὲ τὸ ἀπομάθημα, μαρτύριον δὲ τοῦ ὀρθῶς γεγράφθαι τὰ πρόσθεν γεγράμμενα εἴτε μάλιστα πιεστέα τὰ κατήγματα
45 εἴτε ἥκιστα.

[1] ἑκατέρου.

---

[1] That is, an unhealthy discharge without "purification."

[2] Exposure here cannot mean exposure to cold or even bareness—the foolish surgeons cover the wound with wool or

plained in what has been written above. The treatment, too, is itself evidence; for in a patient so bandaged the swelling necessarily arises in the wound itself, since if even healthy tissue were bandaged on this side and that, and a vacancy left in the middle, it would be especially at the vacant part that swelling and decoloration would occur. How then could a wound fail to be affected in this way? For it necessarily follows that the wound is discoloured with everted edges, and has a watery discharge devoid of pus,[1] and as to the bones, even those which were not going to come away do come away. The wound will become heated and throbbing, and they are obliged to put on an additional plaster because of the swelling; and this too will be harmful to patients bandaged at either side of the wound, for an unprofitable burden is added to the throbbing. They finally take off the dressings, when they find there is aggravation, and treat it for the future without bandaging. Yet none the less, if they get another wound of the same sort, they use the same treatment, for they do not suppose that the outside bandaging and exposure[2] of the wound is to blame, but some mishap. However, I should not have written so much about this had I not known well the harmfulness of this dressing and that many use it; and that it is of vital importance to unlearn the habit. Besides, it is an evidence of the truth of what was written before on the question whether the greatest or least pressure should come at the fracture.[3]

pads—it means absence of due pressure, the proper graduation of which is the main point in Hippocratic bandaging.

[3] According to Adams this warning was still necessary in his time.

XXVI. Χρὴ δέ, ὡς ἐν κεφαλαίῳ εἰρῆσθαι, οἷσιν ἂν μὴ ἐπίδοξος ᾖ ἡ τῶν ὀστέων ἀπόστασις ἔσεσθαι, τὴν αὐτὴν ἰητρείην ἰητρεύειν, ὥσπερ ἂν οἷσιν ὀστέα μὲν κατεηγότα εἴη, ἕλκος δὲ μὴ ἔχοντα· τάς τε γὰρ κατατάσιας καὶ κατορθώσιας τῶν ὀστέων τὸν αὐτὸν τρόπον ποιεῖσθαι, τήν τε ἐπίδεσιν παραπλήσιον τρόπον. ἐπὶ μὲν αὐτὸ τὸ ἕλκος πισσηρὴν κηρωτὴν χρίσαντα, σπλῆνα λεπτὸν διπλόον ἐπιδεθῆναι,[1] τὰ δὲ πέριξ κηρωτῇ
10 λεπτῇ χρίειν. τὰ δὲ ὀθόνια καὶ τὰ ἄλλα πλατύτερά τινι ἐσχισμένα ἔστω, ἢ εἰ μὴ ἕλκος εἶχεν· καὶ ᾧ ἂν πρώτῳ ἐπιδέηται, συχνῷ ἔστω τοῦ ἕλκεος πλατύτερον. τὰ γὰρ στενότερα τοῦ ἕλκεος ζώσαντα ἔχει τὸ ἕλκος· τὸ δὲ οὐ χρή. ἀλλ' ἡ πρώτη περιβολὴ ὅλον κατεχέτω τὸ ἕλκος, καὶ ὑπερεχέτω τὸ ὀθόνιον ἔνθεν τε καὶ ἔνθεν. βάλλεσθαι μὲν οὖν χρὴ τὸ ὀθόνιον κατ' αὐτὴν τὴν ἴξιν τοῦ ἕλκεος, πιέζειν δὲ ὀλίγῳ ἧσσον ἢ εἰ μὴ ἕλκος εἶχεν, ἐπινέμεσθαι δὲ τῇ ἐπιδέσει
20 ὥσπερ καὶ πρόσθεν εἴρηται. τὰ δὲ ὀθόνια αἰεὶ μὲν τοῦ τρόπου τοῦ μαλθακοῦ ἔστωσαν, μᾶλλον δέ τε[2] δεῖ ἐν τοῖσι τοιούτοισιν, ἢ εἰ μὴ ἕλκος εἶχεν. πλῆθος δὲ τῶν ὀθονίων· μὴ ἐλάσσω ἔστω τῶν πρότερον εἰρημένων, ἀλλά τινι καὶ πλείω. ἢν δὲ ἐπιδεθῇ, δοκείτω τῷ ἐπιδεδεμένῳ ἡρμόσθαι[3] μέν, πεπιέχθαι δὲ μή· φάτω δὲ κατὰ τὸ ἕλκος μάλιστα ἡρμόσθαι. τοὺς δὲ χρόνους τοὺς αὐτοὺς μὲν χρὴ εἶναι ἐπὶ τὸ μᾶλλον δοκεῖν ἡρμόσθαι, τοὺς αὐτοὺς δὲ ἐπὶ τὸ μᾶλλον δοκεῖν χαλᾶν,
30 ὥσπερ καὶ ἐν τοῖσι πρόσθεν εἴρηται. μετεπιδεῖν δὲ διὰ τρίτης, πάντα μεταποιέοντα ἐς τοὺς τρόπους τοὺς παραπλησίους, ὥσπερ καὶ πρόσθεν

XXVI.[1] To speak summarily, when there is no likelihood of elimination of bone, one should use the same treatment as in cases of fracture without external wound. The extensions and adjustments of the bones should be made in the same way, and so too with the bandaging. After anointing the wound itself with pitch cerate, bind a thin doubled compress over it, and anoint the surrounding parts with a thin layer of cerate. The bandages and other dressings should be torn in rather broader strips than if there was no wound, and the one first used should be a good deal wider than the wound; for bandages narrower than the wound bind it like a girdle, which should be avoided; rather let the first turn take in the whole wound, and let the bandage extend beyond it on both sides. One should, then, put the bandage just in the line of the wound, make rather less pressure than in cases without a wound, and distribute the dressing as directed above. The bandages should always be of the pliant kind, and more so in these cases than if there was no wound. As to number, let it not be less than those mentioned, before but even a little greater. When the bandaging is finished it should appear to the patient to be firm without pressure, and he should say that the greatest firmness is over the wound. There should be the same periods of a sensation of greater firmness, and greater relaxation as were described in the former cases. Change the dressings every other day, making the changes in similar

[1] Proper treatment of compound fractures.

---

[1] ἐπιθεῖναι. [2] τι. [3] ἡρμάσθαι bis.

εἴρηται, πλὴν ἐς τὸ σύμπαν ἧσσόν τινι πιέζειν ταῦτα ἢ ἐκεῖνα. καὶ ἢν κατὰ λόγον τὰ εἰκότα γένηται, ἰσχνότερον μὲν αἰεὶ εὑρεθήσεται τὸ κατὰ τὸ ἕλκος, ἰσχνὸν δὲ καὶ τὸ ἄλλο πᾶν τὸ ὑπὸ τῆς ἐπιδέσιος κατεχόμενον· καὶ αἵ τε ἐκπυήσιες ἔσονται θάσσους ἢ τῶν ἄλλως ἰητρευμένων ἑλκέων, ὅσα τε σαρκία ἐν τῷ τρώματι ἐμελάνθη
40 καὶ ἐθανατώθη, θᾶσσον περιρρήγνυται καὶ ἐκπίπτει ἐπὶ ταύτῃ τῇ ἰητρείῃ ἢ ἐν τῇσι ἄλλῃσιν, ἐς ὠτειλάς τε θᾶσσον ὁρμᾶται τὸ ἕλκος οὕτως ἢ ἄλλως ἰητρευμένον. πάντων δὲ τούτων αἴτιον ὅτι ἰσχνὸν μὲν τὸ κατὰ τὸ ἕλκος χωρίον γίνεται, ἰσχνὰ δὲ τὰ περιέχοντα. τὰ μὲν οὖν ἄλλα πάντα παραπλησίως χρὴ ἰητρεύειν, ὡς τὰ ἄνευ ἑλκώσιος ὀστέα κατηγνύμενα· τοὺς δὲ νάρθηκας οὐ χρὴ προστιθέναι. διὰ τοῦτο καὶ τὰ ὀθόνια χρὴ τούτοισι πλείω εἶναι ἢ τοῖσιν ἑτέροισιν, ὅτι
50 τε ἧσσον πιέζεται, ὅτι τε οἱ νάρθηκες βραδύτεροι [1] προστιθένται· ἢν μέντοι τοὺς νάρθηκας προστιθῇς, μὴ κατὰ τὴν ἴξιν τοῦ ἕλκεος προστιθέναι, ἄλλως τε καὶ χαλαρῶς προστιθέναι, προμηθεύμενος [2] ὅπως μηδεμίη σφίγξις μεγάλη ἔσται ἀπὸ τῶν ναρθήκων· εἴρηται δὲ τοῦτο καὶ ἐν τοῖσι πρότερον γεγραμμένοισιν. τὴν μέντοι δίαιταν ἀκριβεστέρην καὶ πλείω χρόνον χρὴ ποιεῖσθαι οἷσιν ἐξ ἀρχῆς ἕλκεα γίνεται καὶ οἷσιν ὀστέα ἐξίσχει· καὶ τὸ σύμπαν δὲ εἰρῆσθαι, ἐπὶ τοῖσιν
60 ἰσχυροτάτοισι τρώμασιν ἀκριβεστέρην καὶ
61 πολυχρονιωτέρην εἶναι χρὴ τὴν δίαιταν.

XXVII. Ἡ αὐτὴ ἰητρείη τῶν ἑλκέων καὶ οἷσιν ὀστέα μὲν κατέηγεν, ἕλκος δὲ ἐξ ἀρχῆς μηδὲν ᾖ, ἢν δὲ ἐν τῇ ἰητρείῃ ἕλκος γένηται, ἢ τοῖσιν

fashion except that, on the whole, the pressure should be less in these cases. If the case takes a natural course according to rule, the part about the wound will be found progressively diminished and all the rest of the limb included in the bandage will be slender. Purification[1] will take place more rapidly than in wounds treated otherwise, and all fragments of blackened or dead tissue are more rapidly separated and fall off under this treatment than with other methods. The wound, too, advances more quickly to cicatrisation thus than when treated otherwise. The cause of all this is that the wound and the surrounding parts become free from swelling. In all other respects, then, one should treat these cases like fractures without a wound, but splints should not be used.[2] This is why the bandages should be more numerous than in the other cases both because there is less pressure and because the splints are applied later. But if you do apply splints, do not put them in the line of the wound; especially apply them loosely, taking care that there is no great compression from the splints. This direction was also given above. Diet, however, should be more strict and kept up longer in cases where there is a wound from the first and where the bones protrude, and on the whole, the greater the injury the more strict and prolonged should be the dieting.

XXVII. The same treatment of the wounds applies also to cases of fracture which are at first without wound, but where one occurs during treat-

[1] *i.e.* discharge of laudable pus.
[2] We must evidently understand "so soon."

---

[1] βραδύτερον. [2] προμηθευμένοις codd. Pq.

ὀθονίοισι μᾶλλον πιεχθέντος, ἢ ὑπὸ νάρθηκος
ὠἐνέδρης, ἢ ὑπὸ ἄλλης τινὸς προφάσιος. γινώ-
σκεται μὲν οὖν τὰ τοιαῦτα, ἢν ἕλκος ὑπῇ, τῇ
τε ὀδύνῃ καὶ τοῖσι σφυγμοῖσιν· καὶ τὸ οἴδημα
τὸ ἐν τοῖσι ἄκροισι σκληρότερον γίγνεται τῶν
τοιούτων, καὶ εἰ τὸν δάκτυλον ἐπαγάγοις, τὸ
10 ἔρευθος ἐξαείρεται,[1] ἀτὰρ καὶ αὖθις ἀποτρέχει
ταχέως. ἢν οὖν τι τοιοῦτον ὑποπτεύῃς, λύσαντα
χρή, ἢν μὲν ᾖ κνησμὸς κατὰ τὰς ὑποδεσμίδας
ἢ ἐπὶ[2] τὸ ἄλλο τὸ ἐπιδεδεμένον πισσηρῇ κηρωτῇ
ἀντὶ τῆς ἑτέρης χρῆσθαι· ἢν δὲ τούτων μὲν μηδὲν
ᾖ, αὐτὸ δὲ τὸ ἕλκος ἠρεθισμένον εὑρίσκεται
μέλαν ἐπὶ πολὺ ἢ[3] ἀκάθαρτον, καὶ τῶν μὲν
σαρκῶν ἐκπυησομένων, τῶν δὲ νεύρων προσεκ-
πεσουμένων, τούτους οὐδὲν δεῖ ἀναψύχειν παντάπασιν, οὐδέ τι φοβεῖσθαι τὰς ἐκπυήσιας ταύτας,
20 ἀλλ' ἰητρεύειν τὰ μὲν ἄλλα παραπλήσιον τρόπον,
ὥσπερ καὶ οἷσιν ἐξ ἀρχῆς ἕλκος ἐγένετο. τοῖσι
δὲ ὀθονίοισιν ἄρχεσθαι χρὴ ἐπιδέοντα ἀπὸ τοῦ
οἰδήματος τοῦ ἐν τοῖσιν ἀκρέοισι πάνυ χαλαρῶς,
καὶ ἔπειτα ἐπινέμεσθαι τῇ ἐπιδέσει αἰεὶ ἐς τὸ
ἄνω, καὶ πεπιέχθαι μὲν οὐδαμῇ, ἡρμόσθαι[4] δὲ
μάλιστα κατὰ τὸ ἕλκος, τὰ δὲ ἄλλα ἐπὶ ἧσσον.
τὰ δὲ ὀθόνια τὰ πρῶτα, ταῦτα μὲν καθαρὰ ἔστω
καὶ μὴ στενά· τὸ δὲ πλῆθος τῶν ὀθονίων ἔστω
ὅσον περ καὶ ἐν τοῖσι νάρθηξιν, εἰ ἐπιδέοιντο,[5] ἢ
30 ὀλίγῳ ἔλασσον. ἐπὶ δὲ αὐτὸ τὸ ἕλκος ἱκανὸν
σπληνίον τῇ λευκῇ κηρωτῇ κεχρισμένον· ἤν τε
γὰρ σὰρξ ἤν τε νεῦρον μελανθῇ, προσεκπεσεῖται·
τὰ γὰρ τοιαῦτα οὐ χρὴ δριμέσιν ἰητρεύειν, ἀλλὰ

[1] ἐξείργεται Kw.'s conjecture. Kw.'s note ἐξείργεται scripsi, ἐξαρείαται B[1], ἐξαείρεται B[2] Pq., ἐξαείραται M V, ἐξαιρέεται

ment either through too great compression by bandages or the pressure of a splint or some other cause. In such cases the occurrence of ulceration is recognised by pain and throbbing: also the swelling on the extremities gets harder, and if you apply the finger the redness is removed but quickly returns. So, if you suspect anything of this kind you should undo the dressings, if there is irritation below the under bandages, or in the rest of the bandaged part, and use pitch cerate instead of the other plaster. Should there be none of this, but the sore itself is found to be irritated, extensively blackened or foul with tissues about to suppurate and tendons on the way to be thrown off, it is by no means necessary to leave them exposed, or to be in any way alarmed at these suppurations, but treat them for the future in the same manner as cases in which there is a wound from the first. The bandaging should begin from the swelling at the extremities and be quite slack; then it should be carried right on upwards, avoiding pressure in any place, but giving special support at the wound and decreasing it elsewhere. The first bandages must be clean and not narrow, their number as many as when splints are applied or a little fewer. On the wound itself a compress anointed with white cerate is sufficient; for if flesh or tendon be blackened it will also come away. One should treat such cases not with irritant, but

Litt., *ἐξανίσταται* Wb, *τὸ ἔρευθος ἐξαείραται* Galen *in cit.*, *ἐξαρύαται* : *ἐκκενοῦται ἐκθλίβεται* Galen *in exegesi.* Such is the discord about this word whenever it occurs ; but the meaning seems obvious.

[2] *καὶ* omitting *ἤ*. [3] *ἢ ἐπὶ πολὺ ἀκάθαρτον* omitting *μέλαν*.

[4] *ἡρμάσθαι*. [5] *ἐπιδέοιτο*.

μαλθακοῖσιν, ὥσπερ τὰ περίκαυστα. μετεπιδεῖν δὲ διὰ τρίτης, νάρθηκας δὲ μὴ προστιθέναι· ἀτρεμεῖν δὲ ἐπὶ μᾶλλον ἢ τὸ πρόσθεν, καὶ ὀλιγοσιτεῖν· εἰδέναι δὲ χρὴ εἴ τε σάρξ, εἴ τε νεῦρον τὸ ἐκπεσούμενόν ἐστι, ὅτι οὕτω πολλῷ μὲν ἧσσον νέμεται ἐπὶ πλεῖον, πολλῷ δὲ θᾶσσον
40 ἐκπεσεῖται, πολλῷ δὲ ἰσχνότερα τὰ περιέχοντα ἔσται, ἢ εἴ τις ἀπολύσας τὰ ὀθόνια ἐπιθείη τι τῶν καθαρτικῶν φαρμάκων ἐπὶ τὸ ἕλκος. καίτοι καὶ ἢν ἐκπέσῃ τὸ ἐκπυησόμενον, θᾶσσόν τε σαρκοῦται ἐκείνως ἢ ἑτέρως ἰητρευόμενον, καὶ θᾶσσον ὠτειλοῦται. πάντα μήν ἐστι ταῦτα ὀρθῶς ἐπιδεῖν καὶ μετρίως ἐπίστασθαι. προσσυμβάλλεται δὲ καὶ τὰ σχήματα καὶ οἷα χρὴ εἶναι, καὶ ἡ ἄλλη
48 δίαιτα, καὶ τῶν ὀθονίων ἡ ἐπιτηδειότης.

XXVIII. Ἢν δὲ ἄρα ἐξαπατηθῇς ἐν τοῖσι νεοτρώτοισι, μὴ οἰόμενος ὀστέων ἀπόστασιν ἔσεσθαι, τὰ δ' ἐπίδοξα ᾖ ἀναπλῶσαι, οὐ χρὴ ὀρρωδεῖν τοῦτον τὸν τρόπον τῆς ἰητρείης, οὐδὲν γὰρ ἂν μέγα φλαῦρον γένοιτ' ἄν,[1] ἢν μοῦνον οἷός τε ᾖς τῇ χειρὶ τὰς ἐπιδέσιας ἀγαθὰς καὶ ἀσινέας ποιεῖσθαι. σημεῖον δὲ τόδε, ἢν μέλλῃ ὀστέων ἀπόστασις ἔσεσθαι ἐν τῷ τρόπῳ τούτῳ τῆς ἰητρείης· πῦον γὰρ συχνὸν ῥέει ἐκ τοῦ ἕλκεος
10 καὶ ὀργᾶν φαίνεται. πυκνότερον οὖν μετεπιδεῖσθαι[2] διὰ τὸ πλάδον· ἐπεὶ ἄλλως τε καὶ ἀπύρετοι γίνονται, ἢν μὴ κάρτα πιέζωνται ὑπὸ τῆς ἐπιδέσιος, καὶ τὸ ἕλκος καὶ τὰ περιέχοντα ἰσχνά· ὅσαι μὲν οὖν λεπτῶν πάνυ ὀστέων

[1] γένοιτο.
[2] μετεπιδεῖν.

with mild applications, just like burns. Change the dressing every other day but do not apply splints. Keep the patient at rest and on low diet even more than in the former case. One should know if either flesh and tendon is going to come away that the loss will be much less extensive and will be brought about much quicker, and the surrounding parts will be much less swollen (by this treatment), than if on removing the bandage one applied some detersive plaster to the wound. Besides, when the part that is going to suppurate off does come away, flesh formation and cicatrisation will be more rapid with the former treatment than with any other. The whole point is to know the correct method and due measure in dressing these cases. Correctness of position also contributes to the result, as well as diet and the suitability of the bandages.

XXVIII. If, perchance, you are deceived in fresh cases, and think there will be no elimination of bones, yet they show signs of coming to the surface, the use of the above mode of treatment need not cause alarm, for no great damage will be done if only you have sufficient manual skill to apply the dressings well and in a way that will do no harm. The following is a sign of approaching elimination of bone in a case thus treated. A large amount of pus flows from the wound, which appears turgid. So the dressing should be changed more often because of the soaking,[1] for thus especially they get free from fever, if there is no great compression by the bandages, and the wound and surrounding parts are not engorged. But separations of very small fragments require no great

[1] "Maceration," "abundance of humours."

ἀποστάσιες, οὐδεμίης μεγάλης μεταβολῆς δέονται, ἀλλ' ἢ χαλαρώτερα ἐπιδεῖν, ὡς μὴ ἀπολαμβάνηται τὸ πῦον, ἀλλ' εὐαπόρρυτον, ἢ καὶ πυκνότερον μετεπιδεῖν ἔστ' ἂν ἀποστῇ τὸ ὀστέον, καὶ
19 νάρθηκας μὴ προστιθέναι.

XXIX. Ὁπόσοισι δὲ μείζονος ὀστέου ἀπόστασις ἐπίδοξος γένηται, ἤν τε ἐξ ἀρχῆς προγνῷς, ἤν τε καὶ ἔπειτα μεταγνῷς, οὐκ ἔτι τῆς αὐτῆς ἰητρείης δεῖται,[1] ἀλλὰ τὰς μὲν κατατάσιας καὶ τὰς διορθώσιας οὕτω ποιεῖσθαι ὥσπερ εἴρηται· σπλῆνας δὲ χρὴ διπλοῦς, πλάτος μὲν ἡμισπιθαμιαίους, μὴ ἐλάσσους (ὁποῖον δὲ ἄν τι καὶ τὸ τρῶμα ᾖ, πρὸς τοῦτο τεκμαίρεσθαι), μῆκος δὲ βραχυτέρους μὲν ὀλίγῳ ἢ ὥστε δὶς περιϊκνεῖσθαι
10 περὶ τὸ σῶμα τὸ τετρωμένον, μακροτέρους δὲ συχνῷ ἢ ὥστε ἅπαξ περιϊκνεῖσθαι, πλῆθος δὲ ὁπόσους ἂν συμφέρῃ, ποιησάμενον, τούτους ἐν οἴνῳ μέλανι αὐστηρῷ βρέχοντα, χρὴ ἐκ μέσου ἀρχόμενον, ὡς ἀπὸ δύο ἀρχῶν ὑποδεσμὶς ἐπιδεῖται, περιελίσσειν, κἄπειτα σκεπαρνηδὸν παραλλάσσοντα τὰς ἀρχὰς ἀφιέναι. ταῦτα κατά τε αὐτὸ τὸ ἕλκος ποιεῖν καὶ κατὰ τὸ ἔνθεν καὶ ἔνθεν τοῦ ἕλκεος· καὶ πεπιέχθω μὲν μή, ἀλλ' ὅσον ἑρμασμοῦ ἕνεκεν τοῦ ἕλκεος προσκείσθω. ἐπὶ
20 δὲ αὐτὸ τὸ ἕλκος ἐπιτιθέναι χρὴ πισσηρήν, ἤ τι τῶν ἐναίμων ἤ τι τῶν ἄλλων φαρμάκων, ὅ τι σύντροφόν[2] ἐστιν [ὃ] ἐπιτέγξει.[3] καὶ ἢν μὲν ἡ ὥρη θερινὴ ᾖ, ἐπιτέγγειν τῷ οἴνῳ τοὺς σπλῆνας πυκνά· ἢν δὲ χειμερινὴ ἡ ὥρη ᾖ, εἴρια πολλὰ

alteration of treatment beyond either loose bandaging so as not to intercept the pus but allow it to flow away freely; or even more frequent change of dressing till the bone separates, and no application of splints.

XXIX. But in cases where separation of a rather large bone is probable, whether you prognosticate it from the first, or recognise it later, the treatment should not be the same, but, while the extensions and adjustments should be done as was directed, the compresses should be double, half a span[1] in breadth at least—take the nature of the wound as standard for this—and in length a little less than will go twice round the wounded part, but a good deal more than will go once round. Provide as many of these as may suffice, and after soaking them in dark astringent wine, apply them beginning from their middle as is done with a two headed under bandage; enveloping the part and then leaving the ends crossed obliquely, as with the adze-shaped bandage. Put them both over the wound itself and on either side of it, and though there should be no compression, they should be applied firmly so as to support the wound. On the wound itself one should put pitch cerate or one of the applications for fresh injuries or any other appropriate remedy which will serve as an embrocation. If it is summer time soak the compresses frequently with wine, but if

[1] Adams strangely calls a span a fathom here and elsewhere.

---

[1] δεῖ.

[2] σύντροφόν, as Galen says, means "appropriate," as in *Surgery*, XI.

[3] ἐπιτέγξει Pq. takes as a verb. Kw. apparently takes it as subst., omitting ὃ.

ῥυπαρὰ νενοτισμένα οἴνῳ καὶ ἐλαίῳ[1] ἐπικείσθω. ἰξαλῆν δὲ χρὴ ὑποτετάσθαι, καὶ εὐαπόρρυτα ποιεῖν, φυλάσσοντα τοὺς ὑπορρόους, μεμνημένον ὅτι οἱ τόποι οὗτοι, ἐν τοῖσι αὐτοῖσι σχήμασι πολλὸν χρόνον κειμένοισι, ἐκτρίμματα δυσάκεστα
30 ποιέουσιν.

XXX. Ὅσους δὲ μὴ οἷόν τε ἐπιδεσει ἰήσασθαι διά τινα τούτων τῶν εἰρημένων τρόπων ἢ τῶν ῥηθησομένων, τούτους περὶ πλέονος χρὴ ποιεῖσθαι ὅπως εὐθέτως σχήσουσι τὸ κατεηγὸς τοῦ σώματος κατ' ἰθυωρίην, προσέχοντα τὸν νόον καὶ τῷ ἀνωτέρω δὲ μᾶλλον ἢ τῷ κατωτέρω. εἰ δέ τις μέλλοι καλῶς καὶ εὐχερῶς ἐργάζεσθαι, ἄξιον καὶ μηχανοποιήσασθαι, ὅκως κατάτασιν δικαίην καὶ μὴ βιαίην σχήσῃ[2] τὸ κατεηγὸς τοῦ σώματος·
10 μᾶλλον[3] δὲ ἐν κνήμῃ ἐνδέχεται μηχανοποιεῖν. εἰσὶ μὲν οὖν τινὲς οἳ ἐπὶ πᾶσι τοῖσι τῆς κνήμης κατήγμασι, καὶ τοῖσι ἐπιδεομένοισι καὶ τοῖσι μὴ ἐπιδεομένοισι, τὸν πόδα ἄκρον προσδέουσι πρὸς τὴν κλίνην ἢ πρὸς ἄλλο τι ξύλον παρὰ τὴν κλίνην κατορύξαντες. οὗτοι μὲν οὖν πάντα κακὰ ποιοῦσιν, ἀγαθὸν δὲ οὐδέν· οὔτε γὰρ τοῦ κατατείνεσθαι ἄκος ἐστὶ τὸ προσδεδέσθαι τὸν πόδα, οὐδὲν γὰρ ἧσσον τὸ ἄλλο σῶμα προσχωρήσει πρὸς τὸν πόδα καὶ οὕτως οὐκ ἂν ἔτι τείνοιτο·
20 οὔτ' αὖ[4] ἐς τὴν ἰθυωρίην οὐδὲν ὠφελεῖ, ἀλλὰ καὶ βλάπτει· στρεφομένου γὰρ τοῦ ἄλλου σώματος ἢ τῇ ἢ τῇ, οὐδὲν κωλύσει ὁ δεσμὸς τὸν πόδα καὶ τὰ ὀστέα τὰ τῷ ποδὶ προσηρτημένα ἐπακολουθεῖν τῷ ἄλλῳ σώματι· εἰ δὲ μὴ προσεδέδετο, ἧσσον ἂν διεστρέφετο· ἧσσον γὰρ ἂν ἐγκατελείπετο ἐν τῇ κινήσει τοῦ ἄλλου σώματος. εἰ δέ

winter apply plenty of crude wool moistened with wine and oil. A goat's skin should be spread underneath to make free course for discharges, giving heed to drainage and bearing in mind that these regions (when patients lie a long time in the same posture) develop sores difficult to heal.

XXX. As to cases which cannot be treated by bandaging in one of the ways which have been or will be described, all the more care should be taken that they shall have the fractured limb in good position in accord with its normal lines, seeing to it that the slope is upwards rather than downwards. If one intends to do the work well and skilfully, it is worth while to have recourse to mechanism, that the fractured part may have proper but not violent extension. It is especially convenient to use mechanical treatment for the leg. Now, there are some who in all cases of leg fractures, whether they are bandaged or not, fasten the foot to the bed, or to some post which they fix in the ground by the bed. They do all sorts of harm and no good; for extension is not ensured by fastening the foot, since the rest of the body will none the less move towards the foot, and thus extension will not be kept up. Nor is it of any use for preserving the normal line, but even harmful. For when the rest of the body is turned this way or that, the ligature in no way prevents the foot and the bones connected with it from following the movement: If it were not tied up, there would be less distortion, for it would not be left behind so much in the movement of the rest of the body. Instead of this, one should get two

[1] Cf. the good Samaritan.
[2] σχήσει.
[3] μάλιστα.
[4] αὐτήν.

τις σφαίρας δύο ῥάψαιτο ἐκ σκύτεος Αἰγυπτίου τοιαύτας οἵας φορέουσιν οἱ ἐν τῇσι μεγάλῃσι πέδῃσι πολλὸν χρόνον πεπεδημένοι, αἱ δὲ
30 σφαῖραι ἔχοιεν ἔνθεν καὶ ἔνθεν χιτῶνας τὰ μὲν πρὸς τοῦ τρώματος βαθυτέρους, τὰ δὲ πρὸς τῶν ἄρθρων βραχυτέρους, εἶεν δὲ ὀγκηραὶ μὲν καὶ μαλθακαί, ἁρμόζουσαι δέ, ἡ μὲν ἄνωθεν[1] τῶν σφυρῶν, ἡ δὲ κάτωθεν[2] τοῦ γόνατος· ἐκ δὲ πλαγίης ἑκατέρης[3] δισσὰ ἑκατέρωθεν ἔχοι προσηρτημένα ἢ ἁπλόου ἱμάντος ἢ διπλόου, βραχύτερα[4] ὥσπερ ἀγκύλας, τὰ μέν τι τοῦ σφυροῦ ἑκατέρωθεν, τὰ δέ τι τοῦ γόνατος· [καὶ ἡ ἄνωθεν σφαῖρα ἕτερα τοιαῦτα ἔχοι][5] κατὰ τὴν ἰθυωρίην
40 τὴν αὐτήν. κἄπειτα κρανεΐνας ῥάβδους τέσσαρας λαβών, ἴσας τὸ μέγεθος ἀλλήλῃσιν ἐχούσας, πάχος μὲν ὡς δακτυλιαίας, μῆκος δέ, ὡς κεκαμμέναι ἐναρμόσουσιν ἐς τὰ ἀπαιωρήματα, ἐπιμελόμενος ὅπως τὰ ἄκρα τῶν ῥάβδων μὴ ἐς τὸν χρῶτα, ἀλλ' ἐς τὰ ἄκρα τῶν σφαιρέων ἐγκέλσῃ. εἶναι δὲ χρὴ ζεύγεα τρία τῶν ῥάβδων, καὶ πλέω, καί τινι μακροτέρας τὰς ἑτέρας τῶν ἑτέρων καί τινι καὶ βραχυτέρας καὶ σμικροτέρας, ὡς καὶ μᾶλλον διατείνειν,[6] ἢν βούληται, καὶ ἧσσον·
50 καὶ ἕστωσαν δὲ αἱ ῥάβδοι ἑκάτεραι ἔνθεν καὶ ἔνθεν τῶν σφυρῶν. ταῦτα τοίνυν εἰ καλῶς μηχανοποιηθείη, τήν τε κατάτασιν καὶ δικαίην ἂν παρέχοι καὶ ὁμαλὴν κατὰ τὴν ἰθυωρίην, καὶ τῷ τρώματι πόνος οὐδεὶς ἂν εἴη· τὰ γὰρ ἀποπιέσματα, εἴ τι καὶ ἀποπιέζοιτο, τὰ μὲν ἂν ἐς τὸν πόδα ἀπάγοιτο, τὰ δὲ ἐς τὸν μηρόν· αἵ τε ῥάβδοι εὐθετώτεραι, αἱ μὲν ἔνθεν, αἱ δὲ ἔνθεν τῶν σφυρῶν, ὥστε μὴ κωλύεσθαι τὴν θέσιν τῆς

rounded circlets sewn in Egyptian leather such as are worn by those who are kept a long time shackled in the large fetters. The circlets should have coverings on both sides deeper on the side facing the injury and shallower on that facing the joints. They should be large and soft, fitting the one above the ankle, the other below the knee. They should have on each side two attachments of leather thongs, single or double, short like loops, one set at the ankle on either side, the other on either side of the knee (and the upper circlet should have others like them in the same straight line, *i.e.* just opposite those below). Then take four rods of cornel wood of equal size, the thickness of a finger; and of such length as when bent they fit into the appendices, taking care that the ends of the rods do not press upon the skin but on the projecting edges of the circlet. There should be three or more pairs of rods, some longer than the others and some shorter and more slender, so as to exert greater or less tension at pleasure. Let the rods be placed separately on either side of the ankles. This mechanism if well arranged will make the extension both correct and even in accordance with the normal lines, and cause no pain in the wound, for the outward pressure, if there is any, will be diverted partly to the foot and partly to the thigh. The rods are better placed, some on one side and some on the other side of the ankles, so as not to interfere with the position of the

---

[1] τῷ ἄνωθεν. [2] τῷ κάτωθεν.
[3] ἑκατέρῃ. [4] βραχέα.
[5] Kw. omits; Erm. omits the rest of the sentence also.
[6] διατείνῃς.

κνήμης· τό τε τρῶμα εὐκατάσκεπτον καὶ εὐ-
60 βάστακτον· οὐδὲν γὰρ ἐμποδών, εἴ τις ἐθέλοι τὰς δύο τῶν ῥάβδων τὰς ἀνωτέρω αὐτὰς πρὸς ἀλλήλας ζεῦξαι, καὶ ἤν τις κούφως βούλοιτο ἐπιβάλλειν, ὥστε τὸ ἐπιβαλλόμενον μετέωρον ἀπὸ τοῦ τρώματος εἶναι. εἰ μὲν οὖν αἵ τε σφαῖραι προσηνέες καὶ καλαὶ καὶ μαλθακαὶ καὶ καιναὶ ῥαφεῖεν, καὶ ἡ ἔντασις τῶν ῥάβδων χρηστῶς ἐνταθείη, ὥσπερ ἤδη εἴρηται, εὔχρηστον τὸ μηχάνημα· εἰ δέ τι τούτων μὴ καλῶς ἕξει, βλάπτοι ἂν μᾶλλον ἢ ὠφελέοι. χρὴ δὲ καὶ τὰς
70 ἄλλας μηχανὰς ἢ καλῶς μηχανᾶσθαι, ἢ μὴ μηχανᾶσθαι, αἰσχρὸν γὰρ καὶ ἄτεχνον μηχανο-
72 ποιέοντα ἀμηχανοποιεῖσθαι.

XXXI. Τοῦτο δέ, οἱ πλεῖστοι τῶν ἰητρῶν τὰ κατήγματα καὶ τὰ σὺν ἕλκεσι καὶ τὰ ἄνευ ἑλκέων, τὰς πρώτας τῶν ἡμερέων ἰητρεύουσιν εἰρίοισι ῥυπαροῖσιν· καὶ οὐδέν τι ἄτεχνον δοκέει τοῦτο εἶναι. ὁπόσοι μὲν οὖν ἀναγκάζονται ὑπὸ τῶν αὐτίκα νεοτρώτων ἐόντων, οὐκ[1] ἔχοντες ὀθόνια, εἰρίοισι παρασκευάσασθαι, τούτοισι πλείστη συγγνώμη· οὐ γὰρ ἄν τις ἔχοι ἄνευ ὀθονίων ἄλλο τι πολλῷ βέλτιον εἰρίου ἐπιδῆσαι[2] τοιαῦτα· εἶναι
10 δὲ χρὴ πάμπολλα καὶ πάνυ καλῶς εἰργασμένα καὶ μὴ τρηχέα· τῶν γὰρ ὀλίγων καὶ φλαύρων ὀλίγη καὶ ἡ δύναμις. ὅσοι δὲ ἐπὶ μίην ἢ δύο ἡμέρας εἴρια ἐπιδεῖν δικαιοῦσι, τρίτῃ δὲ καὶ τετάρτῃ ὀθονίοισιν ἐπιδέοντες πιέζουσι, καὶ κατατείνουσι

[1] μή.   [2] ἐπιδῆσαι ἐπί.

leg; and the wound is both easy to examine and easy to handle.[1] For, if one pleases, there is nothing to prevent the two upper rods from being tied together, so that, if one wants to put something lightly over it, the covering is kept up away from the wound. If then the circlets are supple, of good quality, soft and newly sewn, and the extension [2] by the bent rods suitably regulated as just described, the mechanism is of good use, but if any of these things are not well arranged it will harm rather than help. Other mechanisms also should either be well arranged or not used, for it is shameful and contrary to the art to make a machine and get no mechanical effect.

XXXI. Again, most practitioners treat fractures, whether with or without wounds, by applying uncleansed wool during the first days, and this appears in no way contrary to the art. Those who because they have no bandages are obliged to get wool for first-aid treatment [3] are altogether excusable, for in the absence of bandages one would have nothing much better than wool with which to dress such cases; but it should be plentiful, well pulled out and not lumpy; if small in amount and of poor quality its value is also small. Now, those who think it correct to dress with wool for one or two days, and on the third or fourth day use bandages with compression and extension just at this period

[1] "Arrange" (Adams), better than "maintain" (Littré, Petrequin); "sustinere aliquid" (Erm.) suits the context—"easily bears a covering," but see Herod. II. 125.

[2] ἔντασις perhaps connected with use of word in architecture, "slight outward curvature."

[3] Cf. Aristoph. *Acharn.* 12, *Vesp.* 275, *Lysist.* 987 on this use of wool.

τότε μάλιστα, οὗτοι πολύ τι τῆς ἰητρικῆς καὶ κάρτα ἐπίκαιρον ἀσυνετέουσι· ἥκιστα γὰρ χρὴ τῇ τρίτῃ ἡμέρῃ ἢ τῇ τετάρτῃ στυφελίζειν πάντα τὰ τρώματα, ὡς ἐν κεφαλαίῳ εἰρῆσθαι· καὶ μηλώσιας δὲ[1] πάσας φυλάσσεσθαι χρὴ ἐν
20 ταύτῃσι τῇσιν ἡμέρῃσι, καὶ ὁπόσοισιν ἄλλοισι τρώμασι[2] ἠρέθισται. τὸ ἐπίπαν γὰρ ἡ τρίτη καὶ τετάρτη ἡμέρη ἐπὶ τοῖσι πλείστοισι τῶν τρωμάτων τίκτει τὰς παλιγκοτήσιας, καὶ ὅσα ἐς φλεγμονὴν καὶ ἀκαθαρσίην ὁρμᾷ, καὶ ὅσα ἂν ἐς πυρετοὺς ἴῃ· καὶ μάλα πολλοῦ ἄξιον τοῦτο τὸ μάθημα, εἴ πέρ τι καὶ ἄλλο· τίνι γὰρ οὐκ ἐπικοινωνεῖ τῶν ἐπικαιροτάτων ἐν ἰητρικῇ, οὐ κατὰ τὰ ἕλκεα μόνον, ἀλλὰ καὶ κατ' ἄλλα πολλὰ νοσήματα; εἰ μή τις φήσειε καὶ τἄλλα νοσήματα
30 ἕλκεα εἶναι· ἔχει γάρ τινα καὶ οὗτος ὁ λόγος ἐπιείκειαν· πολλαχῇ γὰρ ἠδέλφισται τὰ ἕτερα τοῖσι ἑτέροισι. ὁπόσοι μέντοι δικαιοῦσιν εἰρίοισι χρῆσθαι, ἔστ' ἂν ἑπτὰ ἡμέραι παρέλθωσιν, ἔπειτα κατατείνειν τε καὶ κατορθοῦν καὶ ὀθονίοισιν ἐπιδεῖν, οὗτοι οὐκ ἂν ἀσύνετοι ὁμοίως φανεῖεν· καὶ γὰρ τῆς φλεγμονῆς τὸ ἐπικαιρότατον παρελήλυθε, καὶ τὰ ὀστέα χαλαρὰ [καὶ εὔθετα][3] μετὰ ταύτας τὰς ἡμέρας ἂν εἴη. πολλῷ μέντοι ἥσσηται καὶ αὕτη ἡ μελέτη τῆς ἐξ ἀρχῆς τοῖσιν
40 ὀθονίοισιν ἐπιδέσιος· κεῖνος μὲν γὰρ ὁ τρόπος ἑβδομαίους ἐόντας ἀφλεγμάντους ἀποδείκνυσι, καὶ παρασκευάζει νάρθηξι τελέως ἐπιδεῖν· οὗτος δὲ ὁ τρόπος πολὺ ὑστερεῖ, βλάβας δέ τινας καὶ ἄλλας ἔχει. ἀλλὰ μακρὸν ἂν εἴη πάντα γράφειν.

Ὁπόσοισι δὲ τὰ ὀστέα κατεηγότα καὶ ἐξ-

[1] χρή. [2] τρώματα. [3] Pq. omits.

are very ignorant of the healing art, and that on a most vital point. For, to speak summarily, the third or fourth day is the very last on which any lesion should be actively interfered with; and all probings as well as everything else by which wounds are irritated[1] should be avoided on these days. For, as a rule, the third or fourth day sees the birth of exacerbations in the majority of lesions, both where the tendency is to inflammation and foulness, and in those which turn to fever. And if any instruction is of value this is very much so. For what is there of most vital importance in the healing art to which it does not apply, not only as regards wounds but many other maladies? Unless one calls all maladies wounds, for this doctrine also has reasonableness, since they have affinity one to another in many ways. But those who think it correct to use wool till seven days are completed and then proceed to extension, coaptation and bandaging would appear not so unintelligent, for the most dangerous time for inflammation is past, and the bones after this period will be found loose and easy to put in place. Still, even this treatment is much inferior to the use of bandages from the beginning, for that method results in the patients being without inflammation on the seventh day and ready for complete dressing with splints, while the former one is much slower, and has some other disadvantages; but it would take long to describe everything.

In cases where the fractured and projecting bones

[1] Littré—Adams, "in wounds attended by irritation," seems pleonastic (he has said that no wound is to be interfered with). ὁκόσα ἄλλα οἷσιν ἠρέθισται τρώμασιν (Petrequin). This view is confirmed by Kw.'s reading.

ἴσχοντα μὴ δύνηται ἐς τὴν ἑωυτῶν χώρην καθιδρύεσθαι, ἥδε ἡ κατάστασις·[1] σιδήρια χρὴ ποιεῖσθαι ἐς τοῦτον τὸν τρόπον οὗπερ[2] οἱ μοχλοὶ ἔχουσιν, οἷς οἱ λατύποι χρέονται, τὸ μέν τι
50 πλατύτερον, τὸ δέ τι στενότερον· εἶναι δὲ χρὴ καὶ τρία καὶ ἔτι πλείω, ὡς τοῖσι μάλιστα ἁρμόζουσί τις χρήσαιτο·[3] ἔπειτα τούτοισι χρὴ ἅμα τῇ κατατάσει μοχλεύειν ὑπερβάλλοντα, πρὸς μὲν τὸ κατώτερον[4] τοῦ ὀστέου τὸ κατώτερον ἐρείδοντα, πρὸς δὲ τὸ ἀνώτερον[5] τὸ ἀνώτερον τοῦ σιδηρίου, ἁπλῷ δὲ λόγῳ, ὥσπερ εἰ λίθον τις ἢ ξύλον μοχλεύοι ἰσχυρῶς· ἔστω δὲ σθεναρὰ τὰ σιδήρια ὡς οἷόν τε, ὡς μὴ κάμπτηται. αὕτη μεγάλη τιμωρίη, ἤν τε τὰ σιδήρια ἐπιτήδεια ᾖ
60 καὶ μοχλεύηταί τις ὡς χρή· ὁπόσα γὰρ ἀνθρώποισιν ἄρμενα μεμηχάνηται, πάντων ἰσχυρότατά ἐστι τρία ταῦτα, ὄνου τε περιαγωγὴ καὶ μόχλευσις καὶ σφήνωσις· ἄνευ δὲ τούτων, ἢ ἑνὸς δέ[6] τινος ἢ πάντων, οὐδὲν τῶν ἔργων τῶν ἰσχυροτάτων οἱ ἄνθρωποι ἐπιτελέουσιν. οὔκουν ἀτιμαστέη αὕτη ἡ μόχλευσις· ἢ γὰρ οὕτως ἐμπεσεῖται τὰ ὀστέα, ἢ οὐκ ἄλλως. ἢν δ' ἄρα τοῦ ὀστέου τὸ ἄνω παρηλλαγμένον μὴ ἐπιτήδειον ἔχῃ ἐνέδρην τῷ μοχλῷ, ἀλλὰ πάροξυ ᾖ
70 παραφέρῃ,[7] παραγλύψασα χρὴ τοῦ ὀστέου ἐνέδρην τῷ μοχλῷ ἀσφαλέα ποιήσασθαι· μοχλεύειν δὲ χρὴ καὶ τείνειν αὐθήμερα ἢ δευτεραῖα, τριταῖα δὲ μή, τεταρταῖα δὲ ὡς ἥκιστα καὶ πεμπταῖα. καὶ μὴ ἐμβάλλοντα, ὀχλήσαντι δὲ ἐν ταύτῃσι τῇσιν ἡμέρῃσι, φλεγμονὴν ἂν

[1] καταστῆσαι used by Asiatic Greeks for "put in its place." Galen, XVIII(2). 590.

cannot be settled into their proper place, the following is the method of reduction. One must have iron rods made in fashion like the levers used by stone masons, broader at one end and narrower at the other.[1] There should be three and even more that one may use those most suitable. Then one should use these, while extension is going on, to make leverage, pressing the under side of the iron on the lower bone, and the upper side against the upper bone, in a word just as if one would lever up violently a stone or log. The irons should be as strong as possible so as not to bend. This is a great help, if the irons are suitable and the leverage used properly; for of all the apparatus contrived by men these three are the most powerful in action —the wheel and axle, the lever and the wedge. Without some one, indeed, or all of these, men accomplish no work requiring great force. This lever method, then, is not to be despised, for the bones will be reduced thus or not at all. If, perchance, the upper bone over-riding the other affords no suitable hold for the lever, but being pointed, slips past,[2] one should cut a notch in the bone to form a secure lodgment for the lever. The leverage and extension should be done on the first or second day, but not on the third, and least of all on the fourth and fifth. For to cause disturbance without reduction on these days would set up inflam-

[1] "One rather broader—another narrower," Adams.
[2] "Presents a point which makes the lever slip," Pq.; "the protruding part is sharp," Adams.

---

[2] ὅνπερ. [3] ἁρμόσσουσι . . . χρήσεται.
[4] κατωτέρω. [5] ἀνωτέρω.
[6] τέ. [7] πάροξυν παραφέρῃ. πάροξυ ἐὸν Littré.

ποιήσειε, καὶ ἐμβάλλοντι οὐδὲν ἧσσον· σπασμὸν μέντοι ἐμβάλλοντι πολὺ ἂν μᾶλλον ποιήσειεν ἢ ἀπορήσαντι ἐμβάλλειν. ταῦτα εὖ χρὴ εἰδέναι· καὶ γὰρ εἰ ἐπιγένοιτο σπασμὸς ἐμβάλλοντι,
80 ἐλπίδες μὲν οὐ πολλαὶ σωτηρίης· λυσιτελεῖ δὲ ὀπίσω ἐκβάλλειν τὸ ὀστέον, εἰ οἷόν τε εἴη ἀόχλως. οὐ γὰρ ἐπὶ τοῖσι χαλαρωτέροισι τοῦ καιροῦ σπασμοὶ καὶ τέτανοι ἐπιγίνονται, ἀλλὰ ἐπὶ τοῖσιν ἐντεταμένοισι μᾶλλον. περὶ οὗ οὖν ὁ λόγος, οὐ χρὴ ἐνοχλεῖν ἐν τῇσι προειρημένῃσιν ἡμέρῃσι ταύτῃσι, ἀλλὰ μελετᾶν ὅπως ἥκιστα φλεγμανεῖ τὸ ἕλκος καὶ μάλιστα ἐκπυήσει. ἐπὴν δὲ ἑπτὰ ἡμέραι παρέλθωσιν ἢ ὀλίγῳ πλείους, ἢν ἀπύρετος ᾖ, καὶ μὴ φλεγμαίνῃ τὸ
90 ἕλκος, τότε ἧσσον κωλύει πειρῆσθαι ἐμβάλλειν, ἢν ἐλπίζῃς κρατήσειν, ἢν δὲ μή, οὐδὲν δεῖ μάτην
92 ὀχλεῖν καὶ ὀχλεῖσθαι.

XXXII. Ἢν μὲν οὖν ἐμβάλλῃς τὰ ὀστέα ἐς τὴν ἑωυτῶν χώρην, γεγράφεται ἤδη οἱ τρόποι οἵως[1] χρὴ ἰητρεύειν, ἤν τε ἐλπίζῃς ὀστέα ἀποστήσεσθαι ἤν τε μή. χρὴ δέ, καὶ ἢν μὲν ἐλπίζῃς ὀστέα ἀποστήσεσθαι, [ὡς ἔφην,][2] τῷ τρόπῳ τῶν ὀθονίων ἐπὶ πᾶσι τοῖσι τούτοισι τὴν ἐπίδεσιν ποιεῖσθαι ἐκ μέσου τοῦ ὀθονίου ἀρχόμενον, ὡς ἐπὶ τὸ πολύ, ὡς ἀπὸ δύο ἀρχῶν ὑποδεσμὶς ἐπιδεῖται· τεκμαίρεσθαι δὲ χρὴ πρὸς τὴν μορφὴν τοῦ ἕλκεος,
10 ὅπως ἥκιστα σεσηρὸς καὶ ἐκπεπλιγμένον ἔσται παρὰ τὴν ἐπίδεσιν· τοῖσι μὲν γὰρ ἐπὶ δεξιὰ ἐπιδεῖν συντρόφως[3] ἔχει, τοῖσι δὲ ἐπ' ἀριστερά,
13 τοῖσι δὲ ἀπὸ δύο ἀρχέων.

[1] ὡς.

mation, and no less so if there was reduction; spasm, indeed, would much more likely be caused if reduction succeeded than if it failed. It is well to know this, for if spasm supervenes after reduction there is not much hope of recovery. It is advantageous to reproduce the displacement, if it can be done without disturbance, for it is not when parts are more relaxed than usual that spasms and tetanus supervene, but when they are more on the stretch. As regards our subject, then, one should not disturb the parts on the days above mentioned, but study how best to oppose inflammation in the wound and favour suppuration. At the end of seven days, or rather more, if the patient is free from fever and the wound not inflamed, there is less objection to an attempt at reduction, if you expect to succeed; otherwise you should not give the patient and yourself useless trouble.

XXXII. The proper modes of treatment after you reduce the bones to their place have already been described, both when you expect bones to come away and when you do not. Even when you expect bones to come away you should use in all such cases the method of separate bandages, as I said, beginning generally with the middle of the bandage as when an under-bandage is applied from two heads. Regulate the process with a view to the shape of the wound that it may be as little as possible drawn aside or everted by the bandaging: for in some cases it is appropriate to bandage to the right, in others to the left, in others from two heads.

---

[2] Omit Littré, Erm.

[3] συντρόφως = οἰκείως (Galen). Cf. XXIX.

XXXIII. Ὁπόσα δὲ κατηπορήθη ὀστέα ἐμπεσεῖν, ταῦτα [αὐτὰ][1] εἰδέναι χρὴ ὅτι ἀποστήσεται, καὶ ὅσα τελέως ἐψιλώθη τῶν σαρκῶν. ψιλοῦται δὲ ἐνίων μὲν τὸ ἄνω μέρος, μετεξετέρων δὲ κυκλωθὲν ἀμφιθνήσκουσιν[2] αἱ σάρκες· καὶ τῶν μὲν ἀπὸ τοῦ ἀρχαίου τρώματος σεσάπρισται ἔνια τῶν ὀστέων, τῶν δὲ οὔ· καὶ τῶν μὲν μᾶλλον, τῶν δὲ ἧσσον· καὶ τὰ μὲν σμικρά, τὰ δὲ μεγάλα. διὰ οὖν ταῦτα τὰ εἰρημένα οὐκ ἔστιν ἑνὶ ὀνόματι 10 εἰπεῖν, ὁπότε τὰ ὀστέα ἀποστήσεται· τὰ μὲν γὰρ διὰ σμικρότητα, τὰ δὲ διὰ τὸ ἐπ' ἄκρου ἔχεσθαι, θᾶσσον ἀφίσταται· τὰ δέ, διὰ τὸ μὴ ἀφίστασθαι, ἀλλὰ λεπιδοῦσθαι, καταξηρανθέντα καὶ σαπρὰ γενόμενα· πρὸς δὲ τούτοις, διαφέρει τι καὶ ἰητρείη ἰητρείης. ὡς μὲν οὖν τὸ ἐπίπαν τάχιστα τούτων ὀστέα ἀφίσταται ὧν τάχισται μὲν αἱ ἐκπυήσιες, τάχισται δὲ καὶ κάλλισται αἱ σαρκοφυΐαι, καὶ γὰρ αἱ ὑποφυόμεναι σάρκες κατὰ τὸ σιναρὸν αὗται μετεωρίζουσι τὰ ὀστέα 20 ὡς ἐπὶ τὸ πολύ. ὅλος μὴν ὁ κύκλος τοῦ ὀστέου, ἢν ἐν τεσσαράκοντα ἡμέρῃσιν ἀποστῇ, καλῶς ἀποστήσεται· ἔνια γὰρ ἐς ἑξήκοντα ἡμέρας ἀφικνεῖται [ἢ καὶ πλείους].[3] τὰ μὲν γὰρ ἀραιότερα τῶν ὀστέων θᾶσσον ἀφίσταται, τὰ δὲ στερεώτερα, βραδύτερον· τὰ δὲ ἄλλα τὰ μείω, πολλὸν ἐνδοτέρω, ἄλλα δ' ἄλλως. ἀποπρίειν δ' ὀστέον ἐξέχον ἐπὶ τῶνδε τῶν προφασίων χρή· ἢν μὴ δύνηται ἐμβάλλειν, μικροῦ δέ τινος αὐτῷ δοκῇ δεῖν παρελθεῖν, καὶ οἷόν τε ᾖ παραιρεθῆναι· ἤν 30 τε ἀσηρὸν ᾖ καὶ θραῦόν τι τῶν σαρκίων, καὶ δυσθεσίην παρέχῃ, ψιλόν τε τυγχάνῃ ἐόν, καὶ

[1] Omit B, Pq.

XXXIII. As to bones which cannot be reduced, it should be known that just these will come away, as also will those which are completely denuded. In some cases the upper part of the bones are denuded, in others the soft parts surrounding them perish, and the starting point of the necrosis is, in some of the bones, the old wound, in others not. It is more extensive in some and less so in others, and some bones are small, others large. It follows from the above that one cannot make a single statement as to when the bones will come away, for some separate sooner owing to their small size, others because they come at the end (of the fracture) while others do not come away (as wholes) but are exfoliated after desiccation and corruption. Besides this, the treatment makes a difference. As a general rule, bones are most quickly eliminated in cases where suppuration is quickest, and the growth of new flesh most rapid and good; for it is the growth of new flesh in the lesion that as a rule lifts up the fragments. As to a whole circle of bone, if it comes away in forty days it will be a good separation, for some cases go on to sixty days or even more. The more porous bones come away more quickly, the more solid more slowly; for the rest, the smaller ones take much less time, and so variously. The following are the indications for resection of a protruding bone: if it cannot be reduced, but only some small portion seems to come in the way, and it is possible to remove it; if it is harmful, crushing some of the tissues, and causing wrong position of the part, and if it is denuded, this also should

---

[2] περιθνήσκουσι. [3] Kw. Omits

τὸ τοιοῦτον[1] ἀφαιρεῖν χρή. τὰ δὲ ἄλλα οὐδὲν μέγα διαφέρει, οὔτε ἀποπρῖσαι οὔτε μὴ ἀποπρῖσαι. σαφέως γὰρ εἰδέναι χρὴ ὅτι ὀστέα, ὅσα τελέως στερέεται τῶν σαρκῶν καὶ ἐπιξηραίνεται, ὅτι πάντα τελέως ἀποστήσεται. ὅσα δὲ ἀπολεπιδοῦσθαι μέλλει, ταῦτα οὐ χρὴ ἀποπρίειν· τεκμαίρεσθαι δὲ χρὴ ἀπὸ τῶν τεταγμένων
39 σημείων τὰ τελέως ἀποστησόμενα.

XXXIV. Ἰητρεύειν δὲ τοὺς τοιούτους σπλήνεσι καὶ τῇ οἰνηρῇ ἰητρείῃ, ὥσπερ καὶ πρόσθεν γέγραπται ἐπὶ τῶν ἀποστησομένων ὀστέων. φυλάσσεσθαι δὲ χρὴ μὴ ψυχροῖσι[2] τέγγειν τὸν πρῶτον χρόνον· ῥιγέων γὰρ πυρετώδων κίνδυνος· κίνδυνος δὲ καὶ σπασμῶν· προκαλεῖται γὰρ σπασμὸν τὰ ψυχρά, ποτὶ δὲ καὶ ἕλκη. εἰδέναι δὲ χρὴ ὅτι ἀνάγκη βραχύτερα τὰ σώματα ταύτῃ γίνεσθαι, ὧν ἀμφότερα τὰ ὀστέα κατεηγότα καὶ
10 παρηλλαγμένα ἰητρεύηται, καὶ οἷς ὅλος ὁ κύκλος
11 τοῦ ὀστέου ἀπέστη.

XXXV. Ὅσοισι[3] δὲ μηροῦ ὀστέον ἢ βραχίονος ἐξέσχεν, οὗτοι οὐ μάλα περιγίνονται. τὰ γὰρ ὀστέα μεγάλα καὶ πολυμύελα, καὶ πολλὰ καὶ ἐπίκαιρα τὰ συντιτρωσκόμενα νεῦρα[4] καὶ μύες καὶ φλέβες· καὶ ἢν μὲν ἐμβάλλῃς, σπασμοὶ φιλέουσι ἐπιγίνεσθαι, μὴ ἐμβληθεῖσι δὲ πυρετοὶ ὀξέες καὶ ἐπίχολοι καὶ λυγγώδεες, καὶ ἐπιμελαίνονται· περιγίνονται δὲ οὐχ ἧσσον, οἷσι μὴ ἐμβληθῇ, μὴ πειρηθῇ[5] ἐμβάλλεσθαι· ἔτι δὲ μᾶλλον περι-
10 γίνονται, οἷσι τὸ κάτω μέρος τοῦ ὀστέου ἐξέσχεν,

[1] τοιοῦτο.
[2] καταψυχροῖσι (B M V). Kw. adopts Ermerins's suggestion κάρτα.

be removed. In other cases it makes no great difference whether there is resection or not. For one should bear clearly in mind that when bones are entirely deprived of soft parts and dried up they will all come away completely: and one should not resect those bones which are going to be exfoliated. Draw your conclusion as to bones which will come away completely from the symptoms set forth.

XXXIV. Treat such cases with compresses and vinous applications as described above in the case of bones about to be eliminated. Take care not to moisten with cold fluids at first, for there is risk of feverish rigors and further risk of spasms, for cold substances provoke spasms and sometimes[1] ulcerations. Bear in mind that there must be shortening of the parts in cases where, when both bones are broken, they are treated while over-lapping, also in cases where the circle of bone is eliminated entire.

XXXV. Cases where the bone of the thigh or upper arm protrudes rarely recover; for the bones are large and contain much marrow, while the cords, muscles and blood vessels which share in the injury are numerous and important. Besides, if you reduce the fracture, convulsions are liable to supervene, while in cases not reduced there are acute bilious fevers with hiccough and mortification. Cases where reduction has not been made or even attempted are no less likely to recover, and recovery is more frequent when the lower than when the upper part

[1] This seems the place where ποτὶ means ποτὲ as Galen says in his Lexicon, but ποτὶ καὶ is an expression peculiar to these treatises and means "especially." See Diels *op. cit.*

---

[3] Ὅσων. [4] καὶ νεῦρα.
[5] ἐνεβλήθη . . . ἐπειρήθη.

ἢ οἷσι τὸ ἄνω· περιγίνοιντο δ' ἂν καὶ οἷσιν ἐμβληθείη, σπανίως γε μήν. μελέται γὰρ μελετέων μέγα διαφέρουσι, καὶ φύσιες φυσίων τῶν σωμάτων ἐς εὐφορίην. διαφέρει δὲ μέγα, καὶ ἢν ἔσω τοῦ βραχίονος καὶ τοῦ μηροῦ τὰ ὀστέα ἐξέχῃ· πολλαὶ γὰρ καὶ ἐπίκαιροι κατατάσιες φλεβῶν ἐν τῷ ἔσω μέρει, ὧν ἔνιαι τιτρωσκόμεναι σφάγιαί εἰσιν· εἰσὶ δὲ καὶ ἐν τῷ ἔξω μέρει, ἧσσον δέ. ἐν τοῖσιν οὖν τοιούτοισι τρώμασι
20 τοὺς μὲν κινδύνους οὐ χρὴ λήθειν ὁποῖοί τινές εἰσι, καὶ προλέγειν χρὴ πρὸς τοὺς καιρούς. εἰ δὲ ἀναγκάζοιο μὲν ἐμβάλλειν, ἐλπίζοις δὲ ἐμβάλλειν, καὶ μὴ πολλὴ ἡ παράλλαξις εἴη τοῦ ὀστέου, καὶ μὴ συνδεδραμήκοιεν οἱ μύες—φιλέουσι γὰρ συνθεῖν—ἡ μόχλευσις καὶ τούτοισι
26 μετὰ τῆς κατατάσιος εὖ ἂν συλλαμβάνοιτο.

XXXVI. Ἐμβάλλοντα δέ, ἐλλέβορον μαλθακὸν πιπίσαι χρὴ αὐθήμερον, ἢν αὐθήμερον ἐμβληθῇ, εἰ δὲ μή, οὐδ' ἐγχειρεῖν χρή. τὸ δὲ ἕλκος ἰητρεύειν χρή· οἷσί περ κεφαλῆς ὀστέα κατεηγυίης καὶ ψυχρὸν μηδὲν προσφέρειν, σιτίων δὲ στερῆσαι τελέως· καὶ ἢν μὲν πικρόχολος φύσει ᾖ, ὀξύγλυκυ εὐῶδες ὀλίγον ἐφ' ὕδωρ ἐπιστάζοντα τούτῳ διαιτᾶν· ἢν δὲ μὴ πικρόχολος ᾖ, ὕδατι πόματι χρῆσθαι· καὶ ἢν μὲν πυρεταίνῃ
10 συνεχῶς, τεσσαρακαίδεκα ἡμέρῃσι[1] τὸ ἐλάχιστον οὕτω διαιτᾶν, ἢν δὲ ἀπύρετος ᾖ, ἑπτὰ ἡμέρῃσιν· ἔπειτα ἐκ προσαγωγῆς κατὰ λόγον ἐς φαύλην δίαιταν ἄγειν. καὶ οἷσιν μὴ[2] ἐμβληθῇ τὰ ὀστέα, καὶ τὴν φαρμακείην χρὴ τοιαύτην ποιεῖσθαι, καὶ

[1] ἡμέρας bis. [2] ἂν μή.

of the bone projects. There may be survival even in cases where reduction is made, but it is rare indeed. There are great differences between one way of dealing with the case and another, and between one bodily constitution and another as to power of endurance. It also makes a great difference whether the bone protrudes on the inner or outer side of the arm or thigh, for many important blood vessels stretch along the inner side, and lesions of some of them are fatal; there are also some on the outside, but fewer. In such injuries, then, one must not overlook the dangers or the nature of some of them, but foretell them as suits the occasion. If you have to attempt reduction and expect to succeed and there is no great overriding of the bone, and the muscles are not retracted (for they are wont to retract) leverage combined with extension would be well employed even in these cases.

XXXVI. After reduction one should give a mild dose of hellebore on the first day, if it is reduced on the first day, otherwise one should not even attempt it. The wound should be treated with the remedies used for the bones of a broken head. Apply nothing cold and prescribe entire abstinence from solid food. If he is of a bilious nature give him a little aromatic hydromel[1] sprinkled in water, but if not, use water as beverage. And if he is continuously febrile keep him on this regimen for fourteen days at least, but if there is no fever, for seven days, then return by a regular gradation to ordinary diet. In cases where the bones are not reduced, a similar purgation should be made and so with the management of the wounds

[1] Decoction of honeycomb in water = ἀπόμελι in XI; cf. Galen on its preparation.

τῶν ἑλκέων τὴν μελέτην καὶ τὴν δίαιταν· ὡσαύτως καὶ τὸ ἀπαιωρεύμενον[1] τοῦ σώματος μὴ κατατείνειν, ἀλλὰ καὶ προσάγειν μᾶλλον, ὥστε χαλαρώτερον εἶναι τὸ κατὰ τὸ ἕλκος. τῶν δὲ ὀστέων ἀπόστασις[2] χρονίη, ὥσπερ καὶ πρόσθεν
20 εἴρηται. μάλιστα δὲ χρὴ τὰ τοιαῦτα διαφυγεῖν, ἅμα ἤν τις καλὴν ἔχῃ τὴν ἀποφυγήν. αἵ τε γὰρ ἐλπίδες ὀλίγαι, καὶ οἱ κίνδυνοι πολλοί· καὶ μὴ ἐμβάλλων ἄτεχνος ἂν δοκέοι εἶναι, καὶ ἐμβάλλων ἐγγυτέρω ἂν τοῦ θανάτου ἀγάγοι ἢ
25 σωτηρίης.

XXXVII. Τὰ δὲ ὀλισθήματα τὰ κατὰ τὰ γούνατα καὶ τὰ διακινήματα τῶν ὀστέων εὐηθέστερα πολὺ τῶν κατ' ἀγκῶνα κινημάτων καὶ ὀλισθημάτων· τό τε γὰρ ἄρθρον τοῦ μηροῦ εὐσταλέστερον ὡς ἐπὶ μεγέθει ἢ τὸ τοῦ βραχίονος, καὶ δικαίην φύσιν μοῦνον ἔχον, καὶ ταύτην περιφερέα· τὸ δὲ τοῦ βραχίονος ἄρθρον μέγα τε καὶ βαθμίδας πλείονας ἔχον. πρὸς δὲ τούτοις, τὰ μὲν τῆς κνήμης ὀστέα παραπλήσια μῆκός
10 ἐστι καὶ σμικρόν τε οὐκ ἄξιον λόγου τὸ ἔξω ὀστέον ὑπερέχει, οὐδενὸς μεγάλου κώλυμα ἐόν, ἀφ' οὗ πέφυκεν ὁ ἔξω τένων ὁ παρὰ τὴν ἰγνύην· τὰ δὲ τοῦ πήχεος ὀστέα ἄνισά ἐστιν, καὶ τὸ βραχύτερον παχύτερον συχνῷ, τὸ δὲ λεπτότερον πολλὸν ὑπερβάλλει καὶ ὑπερέχει τὸ ἄρθρον· ἐξήρτηται μέντοι καὶ τούτων[3] τῶν νεύρων κατὰ τὴν κοινὴν σύμφυσιν τῶν ὀστέων· πλεῖον δὲ μέρος ἔχει τῆς ἐξαρτήσιος τῶν νεύρων ἐν τῷ βραχίονι τὸ λεπτὸν ὀστέον ἤπερ τὸ παχύ. ἡ
20 μὲν οὖν φύσις τοιουτότροπος τῶν ἄρθρων τούτων

[1] ἀπορεύμενον. [2] ἡ ἀπόστασις.
[3] τοῦτο.

and the regimen. Likewise do not stretch the unreduced part,[1] but even bring it more together so that the seat of the wound may be more relaxed. Elimination of the bones takes time, as was said before. One should especially avoid such cases if one has a respectable excuse, for the favourable chances are few, and the risks many. Besides, if a man does not reduce the fracture, he will be thought unskilful, while if he does reduce it he will bring the patient nearer to death than to recovery.

XXXVII. Dislocations at the knee and disturbances of the bones are much milder than displacements and dislocations at the elbow; for the articular end of the thigh-bone is more compact in relation to its size than is that of the arm-bone, and it alone has a regular conformation, a rounded one, whereas the articular end of the humerus is extensive, having several cavities. Besides this the leg-bones are about the same size, the outer one overtops the other to some little extent not worth mention,[2] and opposes no hindrance to any large movement though the external tendon of the ham arises from it. But the bones of the forearm are unequal, and the shorter (radius) much the thicker, while the more slender one (ulna) goes far beyond and overtops the joint. This, however, is attached to the ligaments at the common junction of the bones.[3] The slender bone has a larger share than the thicker one of the attachments of ligaments in the arm. Such then is the disposition of these articulations and of

[1] Kw.'s reading is the most suitable.

[2] A curious error, perhaps due to an effort to make the fibula resemble the ulna as far as possible. (The fibula does not reach the top of the tibia.)

[3] The ulna is attached to the ligaments of the elbow joint, at the point where it joins the radius. Galen.

καὶ τῶν ὀστέων τοῦ ἀγκῶνος. καὶ διὰ τὸν τρόπον τῆς φύσιος τὰ κατὰ τὸ γόνυ ὀστέα πολλάκις μὲν ὀλισθάνει, ῥηϊδίως δὲ ἐμπίπτει· φλεγμονὴ δὲ οὐ μεγάλη προσγίνεται, οὐδὲ δεσμὸς τοῦ ἄρθρου. ὀλισθάνει δὲ τὰ πλεῖστα ἐς τὸ ἔσω μέρος, ἔστι δ' ὅτε ἐς τὸ ἔξω, ποτὲ δὲ καὶ ἐς τὴν ἰγνύην. τούτων ἁπάντων αἱ ἐμβολαὶ οὐ χαλεπαί· ἀλλὰ τὰ μὲν ἔξω καὶ ἔσω ὀλισθάνοντα, καθῆσθαι μὲν χρὴ τὸν ἄνθρωπον χαμαὶ ἢ ἐπὶ χαμαιζήλου 30 τινός, τὸ δὲ σκέλος ἀνωτέρω ἔχειν, μὴ μέν πολλῷ. κατάτασις δὲ ὡς ἐπὶ τὸ πολὺ μετρίη ἀρκεῖ, τῇ μὲν κατατείνειν τὴν κνήμην, τῇ δὲ ἀντιτείνειν τὸν 33 μηρόν.[1]

XXXVIII. Τὰ δὲ κατὰ τὸν ἀγκῶνα ὀχλωδέστερά ἐστι τῶν κατὰ τὸ γόνυ, καὶ δυσεμβολώτερα καὶ διὰ τὴν φλεγμονὴν καὶ διὰ τὴν φύσιν, ἢν μή τις αὐτίκα ἐμβάλῃ· ὀλισθάνει μὲν ἧσσον[2] ἢ ἐκεῖνα, δυσεμβολώτερα δὲ καὶ δυσθετώτερα, 6 καὶ ἐπιφλεγμαίνει μᾶλλον καὶ ἐπιπωροῦται.[3]

XXXIX. Ἔστι δὲ καὶ τούτων πλεῖστα[4] σμικραὶ ἐγκλίσιες, ἄλλοτε ἐς τὸ πρὸς τῶν πλευρέων μέρος, ἄλλοτε ἐς τὸ ἔξω, οὐ πᾶν δὲ τὸ ἄρθρον μεταβεβηκός, ἀλλὰ μένον[5] τὸ κατὰ τὸ κοῖλον

[1] End of Galen's Commentary as extant; but later fragments are preserved in Orib. XLVI. 6, XLVII.5, etc.

[2] ἧσσον opposed to πολλάκις above: but not true. Some therefore take it to mean "to a less extent."

[3] ἐπιποροῦται. [4] τὰ μὲν πλεῖστα.

[5] μόνον B, μένοντι τὸ M, μένον τι V, μοῦνον Kw. The reading is important for the writer's account of elbow dislocations. If μένον, the chapter must refer to dislocation of the radius only and "inwards" would imply that the writer looked at the arm and hand as hanging back to front with the bend of the elbow turned inwards, the reverse of our position. Petrequín first noticed this, and showed that

the bones of the elbow. Owing to the way they are disposed the bones at the knee are often dislocated [1] but easily put in, and no great inflammation or fixation of the joint supervenes. Most dislocations are inwards,[2] but some outwards and some into the knee flexure. Reduction is not difficult in any of these cases: as to external and internal dislocations, the patient should be seated on the ground or something low, and have the leg raised, though not greatly. Moderate extension as a rule suffices; make extension on the leg and counter-extension on the thigh.

XXXVIII. Dislocations at the elbow are more troublesome than those at the knee, and harder to put in, both because of the inflammation and because of the conformation of the bones, unless one puts them in at once. It is true that they are more rarely [3] dislocated than the above, but they are harder to put up, and inflammation and excessive formation of callus [4] is more apt to supervene.

XXXIX. (Dislocation of radius.) The majority of these are small displacements sometimes inwards, towards the side and ribs, sometimes outwards (our "forwards" and "backwards"). The joint is not dislocated as a whole, but maintaining the con-

[1] A strange remark, perhaps includes displacement of the kneecap. Displacements of cartilages are not noticed.

[2] Of the thigh-bone.

[3] Pq. says he treated ten times more elbow than knee dislocations.

[4] Cf. Celsus VIII. 16, "callus circumdatur."

---

it explains much. *μόνον* or *μοῦνον* would imply a dislocation of the ulna only, and add another difficulty. It seems clear that the epitomist (M VII, J XVII) read *μένον*; but these chapters have puzzled the scribes as well as the surgeons.

τοῦ ὀστέου τοῦ βραχίονος, ἢ τὸ τοῦ πήχεος ὀστέον τὸ ὑπερέχον ἔχει.[1] τὰ μὲν οὖν τοιαῦτα, κἂν τῇ ἢ τῇ ὀλίσθῃ, ῥηΐδιον ἐμβάλλειν, καὶ ἀποχρὴ ἡ κατάτασις ἡ ἐς τὸ ἰθὺ γινομένη κατ' ἰθυωρίην τοῦ βραχίονος, τὸν μὲν κατὰ τὸν καρπὸν 10 τῆς χειρὸς τείνειν, τὸν δὲ κατὰ τὴν μασχάλην περιβάλλοντα, τὸν δὲ τῇ ἑτέρῃ πρὸς τὸ ἐξεστεὸς ἄρθρον τὸ θέναρ προσβάλλοντα ὠθεῖν, τῇ δὲ 13 ἑτέρῃ ἀντωθεῖν προσβάλλοντα[2] ἐγγὺς τῷ ἄρθρῳ.

XL. Ἐνακούει δὲ οὐ βραδέως ἐμβαλλόμενα τὰ τοιαῦτα ὀλισθήματα, ἢν πρὶν φλεγμήνῃ ἐμβάλλῃ τις. ὀλισθάνει δὲ ὡς ἐπὶ τὸ πολὺ μᾶλλον ἐς τὸ ἔσω μέρος, ὀλισθάνει δὲ καὶ ἐς τὸ ἔξω, εὔδηλα δὲ τῷ σχήματι. καὶ πολλάκις ἐμπίπτει τὰ τοιαῦτα, καὶ ἄνευ ἰσχυρῆς κατατάσιος· χρὴ δὲ τῶν ἔσω ὀλισθανόντων, τὸ μὲν ἄρθρον ἀπωθεῖν ἐς τὴν φύσιν, τὸν δὲ πῆχυν ἐς τὸ καταπρηνὲς μᾶλλον ῥέποντα[3] περιάγειν. τὰ μὲν 10 πλεῖστα ἀγκῶνος τοιαῦτα ὀλισθήματα.

XLI. Ἢν δὲ ὑπερβῇ τὸ ἄρθρον ἢ ἔνθα ἢ ἔνθα ὑπὲρ τὸ ὀστέον τοῦ πήχεος τὸ ἐξέχον ἐς τὸ κοῖλον τοῦ βραχίονος—γίνεται μὲν οὖν ὀλιγάκις τοῦτο, ἢν δὲ γίνηται—οὐκ ἔτι ὁμοίως ἡ κατάτασις ἡ ἐς τὴν ἰθυωρίην γινομένη ἐπιτηδείη τῶν τοιούτων ὀλισθημάτων· κωλύει γὰρ ἐν τῇ τοιαύτῃ κατατάσει τὸ ἀπὸ τοῦ πήχεος ὑπερέχον ὀστέον τὴν ὑπέρβασιν τοῦ βραχίονος. χρὴ τοίνυν τοῖσιν

[1] ἐξέσχεν B, Kw., etc. [2] πρὸς τοῦ πήχεος B, Kw. insert. [3] Pq. omits.

nexion with the cavity of the humerus, where the projecting part of the ulna sticks out. Such cases, then, whether dislocation is to one side or the other, are easy to reduce, and direct extension in the line of the upper arm is quite enough, one person may make traction on the wrist, another does so by clasping the arm at the axilla, while a third presses with the palm of one hand on the projecting part and with the other makes counter-pressure near the joint.

XL. Such dislocations yield readily to reduction if one reduces them before they are inflamed ; the dislocation is usually rather inwards (forwards), but may also be outwards, and is easily recognised by the shape. And they are often reduced even without vigorous extension. In the case of internal dislocations one should push the joint back into its natural place, and turn the forearm rather towards the prone position. Most dislocations of the elbow are of this kind.[1]

XLI. (Complete dislocation of the elbow backwards and forwards). If the articular end of the humerus passes either this way or that[2] over the part of the ulna which projects into its cavity (the latter[3] indeed occurs rarely, if it does occur), extension in the line of the limb is no longer equally suitable, for the projecting part of the ulna prevents the passage of the humerus. In patients with these

[1] Adams agrees that XXXIX is "dislocation of the radius," but has to call XL "incomplete lateral dislocation of the forearm" since the radius alone cannot be dislocated "inwards." The nature of these lesions is discussed on p. 411 ff.

[2] "to either side," Adams.

[3] Refers to "backwards," which can hardly occur without fracture.

οὕτως ἐκβεβληκόσι τὴν κατάτασιν ποιεῖσθαι
10 τοιαύτην, οἵη περ πρόσθεν γέγραπται, ἐπήν τις
ὀστέα βραχίονος κατεηγότα ἐπιδέῃ, ἀπὸ μὲν τῆς
μασχάλης ἐς τὸ ἄνω τείνεσθαι, ἀπὸ δὲ τοῦ
ἀγκῶνος αὐτοῦ ἐς τὸ κάτω ἀναγκάζειν· οὕτω γὰρ
ἂν μάλιστα ὁ βραχίων ὑπεραιωρηθείη ὑπὲρ τῆς
ἑωυτοῦ βαθμίδος, ἢν δὲ ὑπεραιωρηθῇ, ῥηϊδίη ἡ
κατάστασις, τοῖσι θέναρσι τῶν χειρῶν τὸ μὲν
ἐξεστεὸς[1] τοῦ βραχίονος ἐμβάλλοντα ὠθεῖν, τὸ
δὲ ἐς τὸ τοῦ πήχεος ὀστέον τὸ παρὰ τὸ ἄρθρον
ἐμβάλλοντα ἀντωθεῖν, τὸν αὐτὸν τρόπον ἄμφω·
20 ἧσσον μέντοι[2] ἡ τοιαύτη κατάτασις τοῦ τοιούτου
ὀλισθήματος δικαιοτάτη· ἐμβληθείη δ' ἂν καὶ
22 ἀπὸ τῆς ἐς ἰθὺ κατατάσιος, ἧσσον δὲ ἢ οὕτω.

XLII. Ἢν δὲ ἐς τοὔμπροσθεν ὀλίσθῃ ὁ βραχίων, ἐλαχιστάκις μὲν τοῦτο γίνεται, ἀλλὰ τί ἂν
ἐξαπίνης[3] ἐκπάλησις οὐκ ἐμβάλλοι; πολλὰ
γὰρ καὶ παρὰ τὴν οἰκείην[4] φύσιν ἐκπίπτει, καὶ
ἢν μέγα τι ᾖ τὸ κωλῦον· ταύτῃ δὲ τῇ ἐκπαλήσει
μέγα τι τὸ ὑπερβαινόμενον τὸ ὑπὲρ τὸ παχύτερον
τῶν ὀστέων, καὶ τῶν νεύρων συχνὴ κατάτασις·
ὅμως δὲ δή τισιν ἐξεπάλησεν. σημεῖον δὲ τοῖσιν
οὕτως ἐκπαλήσασιν· οὐδὲν γὰρ χρῆμα τοῦ
10 ἀγκῶνος κάμψαι δύνανται, εὔδηλον[5] δὲ καὶ
τὸ ἄρθρον ψαυόμενον. ἢν μὲν οὖν μὴ αὐτίκα
ἐμβληθῇ, ἰσχυραὶ καὶ βίαιαι φλεγμοναὶ καὶ
πυρετώδεες γίνονται· ἢν δὲ δὴ αὐτίκα τις
παρατύχῃ εὐέμβολον, [χρὴ δὲ ὀθόνιον σκληρόν][6]

[1] ἐς τὸ ἐξεστεός.
[2] Kw. ἄμφω, ἧσσον μέντοι . . . He supposes a hiatus.
[3] ἐξαπιναίη.
[4] ἐοικυῖαν.
[5] ἔνδηλον.
[6] Kw. omits.

dislocations, extension should be made after the manner which has been described above for putting up a fractured humerus. Make traction upwards from the armpit, and apply pressure downwards at the elbow itself, for this is the most likely way to get the humerus lifted above its own socket, and if it is so raised, replacement by the palms of hands is easy, using pressure with one hand to put in the projecting part of the humerus, and making counter-pressure on the ulna at the joint to put it back. The same method suits both cases. This has, indeed, less claim to be called the most regular method of extension in such a dislocation and reduction would also be made by direct extension, but less easily.[1]

XLII. (Internal lateral distortion of the forearm, *Petrequin's View*). Suppose the humerus to be dislocated forwards. This happens very rarely; but what might not be dislocated by a sudden violent jerk? For many other bones are displaced from their natural position,[2] though the opposing obstacle may be great. Now, there is a great obstacle to this jerking out, namely the passage over the thicker bone (radius) and the extensive stretching of the ligaments, but nevertheless it is jerked out in some cases. Symptoms in cases of such jerkings out. They cannot bend the elbow at all, and palpation of the joint makes it clear. If, then, it is not reduced at once, violent and grave inflammation occurs with fever, but if one happens to be on the spot it is easily put in. One should take

[1] "Evidently meant as a description of complete lateral dislocation," Adams.
[2] Kw. "beyond what seems natural."

—ὀθόνιον γὰρ σκληρὸν εἰλιγμένον ἀρκεῖ, μὴ μέγα —ἐνθέντα πλάγιον ἐς τὴν καμπὴν τοῦ ἀγκῶνος, ἐξαπίνης συγκάμψαι τὸν ἀγκῶνα καὶ προσαγαγεῖν ὡς μάλιστα τὴν χεῖρα πρὸς τὸν ὦμον. ἱκανὴ μὲν αὕτη ἡ ἐμβολὴ τοῖσιν οὕτως ἐκπαλή-
20 σασιν·[1] ἀτὰρ καὶ ἡ ἐς τὸ ἰθὺ κατάτασις δύναται εὐθετίζειν τοῦτον τὸν τρόπον τῆς ἐμβολῆς· τοῖσι μέντοι θέναρσι τῶν χειρῶν χρή, τὸν μὲν ἐμβάλλοντα ἐς τὸ τοῦ βραχίονος ἐξέχον τὸ παρὰ τὴν καμπὴν ὀπίσω ἀπωθεῖν, τὸν δέ τινα κάτωθεν ἐς τὸ τοῦ ἀγκῶνος ὀξὺ ἐμβάλλοντα ἀντωθεῖν ἐς τὴν ἰθυωρίην τοῦ πήχεος ῥέποντα. δύναται δὲ ἐν τούτῳ τῷ τρόπῳ τῆς ὀλισθήσιος κἀκείνη ἡ κατάτασις ἡ πρόσθεν ἐγγεγραμμένη,[2] ὡς χρὴ κατατείνειν τὰ ὀστέα τοῦ βραχίονος κατεηγότα,
30 ἐπὴν μέλλωσιν ἐπιδεῖσθαι· ἐπὴν δὲ καταταθῇ, οὕτω χρὴ τοῖσι θέναρσι τὰς προσβολὰς ποι-
32 εῖσθαι, ὥσπερ καὶ πρόσθεν γέγραπται.

XLIII. Ἢν δὲ ἐς τὸ ὀπίσω βραχίων ἐκπέσῃ—ὀλιγάκις δὲ τοῦτο γίνεται, ἐπωδυνώτατόν τε τοῦτο πάντων καὶ πυρετωδέστατον, συνεχέων πυρετῶν καὶ ἀκρητοχόλων, θανατωδέων καὶ ὀλιγημέρων—οἱ τοιοῦτοι ἐκτανύειν οὐ δύνανται. ἢν δὲ μὲν οὖν αὐτίκα παρατύχῃς, βιάσασθαι[3] χρὴ ἐκτανύσαντα τὸν ἀγκῶνα, καὶ αὐτομάτως ἐμπίπτει. ἢν δέ σε φθάσῃ πυρεταινήσας, οὐκ ἔτι χρὴ ἐμβάλλειν· κατακτείνειε γὰρ ἂν ἡ ὀδύνη ἀναγκαζομένου. ὡς
10 δ' ἐν κεφαλαίῳ εἰρῆσθαι, οὐδ' ἄλλο χρὴ ἄρθρον
11 πυρεταίνοντι ἐμβάλλειν, ἥκιστα δὲ ἀγκῶνα.

[1] τῷ τοιούτῳ. [2] πρόσθε γεγραμμ'νη. [3] βιάζεσθαι.

a hard bandage (a hard rolled bandage of no great size is sufficient) and put it crosswise in the bend of the elbow, suddenly flex the elbow, and bring the hand as close as possible to the shoulder. This mode of reduction is sufficient for such jerkings out. Direct extension, too, can accomplish this reduction. One must, however, use the palms, putting one on the projecting part of the humerus at the elbow and pushing backwards (our inwards), and with the other making counter-pressure below the point of the elbow, inclining the parts into the line of the ulna.[1] In this form of dislocation, the mode of extension described above as proper to be used in stretching the fractured humerus when it is going to be bandaged is also effective. And when extension is made, application of the palms should be made as described above.

XLIII. (External lateral dislocation of forearm).[2] If the humerus is dislocated backwards (our "inwards")—this occurs rarely, and is the most painful of all, most frequently causing continuous fever with vomiting of pure bile, and fatal in a few days—the patients cannot extend the arm. If you happen to be quickly on the spot, you ought to extend the elbow forcibly, and it goes in of its own accord. But if he is feverish when you arrive, do not reduce, for the pain of a violent operation would kill him. It is a general rule not to reduce any joint when the patient has fever, least of all the elbow.

[1] Adams. "Dislocation of ulna and radius backwards," II. 500, but II. 549, "It would seem to be dislocation of the forearm forwards."

[2] So Petrequin. It seems impossible that this should be dislocation of the forearm backwards, the commonest form, as Adams suggests.

XLIV. Ἔστι δὲ καὶ ἄλλα σίνεα κατ' ἀγκῶνα ὀχλώδεα· τοῦτο μὲν γάρ, τὸ παχύτερον ὀστέον ἔστιν ὅτε ἐκινήθη ἀπὸ τοῦ ἑτέρου, καὶ οὔτε συγκάμπτειν οὔτε καταταν ύειν ὁμοίως δύνανται. δῆλον δὲ γίνεται ψαυόμενον κατὰ τὴν σύγκαμψιν τοῦ ἀγκῶνος παρὰ τὴν διασχίδα τῆς φλεβὸς τὴν ἄνωθεν τοῦ μυὸς τείνουσαν· οἷσι δὲ τὸ τοιοῦτον, οὐκ ἔτι ῥηΐδιον ἐς τὴν ἑωυτοῦ φύσιν ἀγαγεῖν· οὐδὲ γὰρ ἄλλην οὐδεμίην ῥηΐδιον συμφυάδα 10 κοινὴν δύο ὀστέων κινηθεῖσαν ἐς τὴν ἀρχαίην φύσιν ἱδρυνθῆναι, ἀλλ' ἀνάγκη ὄγκον ἴσχειν τὴν διάστασιν. ὡς δὲ ἐπιδεῖν χρὴ ἐν ἄρθρῳ, ἐν τῇ 13 κατὰ σφυρὸν ἐπιδέσει εἴρηται.

XLV. Ἔστι δ' οἷσι κατάγνυται[1] τοῦ πήχεος τὸ ὀστέον τὸ ὑποτεταγμένον τῷ βραχίονι, ὅτε μὲν τὸ χονδρῶδες αὐτοῦ ἀφ' οὗ πέφυκεν ὁ τένων ὁ ὄπισθεν τοῦ βραχίονος <ὅτε δὲ τὰ πρόσω κατὰ τὴν ἀρχὴν τῆς ἐκφύσιος τοῦ προσθίου κορωνοῦ>[2] καί, ἐπὴν τοῦτο κινηθῇ, πυρετῶδες καὶ κακόηθες γίνεται· τὸ μέντοι ἄρθρον μένει ἐν τῇ ἑωυτοῦ χώρῃ· πᾶσα γὰρ ἡ βάσις αὐτοῦ ταύτῃ ὑπερέχει.[3] ὅταν[4] δὲ ἀπαγῇ ταύτῃ ᾗ ὑπερέχει ἡ κεφαλὴ τοῦ 10 βραχίονος, πλανωδέστερον τὸ ἄρθρον γίνεται, ἢν παντάπασιν ἀποκαυλισθῇ. ἀσινέστερα δέ, ὡς ἐν κεφαλαίῳ εἰρῆσθαι, πάντα τὰ καταγνύμενα τῶν ὀστέων ἐστὶν ἢ οἷσιν τὰ μὲν ὀστέα οὐ κατάγνυται, φλέβες δὲ καὶ νεῦρα ἐπίκαιρα ἀμφιφλᾶται ἐν τούτοισι τοῖσι χωρίοισιν· ἐγγυτέρω γὰρ θανάτῳ

[1] ἀπάγνυται.
[2] Omit codd., vulg.; restored by Littré from Galen in Orib. XLVI. 6.
[3] ὑπέχει. [4] ἢν.

XLIV. (Separation of radius). There are also other troublesome lesions of the elbow. Thus the thicker bone is sometimes separated from the other, and they can neither flex nor extend the joint as before. The lesion is made clear by palpation at the bend of the elbow about the bifurcation of the blood vessel[1] which passes upwards along the muscle.[2] In such cases it is not easy to bring the bone into its natural place, for no symphysis of two bones when displaced is permanently settled in its old position, but the diastasis (separation) necessarily remains as a swelling. How a joint ought to be bandaged was described in the case of the ankle.

XLV. (Fractures of olecranon). There are cases in which the bone of the forearm (ulna) is fractured where it is subjacent to the humerus, sometimes the cartilaginous part from which the tendon at the back of the arm arises, sometimes the part in front at the origin of the anterior coronoid process, and when this occurs it is complicated with fever and dangerous, though the joint (articular end of humerus) remains in its place, for its entire base comes above this bone.[3] But when the fracture is in the place on which the articular head of the humerus rests, the joint becomes more mobile if it is a complete cabbage-stalk fracture (*i.e.* right across). Speaking generally, fractures are always less troublesome than cases where no bones are broken, but there is extensive contusion of blood vessels and important cords in these parts. For the latter

[1] Cephalic vein. [2] Biceps.

[3] ὑπερέχει, *supersedet*, "is above," the articular end of the humerus rests entirely on the olecranon, the arm being bent. "Protrudes at this point," Littré-Adams.

πελάζει ταῦτα ἢ ἐκεῖνα, ἣν ἐκπυρωθῇ συνεχεῖ πυρέτῳ· ὀλίγα γε μὴν τὰ τοιαῦτα κατήγματα
18 γίνεται.

XLVI. Ἔστι δὲ ὅτε αὐτὴ ἡ κεφαλὴ τοῦ βραχίονος κατὰ τὴν ἐπίφυσιν κατάγνυται· τοῦτο δὲ δόκεον κακοσινώτατον εἶναι πολλῷ
4 τινὶ[1] εὐηθέστερον τῶν κατ' ἀγκῶνα σινέων ἐστίν.

XLVII. Ὡς μὲν οὖν ἕκαστα τῶν ὀλισθημάτων ἁρμόσσει[2] [ἐμβάλλειν καὶ][3] μάλιστα ἰητρεύειν, γέγραπται, καὶ ὅτι παραχρῆμα ἐμβάλλειν μάλιστα ἄρθρον συμφέρει διὰ τὸ τάχος τῆς φλεγμονῆς τῶν νεύρων. καὶ γὰρ ἢν ἐκπεσόντα αὐτίκα ἐμπέσῃ, ὅμως φιλεῖ τὰ νεῦρα σύντασιν ποιεῖσθαι, καὶ κωλύειν ἐπὶ ποσὸν χρόνον τήν τε ἔκτασιν, ὅσην περ φιλεῖ[4] ποιήσασθαι,[5] τήν τε σύγκαμψιν. ἰητρεύειν δὲ πάντα παραπλησίως τὰ τοιαῦτα
10 συμφέρει καὶ ὁπόσα ἀπάγνυται, καὶ ὁπόσα διΐσταται, καὶ ὁπόσα ὀλισθάνει· πάντα γὰρ χρὴ ὀθονίοισι πολλοῖσι καὶ σπλήνεσι καὶ κηρωτῇ ἰητρεύειν, ὥσπερ καὶ τἄλλα κατήγματα. τὸ δὲ σχῆμα τοῦ ἀγκῶνος ἐν τούτοισι δὴ καὶ παντάπασι χρὴ τοιοῦτον ποιεῖσθαι, οἷόν περ οἷσι βραχίων ἐπεδεῖτο καταγείς, καὶ πῆχυς· κοινότατον μὲν γὰρ πᾶσι τοῖσιν ὀλισθήμασι καὶ τοῖσι κινήμασι καὶ τοῖσι κατήγμασι τοῦτο τὸ σχῆμά ἐστιν· κοινότατον δὲ πρὸς τὴν ἔπειτα διάστασιν,[6]
20 καὶ τὸ ἐκτανύειν ἕκαστα καὶ συγκάμπτειν· ἐντεῦθεν γὰρ ὁδοὶ ἐς ἀμφότερα παραπλήσιοι· εὐοχώτατον καὶ εὐανάληπτον αὐτῷ τῷ κάμνοντι τοῦτο τὸ σχῆμα. ἔτι δὲ πρὸς τούτοισι, εἰ ἄρα κρατηθείη ὑπὸ τοῦ πωρώματος, εἰ μὲν ἐκτετα-

[1] τῷ. [2] ἁρμόσει.

lesions involve greater risk of death than do the former, if one is seized with continued fever. Still, fractures of this kind rarely occur.

XLVI. Sometimes the actual head of the humerus is fractured at the epiphysis, but this, though apparently a very grave lesion, is much milder than injuries of the elbow joint.

XLVII. How, then, each dislocation is most appropriately [reduced and] treated has been described; especially the value of immediate reduction owing to the rapid inflammation of the ligaments. For, even when parts that are put out are put in at once, the tendons are apt to become contracted and to hinder for a considerable time the natural amount of flexion and extension. All such lesions, whether avulsions, separations or dislocations, require similar treatment, for they should all be treated with a quantity of bandages, compresses and cerate, as with fractures. The position of the elbow should in these cases, too, be the same in all respects as in the bandaging of patients with fractured arm or forearm; for this position is most generally used[1] for all the dislocations, displacements and fractures, and is also most useful as regards the future condition, in respect both of extension and flexion in the several cases, since from it the way is equally open in both directions. This attitude is also most easily kept up or returned to by the patient himself. And besides this, if ankylosis should prevail, an arm ankylosed in the

[1] *κοινότατον* almost = "most useful."

---

[3] Omit B, Kw.
[4] *πέφυκε.*
[5] *ποιεῖσθαι.*
[6] *διάτασιν* K.

μένη ἡ χεὶρ κρατηθείη, κρέσσων ἂν εἴη μὴ προσεοῦσα, πολλῷ μὲν γὰρ κώλυμα εἴη, ὀφελείη δὲ ὀλίγῳ, εἰ δ' αὖ συγκεκαμμένη, μᾶλλον εὔχρηστος ἂν εἴη, πολλῷ δὲ εὐχρηστοτέρη, εἰ τὸ διὰ μέσον σχῆμα ἔχουσα πωρωθείη [κρέσσον].[1] τὰ
30 μὲν περὶ τοῦ σχήματος τοιάδε.

XLVIII. Ἐπιδεῖν δὲ χρὴ τήν τε ἀρχὴν τοῦ πρώτου ὀθονίου βαλλόμενον κατὰ τὸ βλαφθέν, ἤν τε καταγῇ, ἤν τε ἐκστῇ, ἤν τε διαστῇ, καὶ τὰς περιβολὰς τὰς πρώτας κατὰ τοῦτο ποιεῖσθαι, καὶ ἐρηρείσθω μάλιστα ταύτῃ, ἔνθεν δὲ καὶ ἔνθεν ἐπὶ ἧσσον. τὴν δὲ ἐπίδεσιν κοινὴν ποιεῖσθαι χρὴ τοῦ τε πήχεος καὶ τοῦ βραχίονος, καὶ ἐπὶ πολὺ πλέον ἑκάτερον ἢ ὡς οἱ πλεῖστοι ποιέουσιν, ὅπως ἐξαρύηται[2] ὡς μάλιστα ἀπὸ τοῦ
10 σίνεος τὸ οἴδημα ἔνθεν καὶ ἔνθεν. προσπεριβαλλέσθω δὲ καὶ τὸ ὀξὺ τοῦ πήχεος, ἢν τὸ σίνος κατὰ τοῦτο ᾖ, ἢν δὲ μή, ἵνα μὴ τὸ οἴδημα ἐνταῦθα περὶ αὐτὰ[3] συλλέγηται. περιφεύγειν δὲ χρὴ ἐν τῇ ἐπιδέσει, ὅπως μὴ κατὰ τὴν καμπὴν πολλὸν τοῦ ὀθονίου ἠθροισμένον ἔσται ἐκ τῶν δυνατῶν· πεπιέχθαι δὲ κατὰ τὸ σίνος ὡς μάλιστα. καὶ τὰ ἄλλα καταλαβέτω αὐτὸν περὶ τῆς πιέξιος καὶ τῆς χαλάσιος ταὐτά, καὶ κατὰ τοὺς αὐτοὺς χρόνους ἕκαστα, ὥσπερ τῶν ὀστέων τῶν κατεηγό-
20 των ἐν τῇ ἰητρείῃ πρόσθεν γέγραπται· καὶ αἱ μετεπιδέσιες διὰ τρίτης ἔστωσαν· χαλᾶν δὲ δοκείτω τῇ τρίτῃ, ὥσπερ καὶ τότε· καὶ νάρθηκας προσπεριβάλλειν ἐν τῷ ἱκνεομένῳ χρόνῳ—οὐδὲν γὰρ ἀπὸ τρόπου, καὶ τοῖσι τὰ ὀστέα κατεηγόσι, καὶ τοῖσι μή, ἢν μὴ πυρεταίνῃ—ὡς χαλαρωτάτους δέ,

[1] κρέσσον or κρέσσων codd. omnes ; but many editors omit.

extended position would be better away, for it would be a great hindrance and little use. If flexed, on the other hand, it would be more useful, and still more useful if the ankylosis occurred in an attitude of semiflexion.[1] So much concerning the attitude.

XLVIII. One should bandage by applying the head of the first roll to the place injured whether it be fractured, dislocated, or separated. The first turns should be made there and the firmest pressure, slackening off towards each side. The bandaging should include both fore and upper arm, and be carried much further each way than most practitioners do, that the oedema may be repelled as far as possible from the lesion to either side. Let the point of the elbow be also included in the bandage, whether the lesion be there or not, that the oedema may not be collected about this part. One should take special care in the dressing that, so far as possible, there shall be no great accumulation of bandage in the bend of the elbow, and that the firmest pressure be made at the lesion. For the rest, let him deal with the case as regards pressure and relaxation, in the same way, and according to the same respective periods, as was previously described in the treatment of fractured bones. Let the change of dressings take place every third day, and he should feel them relaxed on the third day, as in the former case. Apply the splints at the proper time—for their use is not unsuitable whether there is fracture or not, if there is no fever—but they should be applied as loosely as possible, those of

[1] Omit κρέσσον.

[2] ἐξείργηται Kw. [3] αὐτὸ.

τοὺς μὲν ἀπὸ βραχίονος κατατεταγμένους, τοὺς δὲ ἀπὸ τοῦ πήχεος ἀνειμένους· ἔστωσαν δὲ μὴ παχέες οἱ νάρθηκες· ἀναγκαῖον δὲ καὶ ἀνίσους αὐτοὺς εἶναι ἀλλήλοισι, παραλλάσσειν δὲ παρ' ἀλλήλους
30 ᾗ ἂν συμφέρῃ, τεκμαιρόμενον πρὸς τὴν σύγκαμψιν. ἀτὰρ καὶ τῶν σπληνῶν τὴν πρόσθεσιν τοιαύτην χρὴ ποιεῖσθαι, ὥσπερ καὶ τῶν ναρθήκων εἴρηται, ὀγκηροτέρους δὲ ὀλίγῳ κατὰ τὸ σίνος προστιθέναι. τοὺς δὲ χρόνους τοὺς ἀπὸ τῆς φλεγμονῆς τεκμαίρεσθαι χρὴ καὶ ἀπὸ τῶν πρόσθεν
36 γεγραμμένων.

---

[1] Reinhold's emendation, τοὺς μὲν κάτω τεταγμένους, τοὺς δὲ ἄνω κειμένους, seems to give the sense most clearly.

the arm being under and those of the forearm on the top.[1] The splints should not be thick, and must be unequal in length in order to overlap one another where it is convenient, judging by the degree of flexion. So, too, as regards the application of compresses, one should follow the directions for the splints. They should be rather thicker at the point of lesion. The periods are to be estimated by the inflammation and the directions already given.

Hippocrates had no angular splints, and straight ones applied to the bent arm above and below the elbow had to be so arranged that one set overlapped the other at the sides.

ΠΕΡΙ ΑΡΘΡΩΝ[1]

I. Ὦμου δὲ ἄρθρον ἕνα τρόπον οἶδα ὀλίσθανον, τὸν ἐς τὴν μασχάλην· ἄνω δὲ οὐδέποτε εἶδον, οὐδὲ ἐς τὸ ἔξω· οὐ μέντοι διϊσχυριείω ἔγωγε[2] εἰ ὀλισθάνοι ἂν ἢ οὔ, καίπερ ἔχων περὶ αὐτοῦ ὅ τι λέγω. ἀτὰρ οὐδὲ ἐς τὸ ἔμπροσθεν οὐδέπω ὄπωπα ὅ τι ἔδοξέ μοι ὠλισθηκέναι· τοῖσι μέντοι ἰητροῖσι δοκεῖ κάρτα ἐς τοὔμπροσθεν ὀλισθάνειν, καὶ μάλιστα ἐξαπατῶνται ἐν τούτοισιν, ὧν ἂν φθίσις καταλάβῃ τὰς σάρκας τὰς περὶ τὸ ἄρθρον
10 τε καὶ τὸν βραχίονα· φαίνεται γὰρ ἐν τοῖσι τοιούτοισι παντάπασι ἡ κεφαλὴ τοῦ βραχίονος ἐξέχουσα ἐς τοὔμπροσθεν. καὶ ἔγωγέ ποτε τὸ τοιοῦτον οὐ φὰς ἐκπεπτωκέναι ἤκουσα φλαύρως ἀπὸ[3] τῶν ἰητρῶν, ὑπό τε τῶν δημοτέων διὰ τοῦτο τὸ πρῆγμα· ἐδόκεον γὰρ αὐτοῖσιν ἠγνοηκέναι μοῦνος, οἱ δὲ ἄλλοι ἐγνωκέναι, καὶ οὐκ ἠδυνάμην αὐτοὺς ἀναγνῶσαι, εἰ μὴ μόλις,[4] ὅτι τόδ' ἐστὶ τοιόνδε· εἴ τις τοῦ βραχίονος ψιλώσειε μὲν τῶν σαρκῶν τὴν ἐπωμίδα, ψιλώσειε δὲ ᾗ ὁ μῦς
20 ἀνατείνει, ψιλώσειε δὲ τὸν τένοντα τὸν κατὰ τὴν μασχάλην τε καὶ τὴν κληῖδα πρὸς τὸ στῆθος ἔχοντα, φαίνοιτο ἂν ἡ κεφαλὴ τοῦ βραχίονος ἐς τοὔμπροσθεν ἐξέχουσα ἰσχυρῶς, καίπερ οὐκ ἐκπεπτωκυῖα· πέφυκε γὰρ ἐς τοὔμπροσθεν προπετὴς ἡ κεφαλὴ τοῦ βραχίονος· τὸ δ' ἄλλο ὀστέον τοῦ

[1] So Apollonius, Galen and most MSS. B M and Kw. add ΕΜΒΟΛΗΣ.

ON JOINTS

I. As to the shoulder-joint, I know only one dislocation, that into the armpit. I have never observed either the upward or outward form, but do not wish for my part to be positive as to whether such dislocations occur or not, though I can say something on the subject. Nor have I ever seen anything that seemed to me a dislocation forwards. Practitioners, indeed, think forward dislocation often happens, and they are especially deceived in cases where there is wasting of the flesh about the joint and arm, for in all such the head of the humerus has an obvious projection forwards. In such a case I myself once got into disrepute both with practitioners and the public by denying that this appearance was a dislocation. I seemed to them the only person ignorant of what the others recognised, and found it hardly possible to make them understand that the case was as follows:— Suppose one laid bare the point of the shoulder of the fleshy parts from the arm, and also denuded it at the part where the muscle[1] is attached, and laid bare the tendon stretching along the armpit and collar-bone to the chest, the head of the humerus would be seen to have a strongly marked projection forwards, though not dislocated. For the head of the humerus is naturally inclined forwards,

[1] Deltoid.

---

[2] Kw. omits ἔγω. [3] ὑπό τε Pq. [4] μόγις.

βραχιονος ἐς τὸ ἔξω καμπύλον. ὁμιλεῖ δὲ ὁ βραχίων τῷ κοίλῳ τῆς ὠμοπλάτης πλάγιος, ὅταν παρὰ τὰς πλευρὰς παρατεταμένος ᾖ. ὅταν μέντοι ἐς τοὔμπροσθεν ἐκτανυσθῇ ἡ σύμπασα χείρ,
30 τότε ἡ κεφαλὴ τοῦ βραχίονος κατὰ τὴν ἴξιν τῆς ὠμοπλάτης τῷ κοίλῳ γίνεται καὶ οὐκ ἔτι ἐξέχειν ἐς τοὔμπροσθεν φαίνεται. περὶ οὗ οὖν ὁ λόγος, οὐδέποτε εἶδον οὐδὲ ἐς τοὔμροσθεν ἐκπεσόν· οὐ μὴν ἰσχυριείω γε οὐδὲ περὶ τούτου, εἰ μὴ ἐκπέσοι ἂν οὕτως ἢ οὔ· ὅταν οὖν ἐκπέσῃ ὁ βραχίων ἐς τὴν μασχάλην, ἅτε πολλοῖσι ἐκπίπτοντος, πολλοὶ ἐπίστανται ἐμβάλλειν· εὐπαίδευτον δέ ἐστι τὸ εἰδέναι πάντας τοὺς τρόπους, οἷσιν οἱ ἰητροὶ ἐμβάλλουσι, καὶ ὡς ἂν τις αὐτοῖσι τοῖσι τρόποισι
40 τούτοισι κάλλιστα ἂν χρέοιτο·[1] χρῆσθαι δὲ χρὴ τῷ κρατίστῳ τῶν τρόπων, ἢν τὴν ἰσχυροτάτην ἀνάγκην ὁρᾷς· κράτιστος δὲ ὁ ὕστατος γεγραψό-
43 μενος.

II. Ὁκόσοισι μὲν οὖν πυκινὰ[2] ἐκπίπτει ὁ ὦμος, ἱκανοὶ ὡς ἐπὶ τὸ πλεῖστον[3] αὐτοὶ σφίσιν αὐτοῖσιν ἐμβάλλειν εἰσίν· ἐνθέντες γὰρ τῆς ἑτέρης χειρὸς τοὺς κονδύλους ἐς τὴν μασχάλην ἀναγκάζουσιν ἄνω τὸ ἄρθρον, τὸν δὲ ἀγκῶνα παράγουσι παρὰ τὸ στῆθος. τὸν αὐτὸν δὲ τρόπον τοῦτον καὶ ὁ ἰητρὸς ἂν ἐμβάλλοι, εἰ αὐτὸς μὲν ὑπὸ τὴν μασχάλην ἐσωτέρω τοῦ ἄρθρου τοῦ ἐκπεπτωκότος ὑποτείνας τοὺς δακτύλους ἀπαναγκάζοι ἀπὸ
10 τῶν πλευρέων ἐμβάλλων τὴν ἑωυτοῦ κεφαλὴν ἐς τὸ ἀκρώμιον ἀντερείσιος ἕνεκα, τοῖσι δὲ γούνασι παρὰ τὸν ἀγκῶνα ἐς τὸν βραχίονα ἐμβάλλων, ἀντωθέοι πρὸς τὰς πλευράς—συμφέρει δὲ καρτερὰς τὰς χεῖρας ἔχειν τὸν ἐμβάλλοντα—ἢ εἰ

while the rest of the bone is curved outwards. The humerus, when extended along the ribs, meets the cavity of the shoulder-blade obliquely, but when the whole arm is extended to the front, then the head of the humerus comes in line with the cavity of the shoulder-blade, and no longer appears to project forwards. To return to our subject, I never saw a dislocation forwards, but do not want to be positive about this either, whether such dislocation occurs or not. When, then, the humerus is displaced into the axilla, many know how to reduce it since it is a common accident, but expertness[1] includes knowledge of all the methods by which practitioners effect reduction, and the best way of using these methods. You should use the most powerful one when you see the strongest need, and the method that will be described last is the most powerful.

II. Those who have frequent dislocations of the shoulder are usually able to put it in for themselves. For by inserting the fist of the other hand into the armpit they forcibly push up the head of the bone, while they draw the elbow to the chest. And a practitioner would reduce it in the same way if, after putting his fingers under the armpit inside the head of the dislocated bone, he should force it away from the ribs, thrusting his head against the top of the shoulder to get a point of resistance, and with his knees thrusting against the arm at the elbow, should make counter-pressure towards the ribs—it is well for the operator to have strong hands—or, while he

[1] "'Tis a skilful man's part" (Liddell and Scott). "An easy thing to teach" (Adams).

[1] κάλλιστα χρῷτο. [2] οἷσι . . . πυκνὰ. [3] πολὺ.

αὐτὸς μὲν τῇσι χερσὶ καὶ τῇ κεφαλῇ οὕτω ποιοίη, ἄλλος[1] δέ τις τὸν ἀγκῶνα παράγοι παρὰ τὸ στῆθος.

Ἔστι δὲ ἐμβολὴ ὤμου καὶ ἐς τοὐπίσω ὑπερβάλλοντα τὸν πῆχυν ἐπὶ τὴν ῥάχιν, ἔπειτα τῇ
20 μὲν ἑτέρῃ χειρὶ ἀνακλᾶν ἐς τὸ ἄνω τοῦ ἀγκῶνος ἐχόμενον, τῇ δὲ ἑτέρῃ παρὰ τὸ ἄρθρον ὄπισθεν ἐρείδειν. αὕτη ἡ ἐμβολή, καὶ ἡ πρόσθεν εἰρημένη, οὐ κατὰ φύσιν ἐοῦσαι, ὅμως ἀμφισφάλλουσαι τὸ
24 ἄρθρον ἀναγκάζουσιν ἐμπίπτειν.

III. Οἱ δὲ τῇ πτέρνῃ πειρώμενοι ἐμβάλλειν, ἐγγύς τι τοῦ κατὰ φύσιν ἀναγκάζουσιν. χρὴ δὲ τὸν μὲν ἄνθρωπον χαμαὶ κατακλῖναι ὕπτιον, τὸν δὲ ἐμβάλλοντα χαμαὶ ἵζεσθαι ἐφ᾽ ὁπότερα ἂν τὸ ἄρθρον ἐκπεπτώκῃ· ἔπειτα λαβόμενον τῇσι χερσὶ τῇσιν ἑωυτοῦ τῆς χειρὸς τῆς σιναρῆς, κατατείνειν αὐτήν, τήν τε πτέρνην ἐς τὴν μασχάλην ἐμβάλλοντα ἀντωθεῖν, τῇ μὲν δεξιῇ ἐς τὴν δεξιήν, τῇ δὲ ἀριστέρῃ ἐς τὴν ἀριστερήν. δεῖ δὲ ἐς τὸ
10 κοῖλον τῆς μασχάλης ἐνθεῖναι στρογγύλον τι ἐνάρμοσσον· ἐπιτηδειόταται δὲ αἱ πάνυ σμικραὶ σφαῖραι καὶ σκληραί, οἷαι πολλαὶ ἐκ τῶν σκυτέων[2] ῥάπτονται· ἢν γὰρ μή τι τοιοῦτον ἐγκέηται, οὐ δύναται ἡ πτέρνη ἐξικνεῖσθαι πρὸς τὴν κεφαλὴν τοῦ βραχίονος· κατατεινομένης γὰρ τῆς χειρὸς κοιλαίνεται ἡ μασχάλη· οἱ γὰρ τένοντες οἱ ἔνθεν καὶ ἔνθεν τῆς μασχάλης ἀντισφίγγοντες ἐναντίοι εἰσίν. χρὴ δέ τινα ἐπὶ θάτερα τοῦ κατατεινομένου καθήμενον κατέχειν
20 κατὰ τὸν ὑγιέα ὦμον, ὡς μὴ περιέλκηται τὸ σῶμα, τῆς χειρὸς τῆς σιναρῆς ἐπὶ θάτερα τειν-

[1] ἕτερος. [2] ἐκ πολλῶν σκυτέων ποικίλων Weber.

uses his hands and head in this way, an assistant might draw the elbow to the chest.

There is also a way of putting in the shoulder by bringing the forearm backwards on to the spine, then with one hand turn upwards the part at the elbow, and with the other make pressure from behind at the joint. This method and the one described above, though not in conformity with nature,[1] nevertheless, by bringing round the head of the bone, force it into place.

III. Those who attempt to put in the shoulder with the heel, operate in a way nearly conformable with nature. The patient should lie on his back on the ground, and the operator should sit on the ground on whichever side the joint is dislocated. Then grasping the injured arm with both hands he should make extension and exert counter-pressure by putting the heel in the armpit, using the right heel for the right armpit, and the left for the left. In the hollow of the armpit one should put something round fitted to it,—the very small and hard balls such as are commonly sewn up from bits of leather are most suitable. For, unless something of the kind is inserted, the heel cannot reach the head of the humerus, for when extension is made on the arm the axilla becomes hollow and the tendons on either side of it form an obstacle by their contraction. Someone should be seated on the other side of the patient undergoing extension to fix the sound shoulder so that his body is not drawn round when the injured arm is pulled the other way.

[1] "Because without traction," Apollon., referring to *Fract.* I.

ομένης· ἔπειτα ἱμάντος μαλθακοῦ πλάτος ἔχοντος ἱκανόν, ὅταν ἡ σφαίρη ἐντεθῇ ἐς τὴν μασχάλην, περὶ τὴν σφαίραν περιβεβλημένου τοῦ ἱμάντος, καὶ κατέχοντος, λαβόμενον ἀμφοτέρων τῶν ἀρχέων τοῦ ἱμάντος, ἀντικατατείνειν τινά, ὑπὲρ τῆς κεφαλῆς τοῦ κατατεινομένου καθήμενον, τῷ ποδὶ προσβάντα πρὸς τοῦ ἀκρωμίου τὸ ὀστέον. ἡ δὲ σφαῖρα ὡς ἐσωτάτω καὶ ὡς μάλιστα πρὸς
30 τῶν πλευρέων κείσθω, καὶ μὴ ἐπὶ τῇ κεφαλῇ
31 τοῦ βραχίονος.

IV. Ἔστι δὲ καὶ ἄλλη ἐμβολή, ἣ κατωμίζουσιν[1] ἐς ὀρθόν· μείζω μέντοι εἶναι χρὴ τὸν κατωμίζοντα, διαλαβόντα δὲ τὴν χεῖρα ὑποθεῖναι τὸν ὦμον τὸν ἑωυτοῦ ὑπὸ τὴν μασχάλην ὀξύν· κἄπειτα ὑποστρέψαι, ὡς ἂν ἐνίζηται ἕδρῃ, οὕτω στοχασάμενον ὅπως ἀμφὶ τὸν ὦμον τὸν ἑωυτοῦ κρεμάσαι τὸν ἄνθρωπον κατὰ τὴν μασχάλην· αὐτὸς δὲ ἑωυτὸν ὑψηλότερον ἐπὶ τοῦτον τὸν ὦμον ποιείτω ἢ ἐπὶ τὸν ἕτερον· τοῦ δὲ κρεμαμένου τὸν
10 βραχίονα πρὸς τὸ ἑωυτοῦ στῆθος προσαναγκαζέτω ὡς μάλιστα· ἐν τούτῳ δὲ τῷ σχήματι προσανασειέτω, ὁπόταν[2] μετεωρίσῃ τὸν ἄνθρωπον, ὡς ἀντιῤῥέποι τὸ ἄλλο σῶμα αὐτῷ, ἀντίον τοῦ βραχίονος τοῦ κατεχομένου· ἢν δὲ ἄγαν κοῦφος ᾖ ὁ ἄνθρωπος, προσεπικρεμασθήτω[3] τούτου ὄπισθέν τις κοῦφος παῖς. αὗται δὲ ἐμβολαὶ πᾶσαι κατὰ παλαίστρην εὔχρηστοί

[1] ὡς κατωμίζουσιν Galen, Kw. [2] ὅταν—ἀντιῤῥέπῃ.
[3] προσεκκρεμασθήτω.

---

[1] This is the common method of reducing the shoulder-joint, and seems to be that chiefly used in Greek gymnasia. Cf. Galen's account of what happened to him when he dis-

Take, besides, a fairly broad strap of soft leather, and after the ball is put into the armpit, the strap being put round and fixing it, someone, seated at the head of the patient undergoing traction, should make counter-extension by holding the ends of the strap, and pressing his foot against the top of the shoulder-blade. The ball should be put as far into the armpit and as near the ribs as possible, not under the head of the humerus.[1]

IV. There is another mode of reduction in which they put it right by a shoulder lift[2]: but he who does the shoulder lift must be the taller. Grasping the patient's arm, let the operator put the point of his own shoulder under his armpit, then make a turn that it may get seated there, the aim of the manœuvre being to suspend the patient from his shoulder by the armpit. He should hold this shoulder higher than the other, and press in the arm of the suspended patient as far as possible towards his own chest. In this attitude let him proceed to shake the patient when he lifts him up, so that the rest of the body may act as a counterpoise to the arm which is held down. If the patient is very light, a boy of small weight should be suspended to him from behind. All these methods are very useful in the palaestra, since they do not require

located his collar-bone. He rightly remarks that the little ball cannot be put between the ribs and the head of the bone. XVIII(1), 332.

[2] All editors who translate ἐς ὀρθόν make it mean "standing." Föes-Erm: "in erecti et stantis humerum aeger extollitur"; Littré-Adams, "performed by the shoulder of a person standing"; Petrequin alone prefers the patient—"sur le malade debout." But after all the expression seems to go best with the verb.

εἴσιν, ὅτι οὐδὲν ἀλλοίων ἁρμένων δέονται ἐπεισεν-
19 εχθῆναι· χρήσαιτο δ' ἄν τις καὶ ἄλλοθι.

V. Ἀτὰρ καὶ οἱ περὶ τὰ ὕπερα ἀναγκάζοντες ἐγγύς τι τοῦ κατὰ φύσιν ἐμβάλλουσιν. χρὴ δὲ τὸ μὲν ὕπερον κατειλίχθαι ταινίῃ τινὶ μαλθακῇ —ἧσσον γὰρ ἂν ὑπολισθάνοι—ὑπηναγκάσθαι δὲ μεσηγὺ τῶν πλευρέων καὶ τῆς κεφαλῆς τοῦ βραχίονος· καὶ ἢν μὲν βραχὺ ᾖ τὸ ὕπερον, καθῆσθαι χρὴ τὸν ἄνθρωπον ἐπί τινος ὡς μόλις τὸν βραχίονα περιβάλλειν δύνηται περὶ τὸ ὕπερον· μάλιστα δὲ ἔστω μακρότερον τὸ ὕπερον,
10 ὡς ἂν ἑστεὼς ὁ ἄνθρωπος κρέμασθαι μικροῦ δέῃ ἀμφὶ τῷ ξύλῳ. κἄπειτα ὁ μὲν βραχίων καὶ ὁ πῆχυς παρατεταμένος παρὰ τὸ ὕπερον ἔστω, τὸ δὲ ἐπὶ θάτερα τοῦ σώματος καταναγκαζέτω τις, περιβάλλων κατὰ τὸν αὐχένα παρὰ τὴν κληῖδα τὰς χεῖρας. αὕτη ἡ ἐμβολὴ κατὰ φύσιν ἐπιεικέως ἐστί καὶ ἐμβάλλειν δύναται, ἢν χρηστῶς σκευά-
17 σωνται αὐτήν.

VI. Ἀτὰρ καὶ ἡ διὰ τοῦ κλιμακίου ἑτέρη τις τοιαύτη, καὶ ἔτι βελτίων, ὅτι ἀσφαλεστέρως ἂν τὸ σῶμα, τὸ μὲν τῇ, τὸ δὲ τῇ ἀντισηκωθείη μετεωρισθέν· περὶ γὰρ τὸ ὑπεροειδὲς ὁ ὦμος ἢν καὶ καταπεπήγῃ, περισφάλλεσθαι τὸ σῶμα κίνδυνος ἢ τῇ ἢ τῇ. χρὴ μέντοι καὶ ἐπὶ τῷ κλιμακτῆρι ἐπιδεδέσθαι τι ἄνωθεν στρογγύλον ἐνάρμοσσον ἐς τὸ κοῖλον τῆς μασχάλης, ὃ προσδιαναγκάζει τὴν κεφαλὴν τοῦ βραχίονος ἐς
10 τὴν φύσιν ἀπιέναι.

VII. Κρατίστη μέντοι πασέων τῶν ἐμβολῶν ἡ τοιήδε· ξύλον χρὴ εἶναι πλάτος μὲν ὡς πεντεδάκτυλον, ἢ τετραδάκτυλον τὸ ἐπίπαν,

further bringing in of apparatus, and one might also use them elsewhere.

V. Again, those who reduce by a forcible movement round pestles come fairly near the natural method. The pestle should have a soft band wrapped round it (for this will make it less slippery) and be pressed in between the ribs and the head of the humerus. If the pestle is short the patient should be so seated on something that he can just get his arm over it, but as a rule the pestle should be rather long so that the patient when erect is almost suspended on the post. Then let the arm and forearm be pulled down beside the pestle, while an assistant putting his arms round the patient's neck at the collar-bone forces the body down on the other side. This method is tolerably natural and able to reduce the dislocation if they arrange it well.

VI. Again there is another similar method with the ladder, which is still better, since the body when lifted up is more safely kept in equilibrium on either side. For with the pestle, though the shoulder may be fixed, there is danger of the body slipping round to one side or the other. But on the ladder-step also something rounded should be fastened on the upper side, which, fitting into the hollow of the armpit, helps to force the head of the humerus back to its natural place.

VII. The most powerful of all methods of reduction, however, is the following. There should be a piece of wood about five, or four fingers in breadth

πάχος δὲ ὡς διδάκτυλον ἢ καὶ λεπτότερον, μῆκος δὲ δίπηχυ, ἢ καὶ ὀλίγῳ[1] ἔλασσον. ἔστω δὲ ἐπὶ θάτερα τὸ ἄκρον περιφερὲς καὶ στενότατον ταύτῃ καὶ λεπτότατον· ἄμβην δὲ ἐχέτω σμικρὰν ὑπερέχουσαν ἐπὶ τῷ ὑστάτῳ τοῦ περιφερέος, ἐν[2] τῷ μέρει, μὴ τῷ πρὸς τὰς πλευράς, ἀλλὰ τῷ
10 πρὸς τὴν κεφαλὴν τοῦ βραχίονος ἔχοντι, ὡς ὑφαρμώσειε τῇ μασχάλῃ παρὰ τὰς πλευρὰς ὑπὸ τὴν κεφαλὴν τοῦ βραχίονος ὑποτιθέμενον· ὀθονίῳ δὲ ἢ ταινίῃ μαλθακῇ κατακεκολλήσθω ἄκρον τὸ ξύλον, ὅπως προσηνέστερον ᾖ. ἔπειτα χρή, ὑπώσαντα τὴν κεφαλὴν τοῦ ξύλου ὑπὸ τὴν μασχάλην ὡς ἐσωτάτω μεσηγὺ τῶν πλευρέων καὶ τῆς κεφαλῆς τοῦ βραχίονος, τὴν δὲ ὅλην χεῖρα πρὸς τὸ ξύλον κατατείναντα προσκαταδῆσαι κατά τε τὸν βραχίονα, κατά τε τὸν πῆχυν,
20 κατά τε τὸν καρπὸν τῆς χειρός, ὡς ἂν ἀτρεμῇ ὅτι μάλιστα· περὶ παντὸς δὲ χρὴ ποιεῖσθαι, ὅπως τὸ ἄκρον τοῦ ξύλου ὡς ἐσωτάτω τῆς μασχάλης ἔσται, ὑπερβεβηκὸς τὴν κεφαλὴν τοῦ βραχίονος. ἔπειτα χρὴ μεσηγὺ δύο στύλων στρωτῆρα πλάγιον εὖ προσδῆσαι, ἔπειτα ὑπερενεγκεῖν τὴν χεῖρα σὺν τῷ ξύλῳ ὑπὲρ τοῦ στρωτῆρος. ὅπως ἡ μὲν χεὶρ ἐπὶ θάτερα ᾖ, ἐπὶ θάτερα δὲ τὸ σῶμα, κατὰ δὲ τὴν μασχάλην ὁ στρωτήρ· κἄπειτα ἐπὶ μὲν θάτερα τὴν χεῖρα καταναγκάζειν σὺν τῷ
30 ξύλῳ περὶ τὸν στρωτῆρα, ἐπὶ θάτερα δὲ τὸ ἄλλο σῶμα. ὕψος δὲ ἔχων ὁ στρωτὴρ προσδεδέσθω, ὥστε μετέωρον τὸ ἄλλο σῶμα εἶναι ἐπ' ἄκρων τῶν ποδῶν. οὗτος ὁ τρόπος παρὰ πολὺ κράτιστος ἐμβολῆς ὤμου· δικαιότατα μὲν γὰρ μοχλεύει, ἢν καὶ μοῦνον ἐσωτέρω ᾖ τὸ ξύλον τῆς κεφαλῆς

as a rule, about two fingers thick or even thinner, and in length two cubits or a little less. Let it be rounded at one end and be thinnest and narrowest there, and at the extremity of the rounded end let it have a slightly projecting rim (*ambé*) not on the side towards the ribs but on that towards the head of the humerus, so as to fit into the armpit when inserted along the ribs under the head of the humerus, and the end of the wood should have linen or a soft band glued over it that it may be more comfortable. One should then insert the tip of the instrument as far as possible under the armpit between the ribs and the head of the humerus, and extending the whole arm along the wood, fasten it down at the upperarm, forearm and wrist, so as to be as immobile as possible. Above all, one should manage to get the tip of the instrument as far into the armpit as possible, up above the head of the humerus. Then a cross-bar should be firmly fastened between two posts and next one should bring the arm with the instrument over the bar, so that the arm is on one side, the body on the other and the cross-bar at the armpit. Then on one side press down the arm with the instrument round the beam, on the other side the rest of the body. The beam should be fastened at such a height that the rest of the body is suspended on tiptoe. This is by far the most powerful method for reducing the shoulder, for it makes the most correct leverage, if only the instrument is well on

[1] Omit καὶ.
[2] ἐπὶ.

τοῦ βραχίονος· δικαιόταται δὲ αἱ ἀντιῤῥοπαί, ἀσφαλέες δὲ τῷ ὀστέῳ τοῦ βραχίονος. τὰ μὲν οὖν νεαρὰ ἐμπίπτει θᾶσσον ἢ ὡς ἄν τις οἴοιτο, πρὶν ἢ καὶ κατατετάσθαι δοκεῖν· ἀτὰρ καὶ τὰ
40 παλαιὰ μούνη αὕτη τῶν ἐμβολέων οἵη τε ἐμβιβάσαι, ἢν μὴ ἤδη ὑπὸ χρόνου σὰρξ μὲν ἐπελη λύθη ἐπὶ τὴν κοτύλην, ἢν δὲ κεφαλὴ τοῦ βραχίονος ἤδη τρίβον ἑωυτῇ πεποιημένη ᾖ ἐν τῷ χωρίῳ, ἵνα ἐξεκλίθη· οὐ μὴν ἀλλ' ἐμβάλλειν γάρ μοι δοκεῖ[1] καὶ οὕτω πεπαλαιωμένον ἔκπτωμα τοῦ βραχίονος—τί γὰρ ἂν δικαίη μόχλευσις οὐχὶ κινήσειεν;—μένειν μέντοι οὐκ ἄν μοι δοκέοι κατὰ χώρην, ἀλλ' ὀλισθάνειν ἂν ὡς τὸ[2] ἔθος.

Τὸ αὐτὸ δὲ ποιεῖ καὶ περὶ κλιμακτῆρα κατ-
50 αναγκάζειν τοῦτον τὸν τρόπον σκευάσαντα. πάνυ μὴν ἱκανῶς ἔχει καὶ περὶ μέγα ἕδος Θεσσαλικὸν ἀναγκάζειν, ἢν νεαρὸν ᾖ τὸ ὀλίσθημα. ἐσκευάσθαι μέντοι χρὴ τὸ ξύλον οὕτως, ὥσπερ εἴρηται· ἀτὰρ τὸν ἄνθρωπον καθίσαι πλάγιον ἐπὶ τῷ δίφρῳ· κἄπειτα τὸν βραχίονα σὺν τῷ ξύλῳ ὑπερβάλλειν ὑπὲρ τοῦ ἀνακλισμοῦ, καὶ ἐπὶ μὲν θάτερα τὸ σῶμα καταναγκάζειν, ἐπὶ δὲ θάτερα τὸν βραχίονα σὺν τῷ ξύλῳ. τὸ αὐτὸ δὲ ποιεῖ[3] καὶ ὑπὲρ δίκλειδος θύρης ἀναγκάζειν· χρῆσθαι
60 δὲ χρὴ αἰεὶ τούτοισιν, ἃ ἂν τύχῃ παρεόντα.

VIII. Εἰδέναι μὲν οὖν χρὴ ὅτι φύσιες φυσίων

[1] ἄν μοι δοκέοι. [2] ἐς τὸ. [3] ποιεῖν.

---

[1] An old-fashioned straight-backed chair, Galen. Adams is enthusiastic over this method. For the ambé fasten a jack-towel above the patient's elbow: put your foot in the loop and gradually increase the tension. You will do the

the inner side of the head of the humerus. The counterpoise is also most correct and without risk to the bone of the arm. Indeed, recent cases are reduced more rapidly than one would believe, even before any apparent extension has been made, while, as for old standing cases, this method alone is able to reduce them, unless by lapse of time the tissues have already invaded the articular cavity and the head of the humerus has made a friction cavity for itself in the place to which it has slipped. Nevertheless I think it would reduce even so inveterate a dislocation of the arm—for what would not correct leverage move?—but I should not suppose it would stay in position, but slip back to its old place. The same result is obtained by pressure round the rung of a ladder, arranging it in the same way. Also the operation is very effectively done on a large Thessalian chair,[1] if the dislocation is recent. In this case the wooden instrument should be prepared as directed while the patient is seated sideways on the chair. Then put the arm with the instrument over the chair-back, and press down the body on one side, and the arm with the instrument on the other. The same result is obtained by operating over (the lower half of)[2] a double door. One should always make use of what happens to be at hand.

VIII. One should bear in mind that there are

job quickly, safely and almost pleasantly, if the arm and chair top are properly padded.

[2] Apollonius strangely illustrates this by an ordinary vertical (folding) double door. As Galen points out, it refers to doors which open in two halves above and below, usually with a cross-bar between.

μέγα διαφέρουσιν ἐς τὸ ῥηϊδίως ἐμπίπτειν τὰ ἐκπίπτοντα· διενέγκοι μὲν γὰρ ἄν τι καὶ κοτύλη κοτύλης, ἡ μὲν εὐϋπέρβατος ἐοῦσα, ἡ δὲ ἧσσον· πλεῖστον δὲ διαφέρει καὶ τῶν νεύρων ὁ σύνδεσμος, τοῖσι μὲν ἐπιδόσιας ἔχων, τοῖσι δὲ συντεταμένος [ἐών].[1] καὶ γὰρ ἡ ὑγρότης τοῖσι ἀνθρώποισι γίνεται ἡ ἐκ τῶν ἄρθρων, διὰ τῶν νεύρων τὴν ἀπάρτισιν, ἣν χαλαρά τε ᾖ φύσει καὶ τὰς
10 ἐπιτάσιας εὐφόρως φέρῃ· συχνοὺς γὰρ ἄν τις ἴδοι, οἳ οὕτως ὑγροί εἰσιν, ὥστε, ὁπόταν ἐθέλωσι, τότε ἑαυτοῖσι τὰ ἄρθρα ἐξίστανται ἀνωδύνως, καὶ καθίστανται ἀνωδύνως. διαφέρει μέντοι τι καὶ σχέσις τοῦ σώματος· τοῖσι μὲν γὰρ εὖ ἔχουσι τὸ γυῖον καὶ σεσαρκωμένοισιν ἐκπίπτει τε ἧσσον, ἐμπίπτει δὲ χαλεπώτερον· ὅταν δὲ αὐτοὶ σφέων αὐτῶν λεπτότεροι καὶ ἀσαρκότεροι ἔωσι, τότε ἐκπίπτει τε μᾶλλον, ἐμπίπτει δὲ ῥᾷον. σημεῖον δέ, ὅτι ταῦτα οὕτως ἔχει, καὶ τόδε· τοῖσι γὰρ
20 βουσὶ τότε ἐκπίπτουσι μᾶλλον οἱ μηροὶ ἐκ τῆς κοτύλης, ἡνίκα ἂν αὐτοὶ σφέων αὐτῶν λεπτότατοι ἔωσιν· γίνονται δὲ βόες λεπτότατοι, τοῦ χειμῶνος τελευτῶντος· τότε οὖν καὶ ἐξαρθρέουσι μάλιστα, εἰ δή τι καὶ τοιοῦτο δεῖ ἐν ἰητρικῇ γράψαι· δεῖ δέ· καλῶς γὰρ Ὅμηρος καταμεμαθήκει, ὅτι πάντων τῶν προβάτων βόες μάλιστα πονέουσι[2] ταύτην τὴν ὥρην, καὶ βοῶν οἱ ἀρόται, ὅτι [κατὰ][3] τὸν χειμῶνα ἐργάζονται. τούτοισι τοίνυν καὶ ἐκπίπτει μάλιστα· οὗτοι γὰρ μάλιστα λεπτύνονται·
30 τὰ μὲν γὰρ ἄλλα βοσκήματα δύναται βραχείην τὴν ποίην βόσκεσθαι· βοῦς δὲ οὐ μάλα, πρὶν βαθεῖα γένηται· τοῖσι μὲν γὰρ ἄλλοισίν ἐστι λεπτὴ ἡ προβολὴ τοῦ χείλεος, λεπτὴ δὲ ἡ ἄνω

great natural diversities as to the easy reduction of dislocations. There may be some difference in the sockets, one having a rim easy to cross, the other one less so; but the greatest diversity is the attachment of the ligaments, which in some cases is yielding, in others constricted. For the humidity in individuals as regards the joints comes from the disposition of the ligaments which may be slack by nature and easily lend themselves to extensions. In fact one may see many persons of so humid a temperament that when they choose they can dislocate and reduce their joints without pain. The state of the body makes a further difference, for in those who are muscular and have the limb in good condition dislocation is rarer and reduction more difficult, but when they are thinner and less muscular than usual dislocation is more frequent and reduction easier. The following also shows that this is so. In the case of cattle the thigh bones get dislocated from the socket when they are at their thinnest. Now cattle are thinnest at the end of winter, and it is then especially that they have dislocations, if indeed such a matter should be cited in a medical work. And it should be, for Homer has well observed that of all farm beasts cattle suffer most during this season, and among cattle the ploughing oxen because they work in the winter. It is in these, then, that dislocation especially occurs, for they are especially attenuated. For other farm animals can graze on herbage while short, but cattle can hardly do so till it is long, since in the others the projection of the lip is thin,

---

[1] Omit Erm., Kw. [2] ἀτονέουσι.
[3] Omit Erm., Kw.

γνάθος· βοῒ δὲ παχείη μὲν ἡ προβολὴ τοῦ χείλεος, παχείη δὲ καὶ ἀμβλεῖα ἡ ἄνω γνάθος· διὰ ταῦτα ὑποβάλλειν ὑπὸ τὰς βραχείας ποίας οὐ δύναται. τὰ δὲ αὖ μώνυχα τῶν ζώων, ἅτε ἀμφώδοντα ἐόντα, δύναται μὲν σαρκάζειν, δύναται δὲ ὑπὸ τὴν βραχείην ποίην ὑποβάλλειν τοὺς 40 ὀδόντας, καὶ ἥδεται τῇ οὕτως ἐχούσῃ ποίῃ μᾶλλον ἢ τῇ βαθείῃ· καὶ γὰρ τὸ ἐπίπαν ἀμείνων καὶ στερεωτέρη ἡ βραχείη ποίη τῆς βαθείης ποτὶ καὶ πρὶν ἐκκαρπεῖν τὴν βαθείην. διὰ τοῦτο οὖν ἐποίησεν ὧδε τάδε τὰ ἔπη—Ὡς δ' ὁπότ' ἀσπάσιον ἔαρ ἤλυθε βουσὶν ἕλιξιν—ὅτι ἀσμενωτάτη [τοῖσιν][1] αὐτοῖσιν ἡ βαθείη ποίη φαίνεται. ἀτὰρ καὶ ἄλλως ὁ βοῦς χαλαρὸν φύσει τὸ ἄρθρον τοῦτο ἔχει μᾶλλον τῶν ἄλλων ζώων· διὰ τοῦτο καὶ εἰλίπουν[2] ἐστὶ μᾶλλον τῶν ἄλλων ζώων, καὶ 50 μάλιστα ὅταν λεπτὸν[3] καὶ γηραλέον[4] ᾖ. διὰ ταῦτα πάντα καὶ ἐκπίπτει βοῒ μάλιστα. πλείω δὲ γέγραπται περὶ αὐτοῦ, ὅτι πάντων τῶν προειρημένων ταῦτα μαρτύριά ἐστιν.

Περὶ οὗ οὖν ὁ λόγος, τοῖσιν[5] ἀσάρκοισι μᾶλλον ἐκπίπτει καὶ θᾶσσον ἐμπίπτει ἢ τοῖσιν εὖ σεσαρκωμένοισι· καὶ ἧσσον ἐπιφλεγμαίνει τοῖσι ὑγροῖσι καὶ τοῖσιν ἀσάρκοισιν ἢ τοῖσι σκελιφροῖσι[6] καὶ σεσαρκωμένοισι, καὶ ἧσσόν γε δέδεται ἐς τὸν ἔπειτα χρόνον· ἀτὰρ καὶ εἰ μύξα 60 πλείων ὑπείη τοῦ μετρίου μὴ σὺν φλεγμονῇ, καὶ οὕτως ἂν ὀλισθηρὸν εἴη, μυξωδέσ-

[1] Omit Littré, Erm. Kw.

[2] εἰλίπους: Erm.'s correction which Kw. follows as with the other adjectives, but they surely go with ζώων.

[3] λεπτός. [4] γέρων.

as is also the upper jaw, but in the ox the projection of the lip is thick and the upper jaw thick and blunt, wherefore he cannot grasp the short herbage. But the solid-hoofed animals, having a double row of teeth, can not only browse but can also grasp the short herbage with their teeth, and they prefer this kind to the long grass. In fact the short grass is on the whole better and of more substance than the long, especially when the long is just going to seed. It is in allusion to this that he wrote the following verse:—

"As when the season of spring arrives welcome to crumple-horned cattle,"[1]

because the long grass appears most welcome to them. Moreover in the ox this joint is generally more lax than in other animals, and for this reason it has a more shambling gait than other animals, especially when it is thin and old. For all these reasons the joint is especially liable to dislocation in the ox, and more has been written about it because these facts testify to all the preceding statements.

To return to the subject, dislocation occurs more easily and is more quickly reduced in emaciated than in muscular persons, and inflammation more rarely supervenes in the moist and thin than in muscular subjects of a dry habit, but the joint is not so firm afterwards. Further, if an excess of mucous substance is engendered without inflammation, this too will make it liable to slip, and, on

[1] Not in our Homer.

---

[5] ὅτι τοῖσι. [6] σκληροῖσι.

τερα γὰρ τοὐπίπαν τὰ ἄρθρα τοῖσι ἀσάρκοισι ἢ τοῖσι σεσαρκωμένοισίν ἐστιν· καὶ γὰρ αὗται αἱ σάρκες τῶν μὴ ἀπὸ τέχνης ὀρθῶς[1] λελιμαγχημένων, αἱ τῶν λεπτῶν μυξωδέστεραί εἰσιν ἢ αἱ τῶν παχέων. ὅσοισι μέντοι σὺν φλεγμονῇ μύξα ὑπογίνεται, ἡ φλεγμονὴ δήσασα ἔχει τὸ ἄρθρον· διὰ τοῦτο οὐ μάλα ἐκπίπτει τὰ ὑπόμυξα, ἐκπίπτοντα ἄν, εἰ μή τι ἢ πλέον ἢ ἔλασσον
70 φλεγμονῆς ὑπεγένετο.

IX. Οἷσι μὲν οὖν ὅταν[2] ἐμπέσῃ τὸ ἄρθρον καὶ μὴ ἐπιφλεγμήνῃ τὰ περιέχοντα, χρῆσθαί τε ἀνωδύνως αὐτίκα τῷ ὤμῳ δύνανται, οὗτοι μὲν οὐδὲν νομίζουσι δεῖν ἑωυτῶν ἐπιμελεῖσθαι· ἰητροῦ μήν ἐστι καταμαντεύσασθαι τῶν τοιούτων· τοῖσι τοιούτοισι γὰρ ἐκπίπτει καὶ αὖθις μᾶλλον ἢ οἷσιν ἂν ἐπιφλεγμήνῃ τὰ νεῦρα. τοῦτο κατὰ πάντα τὰ ἄρθρα οὕτως ἔχει, καὶ μάλιστα κατ' ὦμον καὶ κατὰ γόνυ· μάλιστα γὰρ οὖν καὶ
10 ὀλισθάνει ταῦτα. οἷσι δ' ἂν ἐπιφλεγμήνῃ [τὰ νεῦρα],[3] οὐ δύνανται χρῆσθαι τῷ ὤμῳ· κωλύει γὰρ ἡ ὀδύνη καὶ ἡ σύντασις τῆς φλεγμονῆς. τοὺς οὖν τοιούτους ἰῆσθαι χρὴ κηρωτῇ καὶ σπλήνεσι καὶ ὀθονίοισι πολλοῖσι ἐπιδέοντα· ὑποτιθέναι δὲ ἐς τὴν μασχάλην εἴριον μαλθακὸν καθαρὸν συνειλίσσοντα ἐκπλήρωμα τοῦ κοίλου ποιοῦντα ἵνα ἀντιστήριγμα μὲν τῇ ἐπιδέσει ᾖ, ἀνακωχῇ δὲ τὸ ἄρθρον· τὸν δὲ βραχίονα χρὴ ἐς τὸ ἄνω ῥέποντα ἴσχειν τὰ πλεῖστα· οὕτω γὰρ
20 ἂν ἑκαστάτω εἴη τοῦ χωρίου ἐς ὃ ὤλισθεν ἡ κεφαλὴ τοῦ ὤμου· χρὴ δέ, ὅταν ἐπιδήσῃς τὸν

[1] ὀρθῆς. [2] ἄν, Littré's suggestion.
[3] Omit B, Kw.

the whole, the joints of emaciated persons contain more mucus than those of muscular individuals. One sees, in fact, that these tissues in emaciated persons, who have not been normally reduced according to the principles of the art, have more mucosity than those of stout people. But in those in whom mucus develops along with inflammation, the inflammation keeps the joint firm. This is why the joints do not often get dislocated from a slight excess of mucus, though they would do so were there not more or less inflammation at the bottom of it.

IX. Should, however, no inflammation of the surrounding parts supervene after the reduction of the joint, patients can at once use the shoulder without pain, and these persons think there is no further necessity to take care of themselves. It is, then, the practitioner's business to act the prophet for such, for it is in such that dislocation occurs again, rather than in cases where inflammation of the ligaments may have supervened. This is the case with all joints and especially those at the shoulder and knee, for they are specially liable to dislocation. Those in whom inflammation may have supervened cannot use the shoulder, for the pain and inflammatory tension prevents it. One should treat such cases with cerate, compresses, and plenty of bandages, also put a soft roll of cleansed wool under the armpit, making a plug for the cavity that it may form a fulcrum for the bandage and prop up the head of the bone. The arm should be kept as far as possible pressed upwards, for so the head of the humerus will be furthest from the place into which it was dislocated. After bandaging the shoulder you should proceed to fasten

ὦμον, ἔπειτα προσκαταδεῖν τὸν βραχίονα πρὸς
τὰς πλευρὰς ταινίῃ τινὶ κύκλῳ περὶ τὸ σῶμα
περιβάλλοντα. χρὴ δὲ καὶ ἀνατρίβειν τὸν ὦμον
ἡσυχαίως καὶ λιπαρῶς· πολλῶν ἔμπειρον δεῖ
εἶναι τὸν ἰητρόν, ἀτὰρ δὴ καὶ ἀνατρίψιος· ἀπὸ
τοῦ αὐτοῦ ὀνόματος οὐ τωὐτὸ ἀποβαίνει· καὶ
γὰρ ἂν δήσειεν ἄρθρον ἀνάτριψις, χαλαρώτερον
τοῦ καιροῦ ἐόν, καὶ λύσειεν ἄρθρον σκληρότερον
30 τοῦ καιροῦ ἐόν· ἀλλὰ διοριεῖται ἡμῖν περὶ
ἀνατρίψιος ἐν ἄλλῳ λόγῳ. τὸν γοῦν τοιοῦτον
ὦμον μαλθακῇσί τε χερσὶν ἀνατρίβειν συμφέρει,
καὶ ἄλλως πρηέως· τὸ δὲ ἄρθρον διακινεῖν, μὴ
βίῃ, ἀλλὰ τοσοῦτον ὅσον ἀνωδύνως κινήσεται.
καθίσταται δὲ πάντα, τὰ μὲν ἐν πλέονι χρόνῳ, τὰ
36 δ' ἐν ἐλάσσονι.

X. Γιγνώσκειν δὲ εἰ ἐκπέπτωκεν ὁ βραχίων
τοιοῖσδε χρὴ τοῖς σημείοισι· τοῦτο μέν, ἐπειδὴ
δίκαιον ἔχουσι τὸ σῶμα οἱ ἄνθρωποι, καὶ τὰς
χεῖρας καὶ τὰ σκέλεα, παραδείγματι χρῆσθαι
δεῖ τῷ ὑγιεῖ πρὸς τὸ μὴ ὑγιές, καὶ τῷ μὴ
ὑγιεῖ πρὸς τὸ ὑγιές, μὴ τὰ ἀλλότρια ἄρθρα
καθορῶντα—ἄλλοι γὰρ ἄλλων μᾶλλον ἔξαρθροι
πεφύκασιν—ἀλλὰ τοῦ αὐτοῦ τοῦ κάμνοντος, ἢν
ἀνόμοιον ᾖ τὸ ὑγιὲς τῷ κάμνοντι. καὶ τοῦτο
10 εἴρηται μὲν ὀρθῶς, παρασύνεσιν δὲ ἔχει πάνυ
πολλὴν διὰ τὰ τοιαῦτα, καὶ οὐκ ἀρκεῖ μοῦνον
λόγῳ εἰδέναι τὴν τέχνην ταύτην, ἀλλὰ καὶ
ὁμιλίῃ ὁμιλεῖν· πολλοὶ γάρ, ὑπὸ ὀδύνης, ἢ καὶ
ὑπ' ἀλλοίης προφάσιος, οὐκ ἐξεστεώτων αὐτοῖσι
τῶν ἄρθρων, ὅμως οὐ δύνανται ἐς τὰ ὅμοια
σχήματα καθεστάναι ἐς οἷά περ τὸ ὑγιαῖνον[1]
σῶμα σχηματίζεται· προσσυνιέναι μὲν οὖν καὶ

the arm to the side with some sort of band, passing it horizontally round the body, and the shoulder should be gently and perseveringly rubbed. The practitioner must be skilled in many things and particularly in friction (massage). Though called by one name it has not one and the same effect, for friction will make a joint firm when looser than it should be, and relax it when too stiff. But we shall define the rules for friction in another treatise. Now, for such a shoulder the proper friction is that with soft hands, and always gently. Move the joint about, without force, but so far as it can be moved without pain. All symptoms subside,[1] some in a longer, others in a shorter time.

X. A dislocation of the humerus may be recognised by the following signs. First, since men's bodies are symmetrical as to arms and legs, one should use the sound in comparison with the unsound, and the unsound with the sound; not observing other people's joints (for some have more projecting joints than others), but those of the patient himself, to see if the sound one is dissimilar to the one affected. And though this is correct advice there is a good deal of fallacy about it.[2] This is why it is not enough to know the art in theory only, but by familiar practice. For many persons owing to pain or some other cause, though their joints are not dislocated, cannot hold themselves in the attitude which the healthy body assumes. One must, therefore, take this also into

[1] "All joints re-establish themselves." Pq.; "Things get restored," Adams.

[2] Kw. punctuates after τοιαῦτα.

---

[1] ὑγιηρὸν.

ἐννοεῖν καὶ τὸ τοιόνδε σχῆμα χρή. ἀτὰρ καὶ[1] ἐν τῇ μασχάλῃ ἡ κεφαλὴ τοῦ βραχίονος φαίνεται 20 ἐγκειμένη πολλῷ μᾶλλον τοῦ ἐκπεπτωκότος ἢ τοῦ ὑγιέος· τοῦτο δέ, ἄνωθεν κατὰ τὴν ἐπωμίδα κοῖλον φαίνεται τὸ χωρίον· καὶ τὸ τοῦ ἀκρωμίου ὀστέον ἐξέχον[2] φαίνεται, ἅτε ὑποδεδυκότος τοῦ ἄρθρου ἐς τὸ κάτω τοῦ χωρίου—παρασύνεσιν μὴν καὶ ἐν τούτῳ ἔχει τινά, ἀλλὰ ὕστερον περὶ αὐτοῦ γεγράψεται, ἄξιον γὰρ γραφῆς ἐστί— τοῦτο δέ, τοῦ ἐκπεπτωκότος ὁ ἀγκὼν φαίνεται ἀφεστεὼς μᾶλλον ἀπὸ τῶν πλευρέων ἢ τοῦ ἑτέρου· εἰ μέντοι τις προσαναγκάζοι, προσάγεται 30 μέν, ἐπιπόνως δέ· τοῦτο δέ, ἄνω τὴν χεῖρα ἆραι εὐθεῖαν παρὰ τὸ οὖς, ἐκτεταμένου τοῦ ἀγκῶνος, οὐ μάλα δύνανται, ὥσπερ τὴν ὑγιέα, οὐδὲ παράγειν ἔνθα καὶ ἔνθα ὁμοίως. τά τε οὖν σημεῖα ταῦτά ἐστιν, ὤμου ἐκπεπτωκότος· αἱ δὲ ἐμβολαὶ αἱ γεγραμμέναι αἵ τε ἰατρεῖαι 36 αὗται.

XI. Ἐπάξιον δὲ τὸ μάθημα ὡς χρὴ ἰητρεύειν τοὺς πυκινὰ ἐκπίπτοντας ὤμους· πολλοὶ μὲν γὰρ ἤδη ἀγωνίης ἐκωλύθησαν διὰ ταύτην τὴν συμφορήν, τἄλλα πάντα ἀξιοχρήϊοι ἐόντες· πολλοὶ δὲ ἐν πολεμικοῖσιν ἀχρήϊοι[3] ἐγένοντο καὶ διεφθάρησαν διὰ ταύτην τὴν συμφορήν· ἅμα τε ἐπάξιον καὶ διὰ τοῦτο, ὅτι οὐδένα οἶδα ὀρθῶς ἰητρεύοντα, ἀλλὰ τοὺς μὲν μηδὲ ἐγχειρέοντας, τοὺς δὲ τἀναντία τοῦ συμφέροντος 10 φρονέοντάς τε καὶ ποιέοντας. συχνοὶ γὰρ ἤδη ἰητροὶ ἔκαυσαν ὤμους ἐκπίπτοντας, κατά τε τὴν

[1] τοῦτο μὲν Apoll. B.Kw. [2] ἔξοχον.
[3] πολέμοις ἀχρεῖοι.

consideration and have such a position in mind. Now, first,[1] the head of the humerus is much more obvious in the armpit on the injured than on the sound side. Again, towards the top of the shoulder the part appears hollow, while the bone at the shoulder-point (acromion) is seen to project, since the articular end of the humerus has sunk to the lower part of the region. Yet there is some fallacy in this too, but it will be described later, for it merits description. Again the elbow of the dislocated limb obviously stands out more from the ribs than that of the other. If, indeed, one should forcibly adduct it, it yields, but with much pain. Further, the patient is quite unable to raise the arm straight alongside the ear, with the elbow extended, as he does with the sound one, or move it about in the same way. These, then, are the signs of a dislocated shoulder, the modes of reduction are the ones described, and these the methods of treatment.

XI. The proper treatment of those whose shoulders are often being dislocated is a thing worth learning. For many have been debarred from gymnastic contests, though well fitted in all other respects, and many have become worthless in warfare and have perished through this misfortune.[2] Another reason for its importance is the fact that I know of no one who uses the correct treatment, some not even attempting to take it in hand, while others have theories and practices the reverse of what is appropriate. For many practitioners cauterize shoulders

[1] Reading τοῦτο μὲν.
[2] Cf. *Airs Waters*, XX. on flabby joints of Scythians and their use of cautery.

ἐπωμίδα, κατά τε ἔμπροσθεν, ᾗ ἡ κεφαλὴ τοῦ βραχίονος ἐξογκεῖ, κατά τε τὸ ὄπισθεν ὀλίγον τῆς ἐπωμίδος. αὗται οὖν αἱ καύσεις, εἰ μὲν ἐς τὸ ἄνω ἐξέπιπτεν ὁ βραχίων, ἢ ἐς τὸ ἔμπροσθεν ἢ ἐς τὸ ὄπισθεν, ὀρθῶς ἂν ἔκαιον· νῦν δὲ δή, ὅτε ἐς τὸ κάτω ἐκπίπτει, ἐκβάλλουσιν αὗται αἱ καύσεις μᾶλλον ἢ κωλύουσιν· ἀποκλείουσι γὰρ
20 τῆς ἄνω εὐρυχωρίης τὴν κεφαλὴν τοῦ βραχίονος.

Χρὴ δὲ ὧδε καίειν ταῦτα· ἀπολαβόντα τοῖσι δακτύλοισι κατὰ τὴν μασχάλην τὸ δέρμα, ἀφελκύσαι κατ᾽ αὐτὴν τὴν ἴξιν μάλιστα, καθ᾽ ἣν ἡ κεφαλὴ τοῦ βραχίονος ἐκπίπτει· ἔπειτα οὕτως ἀφειλκυσμένον τὸ δέρμα, διακαῦσαι ἐς τὸ πέρην. σιδηρίοισι δὲ χρὴ ταῦτα[1] καίειν, μὴ παχέσι, μηδὲ λίην φαλακροῖσιν, ἀλλὰ προμήκεσι—ταχυπορώτερα γάρ—καὶ τῇ χειρὶ ἐπερείδειν· χρὴ δὲ καὶ διαφανέσι καίειν, ὡς ὅτι τάχιστα περαιωθῇ
30 κατὰ δύναμιν· τὰ γὰρ παχέα βραδέως περαιούμενα πλατυτέρας τὰς ἐκπτώσιας τῶν ἐσχαρέων ποιεῖται, καὶ κίνδυνος ἂν εἴη συρραγῆναι τὰς ὠτειλάς· καὶ κάκιον μὲν οὐδὲν ἂν εἴη, αἴσχιον δὲ καὶ ἀτεχνότερον. ὅταν διακαύσῃς ἐς τὸ πέρην, τῶν μὲν πλείστων ἱκανῶς ἂν ἔχοι ἐν τῷ κάτω μέρει τὰς ἐσχάρας ταύτας μούνας θεῖναι· ἢν δὲ μὴ κίνδυνος φαίνηται εἶναι συρραγῆναι τὰς ὠτειλάς, ἀλλὰ πολὺ τὸ διὰ μέσου ᾖ, ὑπάλειπτρον χρὴ λεπτὸν διέρσαι διὰ τῶν καυμάτων, ἔτι
40 ἀναλελημμένου τοῦ δέρματος, οὐ γὰρ ἂν ἄλλως δύναιο διέρσαι· ἐπὴν δὲ διέρσῃς, ἀφεῖναι τὸ δέρμα, ἔπειτα μεσηγὺ τῶν ἐσχαρῶν ἄλλην

[1] τὰ τοιαῦτα.

liable to dislocation at the top and in front where the head of the humerus forms a prominence, and behind a little away from the top of the shoulder. Now these cauterizations would be properly done if the dislocations of the arm were upwards, forwards or backwards, but, as it is, since the dislocation is downwards, these cauterizations rather bring it about than prevent it, for they shut out the head of the humerus from the space above it.

One should cauterize these cases thus:—Grasp the skin at the armpit between the fingers and draw it in the direction towards which the head of the humerus gets dislocated (*i.e.* downwards), then pass the cautery right through the skin thus drawn away. The cautery irons for this operation should not be thick nor very rounded, but elongated (for so they pass through more quickly), and pressure should be made with the hand. They should be white hot, so that the operation may be completed with all possible speed. For thick irons, since they pass through slowly, leave larger eschars to come away, and there is risk of the cicatrices breaking into one another. This indeed is no great evil, but looks rather bad and shows want of skill. When your cautery has gone right through, these two eschars in the part below will in most cases be sufficient by themselves. But if there seems no risk of the cicatrices breaking into one another, and there is a good interval between them, one should pass a thin spatula through the cautery holes, the skin being still held up, for otherwise you could not pass it. After passing it, let go the skin and then make another eschar between the others with a thin

ἐσχάρην ἐμβάλλειν λεπτῷ σιδηρίῳ, καὶ διακαῦσαι ἄχρις ἂν τῷ ὑπαλείπτρῳ ἐγκύρσῃ. ὁπόσον δέ τι χρὴ τὸ δέρμα τὸ ἀπὸ τῆς μασχάλης ἀπολαμβάνειν, τοισίδε χρὴ τεκμαίρεσθαι· ἀδένες ὕπεισιν ἢ ἐλάσσους ἢ μείζους πᾶσιν ὑπὸ τῇ μασχάλῃ, πολλαχῇ δὲ καὶ ἄλλῃ τοῦ σώματος. ἀλλὰ ἐν ἄλλῳ λόγῳ περὶ ἀδένων οὐλομελίης γεγράψεται,
50 ὅ τι τέ εἰσι, καὶ οἷα ἐν οἵοισι σημαίνουσί τε καὶ δύνανται. τοὺς μὲν οὖν ἀδένας οὐ χρὴ προσαπολαμβάνειν, οὐδ' ὅσα ἐσωτέρω τῶν ἀδένων· μέγας γὰρ ὁ κίνδυνος· τοῖσι γὰρ ἐπικαιροτάτοισι τόνοισι γειτονεύονται· ὅσον δὲ ἐξωτέρω τῶν ἀδένων ἐπὶ[1] πλεῖστον ἀπολαμβάνειν· ἀσινέα γάρ. γινώσκειν δὲ χρὴ καὶ τάδε, ὅτι ἢν μὲν ἰσχυρῶς τὸν βραχίονα ἀνατείνῃς, οὐ δυνήσῃ τοῦ δέρματος ἀπολαβεῖν οὐδὲν τοῦ ὑπὸ τῇ μασχάλῃ, ὅ τι καὶ ἄξιον λόγου· καταναισιμοῦται γὰρ ἐν
60 τῇ ἀνατάσει· οἱ δὲ αὖ τόνοι, οὓς οὐδεμιῇ μηχανῇ δεῖ τιτρώσκειν, οὗτοι πρόχειροι γίνονται καὶ κατατεταμένοι ἐν τούτῳ τῷ σχήματι· ἢν δὲ σμικρὸν ἐπάρῃς τὸν βραχίονα, πολὺ μὲν τοῦ δέρματος ἀπολήψῃ, οἱ δὲ τόνοι ὧν δεῖ προμηθεῖσθαι, ἔσω καὶ πρόσω τοῦ χειρίσματος γίνονται. ἆρ' οὖν οὐκ ἐν πάσῃ τῇ τέχνῃ περὶ παντὸς χρὴ ποιεῖσθαι, τὰ δίκαια σχήματα ἐξευρίσκειν ἐφ' ἑκάστοισι; ταῦτα μὲν τὰ κατὰ τὴν μασχάλην, καὶ ἱκαναὶ αὗται αἱ καταλήψεις, ἢν ὀρθῶς τεθῶσιν αἱ
70 ἐσχάραι. ἔκτοσθεν δὲ τῆς μασχάλης δισσὰ μοῦνά ἐστι χωρία, ἵνα ἄν τις ἐσχάρας θείη τιμωρεούσας τῷ παθήματι, μίαν μὲν ἐν τῷ ἔμπροσθεν μεσηγὺ τῆς τε κεφαλῆς τοῦ βραχίονος

[1] ὡς.

cautery, and burn through till you come on to the spatula. The amount of skin that one should take up from the armpit should be estimated thus:—All men have glands, smaller or larger, in the armpit and many other parts of the body.—But the whole structure of glands will be described in another treatise, both what they are, and their signification and function in the parts they occupy.[1]—The glands, then, must not be caught up with the skin, nor any parts internal to the glands. The danger, indeed, is great, for they lie close to cords of the utmost importance. But take up as much as possible of what is superficial to the glands, for that is not dangerous. One should also know the following, namely that if you stretch the arm strongly upwards you cannot take up any part of the skin under the armpit worth mentioning, for it is used up for the extension. The cords, again, which must by no means be wounded, come close to the surface and are on the stretch in this attitude; but if you raise the arm slightly you can take up a good deal of skin, while the cords which are to be guarded lie within, and far from the field of operation. Ought we not then, in all our practice, to consider it of the highest importance to discover the proper attitudes in each case? So much for the parts about the armpit, and these gathers (lit. interceptions) suffice if the eschars are properly placed. Outside the armpit there are only two places where one might put eschars efficacious against the malady; one in front between the head of the humerus and the

[1] The extant treatise on glands is an attempt by a later writer to supply this vacancy. Galen XVIII (1), 379.

καὶ τοῦ τένοντος τοῦ κατὰ τὴν μασχάλην· καὶ ταύτῃ τὸ δέρμα τελέως διακαίειν χρή, βαθύτερον δὲ οὐ χρή· φλέψ τε γὰρ παχείη πλησίη καὶ νεῦρα, ὧν οὐδέτερα θερμαντέα. ὄπισθέν τε αὖ ἄλλην ἐσχάρην ἐνδέχεται ἐνθεῖναι ἀνωτέρω μὲν συχνῷ τοῦ τένοντος τοῦ κατὰ τὴν μασχάλην,
80 κατωτέρω δὲ ὀλίγῳ τῆς κεφαλῆς τοῦ βραχίονος· καὶ τὸ μὲν δέρμα τελέως χρὴ διακαίειν, βαθείην δὲ μηδὲ κάρτα ταύτην ποιεῖν· πολέμιον γὰρ τὸ πῦρ νεύροισιν. ἰητρεύειν μὲν οὖν χρὴ διὰ πάσης τῆς ἰητρείης τὰ ἕλκεα, μηδέποτε ἰσχυρῶς ἀνατείνοντα τὸν βραχίονα, ἀλλὰ μετρίως, ὅσον τῶν ἑλκέων ἐπιμελείης εἵνεκα·—ἧσσον μὲν γὰρ ἂν διαψύχοιτο—συμφέρει γὰρ πάντα τὰ καύματα σκέπειν, ὡς[1] ἐπιεικέως ἰητρεύειν—ἧσσον δ' ἂν ἐκπλίσσοιτο· ἧσσον δ' ἂν αἱμοῤῥαγοίη· ἧσσον δ'
90 ἂν σπασμὸς ἐπιγένοιτο. ὁπόταν δὲ δὴ καθαρὰ γένηται τὰ ἕλκεα, ἐς ὠτειλάς τε ἴῃ, τότε δὴ καὶ παντάπασι χρὴ αἰεὶ τὸν βραχίονα πρὸς τῇσι πλευρῇσι προσδεδέσθαι, καὶ νύκτα καὶ ἡμέρην· ἀτὰρ καὶ ὁπόταν ὑγιέα γένηται τὰ ἕλκεα, ὁμοίως ἐπὶ πολὺν χρόνον χρὴ προσδεῖν τὸν βραχίονα πρὸς τὰς πλευράς· οὕτω γὰρ ἂν μάλιστα ἐπουλωθείη καὶ ἀποληφθείη ἡ εὐρυχωρίη, καθ' ἣν
98 μάλιστα ὀλισθάνει ὁ βραχίων.

XII. Ὅσοισι δ' ἂν ὦμος κατηπορηθῇ ἐμβληθῆναι, ἢν μὲν ἔτι ἐν αὐξήσει ἔωσιν, οὐκ ἐθέλει συναύξεσθαι τὸ ὀστέον τοῦ βραχίονος ὁμοίως τῷ ὑγιεῖ, ἀλλὰ αὔξεται μὲν ἐπί τι, βραχύτερον δὲ τοῦ ἑτέρου γίνεται· καὶ οἱ καλούμενοι δὲ ἐκ γενεῆς γαλιάγκωνες, διὰ δισσὰς συμφορὰς ταύτας

[1] ὡς καί.

tendon at the armpit,[1] and here the cautery should go right through the skin, but no deeper, for there is a large blood vessel in the neighbourhood, and cords, none of which must be heated. Again, another eschar may be placed behind, well above the tendon at the armpit, but a little below the head of the humerus. Burn through the skin completely but do not make this cauterization very deep either, for fire is hostile to nerves. During the whole treatment, the wounds must be dressed without ever lifting the arm up strongly, but only such moderate distance as the care of the wounds requires. They will thus be less exposed to cold—(it is well to cover all burns if they are to be treated properly)—less drawn apart, less liable to haemorrhage, and spasm will be less likely to supervene. When, finally, the wounds get cleansed and begin to cicatrize, then above all should the arm be kept continually bound to the side both night and day, nay, even when the wounds get healed, one should bind the arm to the side in the same way for a long time; for so would the cavity into which the humerus is mostly displaced be best cicatrized up and cut off.

XII. In cases where reduction of the shoulder has failed, if the patients are still adolescent, the bone of the arm will not grow like the sound one. It grows a little indeed, but gets shorter than the other. As to those who are called congenitally weasel-armed[2], they owe this infirmity to two

[1] Pectoralis major tendon.

[2] Strictly weasel-elbowed. Galen in his Lexicon says they have shrivelled upper arms and swollen elbows "like the weasels," but he doubts the derivation. In his Commentary he is still more doubtful, but leaves "those who study such matters" to clear it up, which they have not yet done.

γινονται, ἤν γέ τι τοιοῦτον αὐτοὺς ἐξάρθρημα καταλάβῃ ἐν τῇ γαστρὶ ἐόντας, διά τε ἄλλην[1] συμφορήν, περὶ ἧς ὕστερόν ποτε γεγράψεται·
10 ἀτὰρ καὶ οἷσιν ἔτι νηπίοισιν ἐοῦσι κατὰ τὴν κεφαλὴν τοῦ βραχίονος βαθεῖαι καὶ ὑποβρύχιοι ἐκπυήσιες γίνονται, καὶ οὗτοι πάντες γαλιάγκωνες γίνονται· καὶ ἤν τε τμηθῶσιν, ἤν τε καυθῶσιν, ἤν τε αὐτόματόν σφιν ἐκραγῇ, εὖ εἰδέναι χρὴ ὅτι ταῦτα οὕτως ἔχει. χρῆσθαι μέντοι τῇ χειρὶ δυνατώτατοί[2] εἰσιν οἱ ἐκ γενεῆς γαλιάγκωνες, οὐ μὴν οὐδὲ ἐκεῖνοί γε ἀνατεῖναι παρὰ τὸ οὖς τὸν βραχίονα ἐκτανύσαντες τὸν ἀγκῶνα δύνανται, ἀλλὰ πολὺ ἐνδεεστέρως ἢ τὴν ὑγιέα χεῖρα. οἷσι
20 δ' ἂν ἤδη ἀνδράσιν ἐοῦσιν ἐκπέσῃ ὁ ὦμος καὶ μὴ ἐμβληθῇ, ἡ ἐπωμὶς ἀσαρκοτέρη γίνεται, καὶ ἡ ἕξις λεπτὴ ἡ κατὰ τοῦτο τὸ μέρος· ὅταν μέντοι ὀδυνώμενοι παύσωνται, ὁπόσα μὲν δεῖ ἐργάζεσθαι ἐπαίροντας τὸν ἀγκῶνα ἀπὸ τῶν πλευρέων ἐς τὸ πλάγιον, ταῦτα μὲν οὐ δύνανται ἅπαντα ὁμοίως ἐργάζεσθαι· ὁπόσα δὲ δεῖ ἐργάζεσθαι, παραφέροντας τὸν βραχίονα παρὰ τὰς πλευράς, ἢ ἐς τοὐπίσω ἢ ἐς τοὔμπροσθεν, ταῦτα δὲ δύνανται ἐργάζεσθαι· καὶ γὰρ ἂν ἀρίδα ἑλκύσαιεν[3] καὶ
30 πρίονα, καὶ πελεκήσαιεν ἄν, καὶ σκάψαιεν ἄν, μὴ κάρτα ἄνω αἴροντες τὸν ἀγκῶνα, καὶ τἄλλα ὅσα ἐκ τῶν τοιούτων σχημάτων ἐργάζονται.

XIII. Ὅσοισι δ' ἂν τὸ ἀκρώμιον ἀποσπασθῇ, τούτοισι φαίνεται ἐξέχον τὸ ὀστέον τὸ ἀπεσπασμένον· ἔστι δὲ τοῦτο ὁ σύνδεσμος τῆς κληῖδος καὶ τῆς ὠμοπλάτης· ἑτεροίη γὰρ ἡ φύσις

[1] ἑτέρην. [2] δυνατώτεροι.

separate causes. Either a dislocation of this kind has befallen them in the womb, or another accident which will be described somewhat later;[1] so, too, those in whom deep suppuration bathing the head of the humerus occurs while they are still children all become weasel-armed. And whether they are operated on by the knife or cautery, or the abscess breaks of itself, be sure that this will be the result. Still, those who are congenitally weasel-armed are quite able to use the arm, though they, too, cannot stretch the arm up by the ear with the elbow extended, but to a much less extent than the sound one. In adults, when the shoulder is dislocated and not reduced, its point is less fleshy than usual and this part assumes a lean habit. Still, when they cease to suffer pain, though as regards all such work as requires raising the elbow outwards from the side they are unable to do it as before, any work such as involves moving the arm either backwards or forwards along the side they can execute. For they might work a bow-drill[2] or saw,—and might use pick or spade without much raising of the elbow, and so with all other works which are done in such attitudes.

XIII. In cases of avulsion of the acromion, the bone torn off makes an obvious projection. This bone is the bond between the clavicle and the shoulder-blade, for man's structure is here diverse

[1] As Galen remarks, if we deduct the dislocation and the disease from the two causes, it is difficult to see what remains.

[2] "File" most translators, "auger" Adams, but the ἀρίς was used to work the trephine. See Oribasius, XLVI. ii.

---

[3] ἑλκύσειαν . . . πελεκήσειαν . . . σκάψειαν. Κω.

ἀνθρώπου ταύτῃ ἢ τῶν ἄλλων ζώων· οἱ οὖν ἰητροὶ μάλιστα ἐξαπατῶνται ἐν τούτῳ τῷ τρώματι—ἅτε γὰρ ἀνασχόντος τοῦ ὀστέου τοῦ ἀποσπασθέντος, ἡ ἐπωμὶς φαίνεται χαμαιζήλη καὶ κοίλη—ὥστε[1] καὶ προμηθεῖσθαι τῶν ὤμων τῶν
10 ἐκπεπτωκότων. πολλοὺς οὖν οἶδα ἰητροὺς τἄλλα οὐ φλαύρους ἐόντας, οἳ πολλὰ ἤδη ἐλυμήναντο, ἐμβάλλειν πειρώμενοι τοὺς τοιούτους ὤμους, οὕτως οἰόμενοι ἐκπεπτωκέναι, καὶ οὐ πρόσθεν παύονται πρὶν ἢ ἀπογνῶναι ἢ ἀπορῆσαι, δοκοῦντες αὐτοὶ σφέας αὐτοὺς ἐμβάλλειν τὸν ὦμον. τούτοισιν ἰητρείη μέν, ἥπερ καὶ τοῖσιν ἄλλοισιν τοῖσι τοιούτοισι, κηρωτὴ καὶ σπλῆνες καὶ ὀθόνια, καὶ ἐπίδεσις τοιαύτη. καταναγκάζειν μέντοι τὸ ὑπερέχον χρή, καὶ τοὺς σπλῆνας κατὰ τοῦτο
20 τιθέναι πλείστους, καὶ πιέζειν ταύτῃ μάλιστα, καὶ τὸν βραχίονα πρὸς τῇσι πλευρῇσι προσηρτημένον ἐς τὸ ἄνω μέρος ἔχειν, οὕτω γὰρ ἂν μάλιστα πλησιάζοι τὸ ἀπεσπασμένον. τάδε μὲν εὖ εἰδέναι χρή, καὶ προλέγειν ὡς ἀσφαλέα, εἰ ἄλλως ἐθέλεις, ὅτι βλάβη μὲν οὐδεμίη, οὔτε σμικρὴ οὔτε μεγάλη, τῷ ὤμῳ γίνεται ἀπὸ τούτου τοῦ τρώματος, αἴσχιον δὲ τὸ χωρίον· οὐδὲ γὰρ τοῦτο τὸ ὀστέον ἐς τὴν ἀρχαίην ἕδρην ὁμοίως ἂν ἱδρυνθείη, ὥσπερ ἐπιπέφυκεν,[2] ἀλλ' ἀνάγκη
30 πλέον ἢ ἔλασσον ὀγκηρότερον εἶναι ἐς τὸ ἄνω. οὐδὲ γὰρ ἄλλο ὀστέον οὐδὲν ἐς τωὐτὸ καθίσταται ὅ τι ἂν κοινωνέον ᾖ ἑτέρῳ ὀστέῳ καὶ προσπεφυκὸς ἀποσπασθῇ ἀπὸ τῆς ἀρχαίης φύσιος. ἀνώδυνόν

[1] ὥσπερ τῶν ὤμων.
[2] ὡς ἐπεφύκει.

from that of animals. Thus practitioners are especially deceived by this injury—since, the detached bone being raised up, the point of the shoulder looks depressed and hollow—even to the extent of treating the patients for dislocated shoulders.[1] I know many otherwise excellent practitioners who have done much damage in attempting to reduce shoulders of this kind, which they thought were dislocated: and who did not cease their efforts till they recognised either their error or their impotence if they still supposed they were reducing the shoulder-joint. The treatment in these, as in other like cases, consists of cerate, compresses, bandages and the like mode of dressing. The projecting part however should be forced down, the bulk of the compresses placed over it and strongest pressure made here. Also the arm should be fixed to the ribs and kept up, for so it will best be brought near the part torn off. For the rest, keep well in mind and predict with assurance, if you think proper, that no harm, small or great, happens to the shoulder from this injury, but the part will be deformed. This bone, in fact, cannot be fixed in its old natural position as it was, but there will necessarily be more or less of a tuberosity on the top. Nor, indeed, is any bone brought back to the same place, if, after forming an annex or outgrowth of another bone, it has been torn away from its old natural position.

[1] "Looks hollow" as when the shoulders are dislocated, (Kw.'s reading).

τε τὸ ἀκρώμιον ἐν ὀλίγῃσιν ἡμέρῃσι γίνεται, ἢν
35 χρηστῶς ἐπιδέηται.

XIV. Κληὶς δὲ κατεαγεῖσα, ἢν μὲν ἀτρεκέως ἀποκαυλισθῇ, εὐϊητοτέρη ἐστίν· ἢν δὲ παραμηκέως, δυσιητοτέρη. τἀναντία δὲ τούτοισίν ἐστιν ἢ ὡς ἄν τις οἴοιτο, τὴν μὲν γὰρ ἀτρεκέως ἀποκαυλισθεῖσαν προσαναγκάσειεν[1] ἄν τις μᾶλλον ἐς τὴν φύσιν ἐλθεῖν· καὶ γὰρ εἰ πάνυ προμηθηθείη, τὸ ἀνωτέρω κατωτέρω ἂν ποιήσειε σχήμασί τε ἐπιτηδείοισι καὶ ἐπιδέσει ἁρμοζούσῃ· εἰ δὲ μὴ τελέως ἱδρυνθείη, ἀλλ' οὖν τὸ ὑπερέχον γε τοῦ ὀστέου
10 οὐ κάρτα ὀξὺ γίνεται· ὧν δ' ἂν παραμηκὲς τὸ ὀστέον κατεαγῇ, ἰκέλη ἡ συμφορὴ γίνεται τοῖσιν ὀστέοισι τοῖσι ἀπεσπασμένοισι, περὶ ὧν πρόσθεν γέγραπται· οὔτε γὰρ ἱδρυνθῆναι αὐτὸ πρὸς ἑωυτὸ κάρτα ἐθέλει, ἥ τε ὑπερέχουσα ὄκρις τοῦ ὀστέου ὀξείη γίνεται κάρτα. τὸ μὲν οὖν σύμπαν, εἰδέναι χρὴ ὅτι βλάβη οὐδεμίη τῷ ὤμῳ οὐδὲ τῷ ἄλλῳ σώματι γίνεται διὰ τὴν κάτηξιν τῆς κληῖδος, ἢν μὴ ἐπισφακελίσῃ· ὀλιγάκις δὲ τοῦτο γίνεται. αἶσχός γε μὴν προσγίνεται περὶ τὴν κάτηξιν τῆς
20 κληῖδος, καὶ τούτοισι τὸ πρῶτον αἴσχιστον, ἔπειτα μὴν ἐπὶ ἧσσον γίνεται. συμφύεται δὲ ταχέως κληὶς καὶ τἄλλα πάντα ὅσα χαῦνα ὀστέα· ταχείην γὰρ τὴν ἐπιπώρωσιν ποιεῖται τὰ τοιαῦτα. ὅταν μὲν οὖν νεωστὶ κατεαγῇ, οἱ τετρωμένοι σπουδάζουσι, οἰόμενοι μέζον τὸ κακὸν εἶναι ἢ ὅσον ἐστίν· οἵ τε ἰητροὶ προθυμέονται δῆθεν

[1] προσαναγκάζοι.

---

[1] This is probably dislocation of the clavicle at the outer end. The anatomy of the part was imperfectly understood

The acromion becomes painless in a few days, if it is properly bandaged.[1]

XIV. A fractured collar-bone is more easily treated if broken straight across; but if fractured obliquely, treatment is more difficult. In these cases matters are the reverse of what one would expect. For one will more readily force a collar-bone fractured straight across into its natural position, and by thoroughly careful treatment will succeed in adjusting the upper to the lower fragment by appropriate attitudes and suitable bandaging. And should it not be completely reduced, at least the projection of bone will not be very pointed. But those in whom the bone is fractured obliquely suffer an accident like the avulsions of bones described above; for the fracture hardly lends itself to reduction, and the projecting ridge of bone becomes very sharp. Still, when all is said, one must bear in mind that no harm happens to the shoulder, or body generally, from a fractured collar-bone, unless necrosis supervenes, and this rarely happens. Deformity, it is true, accompanies fracture of the clavicle, and this is very marked at first, but afterwards gets less. The collar-bone unites quickly, as do all spongy bones, for with such the formation of callus is rapid. Thus, when the fracture is recent, patients take it seriously, thinking the damage is worse than it is, and practitioners on their side are careful in applying proper treatment;

even in Galen's time, some saying that the acromion was a distinct bone found only in man; while others thought there was a third bone or cartilage between the clavicle and acromion. The accident occurred to Galen when 35 years old, and he relates vividly how it was first mistaken for a dislocated shoulder, and how, by forty days' endurance of tight bandaging, he recovered without any deformity.

ὀρθῶς ἰῆσθαι· προϊόντος δὲ τοῦ χρόνου οἱ τετρωμένοι, ἅτε οὐκ ὀδυνώμενοι οὐδὲ κωλυόμενοι οὔτε ὁδοιπορίης οὔτε ἐδωδῆς, καταμελέουσι· οἵ τε αὖ
30 ἰητροί, ἅτε οὐ δυνάμενοι καλὰ τὰ χωρία ἀποδεικνύναι, ὑποδιδράσκουσι, καὶ οὐκ ἄχθονται τῇ ἀμελείῃ τῶν τετρωμένων· ἐν τούτῳ τε ἡ ἐπιπώρωσις συνταχύνεται.

Ἐπιδέσιος μὲν οὖν τρόπος καθέστηκε παραπλήσιος τοῖσι πλείστοισι κηρωτῇ καὶ σπλήνεσι καὶ ὀθονίοισι μαλθακοῖσιν ἰητρεύειν· καὶ τάδε δεῖ προσιητρεύειν, καὶ τάδε δεῖ προσσυνιέναι καὶ μάλιστα ἐν τούτῳ τῷ χειρίσματι, ὅτι τούς τε σπλῆνας πλείστους κατὰ τὸ ἐξέχον χρὴ τιθέναι,
40 καὶ τοῖσι ἐπιδέσμοισι πλείστοισι καὶ μάλιστα κατὰ τοῦτο χρὴ πιέζειν. εἰσὶ δὲ δή τινες, οἳ ἐπεσοφίσαντο ἤδη μολύβδιον βαρὺ προσεπικαταδεῖν, ὡς καταναγκάζοι[1] τὸ ὑπερέχον· συνιᾶσι μὲν οὖν ἴσως οὐδὲ οἱ ἁπλῶς ἐπιδέοντες· ἀτὰρ δὴ οὐδ' οὗτος ὁ τρόπος κληῖδος κατήξιός ἐστιν· οὐ γὰρ δυνατὸν τὸ ὑπερέχον καταναγκάζεσθαι οὐδὲν ὅ τι ἄξιον λόγου. ἄλλοι δ' αὖ τινές εἰσιν, οἵτινες, καταμαθόντες τοῦτο, ὅτι αὗται αἱ ἐπιδέσιες παράφοροί εἰσι καὶ οὐ κατὰ φύσιν καταναγκά-
50 ζουσι τὰ ὑπερέχοντα, ἐπιδέουσι μὲν οὖν αὐτοὺς σπλήνεσι καὶ ὀθονίοισι χρεώμενοι, ὥσπερ καὶ οἱ ἄλλοι· ζώσαντες δὲ τὸν ἄνθρωπον ταινίῃ τινί, ᾗ εὐζωστότατος αὐτὸς ἑωυτοῦ ἐστίν, ὅταν ἐπιθέωσι τοὺς σπλῆνας ἐπὶ τὰ ὑπερέχοντα τοῦ κατήγματος, ἐξογκώσαντες ἐπὶ τὰ ἐξέχοντα, τὴν ἀρχὴν τοῦ ὀθονίου προσέδησαν πρὸς τὸ ζῶσμα ἐκ τοῦ ἔμπροσθεν, καὶ οὕτως ἐπιδέουσιν, ἐπὶ τὴν ἴξιν τῆς κληῖδος ἐπιτανύοντες, ἐς τοὔπισθεν ἄγοντες·

but as time goes on the patients, since they feel no pain and are not hindered either in getting about or eating, neglect the matter, and physicians too, since they cannot make the parts look well, withdraw gradually, and are not displeased by the patients' carelessness, and meanwhile the callus formation quickly develops.

Now, the established mode of treatment is like that used for most fractures, cerate, compresses, and soft bandages; also the following extra treatment is required, and it must be kept in mind especially in handling this injury that one should put the bulk of the compresses on the projecting part and apply pressure with most of the bandages, especially at this point. There are some, indeed, who in their wisdom have contrived something further and bind on a heavy piece of lead as well, so as to press down the projection. Perhaps those who use a simple bandage are no wiser, yet after all, this is not a suitable plan for a fractured collar-bone, for the projecting part cannot be pressed down to any extent worth mentioning. Again, there are certain others, who, recognizing a tendency to slip in these dressings and their inability to press down the projecting parts in a natural way, use compresses and bandages like the rest, but gird the patient with a belt at the most suitable part of his body. Then they put compresses on the part of the fracture that sticks up, piling them on to the projection, fix the end of the bandage to the belt in front and apply by stretching it vertically over the collar-bone and bringing it to the back. Then,

---

[1] καταναγκάζειν.

κἄπειτα περιβάλλοντες περὶ τὸ ζῶσμα, ἐς τοὔμ-
60 προσθεν ἄγουσι, καὶ αὖθις ἐς τοὔπισθεν. οἱ δέ
τινες οὐχὶ περὶ τὸ ζῶσμα περιβάλλουσι τὸ
ὀθόνιον, ἀλλὰ περὶ τὸν περίναιόν τε καὶ παρ'
αὐτὴν τὴν ἕδρην καὶ παρὰ τὴν ἄκανθαν κυκλεύ-
οντες τὸ ὀθόνιον, οὕτω πιέζουσι τὸ κάτηγμα.
ταῦτα γοῦν ἀπείρῳ μὲν ἀκοῦσαι φαίνεται ἐγγύς
τι τοῦ κατὰ φύσιν εἶναι, χρεομένῳ δὲ ἄχρηστα·
οὔτε γὰρ μόνιμα οὐδένα χρόνον, οὐδ' εἰ κατα-
κέοιτό τις—καίτοι ἐγγυτάτω ἂν οὕτως—ἀλλ'
ὅμως, εἰ καὶ κατακείμενος ἢ τὸ σκέλος συγκάμ-
70 ψειεν ἢ αὐτὸς καμφθείη, πάντα ἂν τὰ ἐπιδέσματα κινέοιτο· ἄλλως τε ἀσηρὴ ἡ ἐπίδεσις· ἥ τε
γὰρ ἕδρη ἀπολαμβάνεται, ἀθρόα τε τὰ ὀθόνια ἐν
ταύτῃ τῇ στενοχωρίῃ γίνεται· τά τε αὖ περὶ τὴν
ζώνην περιβαλλόμενα οὐχ οὕτως ἰσχυρῶς ἔζωσ-
ται, ὡς οὐκ ἀναγκάσαι ἐς τὸ ἄνω τὴν ζώνην
ἐπανιέναι, καὶ οὕτως ἀνάγκη ἂν εἴη χαλᾶν [1] τὰ
ἐπιδέσματα. ἄγχιστα δ' ἄν τις δοκέοι ποιεῖν,
καίπερ οὐ μεγάλα ποιῶν, εἰ τοῖσι μέν τισι τῶν
ὀθονίων περὶ τὴν ζώνην περιβάλλοι, τοῖσι δὲ
80 πλείστοισι τῶν ὀθονίων τὴν ἀρχαίην ἐπίδεσιν
ἐπιδέοι· οὕτω γὰρ ἂν μάλιστα τὰ ἐπιδέσματα
μόνιμά τε εἴη καὶ ἀλλήλοισι τιμωρέοι.

Τὰ μὲν οὖν πλεῖστα εἴρηται, ὅσσα καταλαμβάνει τοὺς τὴν κληῖδα καταγνυμένους. προσσυνιέναι δὲ τόδε χρή, ὅτι κληὶς ὡς ἐπιτοπολὺ
κατάγνυται, ὥστε τὸ μὲν ἀπὸ τοῦ στήθεος
πεφυκὸς ὀστέον ἐς τὸ ἄνω μέρος ὑπερέχειν, τὸ δὲ
ἀπὸ τῆς ἀκρωμίης ἐν τῷ κάτω μέρει εἶναι. αἴτια
δὲ τούτων τάδε, ὅτι τὸ μὲν στῆθος οὔτε κατωτέρω
90 ἂν πολὺ οὔτε ἀνωτέρω χωρήσειεν· σμικρὸς γὰρ ὁ

passing it through the belt, they bring it to the front and again to the back. There are others who pass the bandage, not through a belt, but round the perineum near the fundament itself, and, completing the circle along the spine, thus make pressure on the fracture. To an inexperienced person these methods seem to come near the natural, but to one who uses them useless; for they have no permanent stability, not even if the patient keeps his bed, though this would come nearest. Yet even if, when recumbent, he bends his leg or curves his body all the bandages will be deranged. Besides the dressing is troublesome, for the fundament is included, and all the bandages accumulate in this narrow part, while, as for those passed through the belt, it is impossible to gird it so tightly as not to yield to the force pulling upwards, and so the bandages will necessarily become lax. One would appear to be most effective, though without effecting much, by making some turns of bandage through the belt while applying most in the old fashion,[1] for so the bandages would best keep in place and support one another.

Almost all then has been said on the subject of patients with broken collar-bones; but the following should also be borne in mind, namely, that the clavicle as a rule is so fractured that the part arising from the breast-bone is on the top and that from the shoulder-point (acromion) below. The reason of this is as follows: the breast-bone does not move much either downwards or upwards, for the range of the joint at

[1] Some make ἀρχαίην ἐπίδεσιν = the under bandage, first applied, but cf. ἀρχαίη φύσις = νομίμη, XIII. 33.

---

[1] πάντα χαλᾶν.

κιγκλισμὸς τοῦ ἄρθρου τοῦ ἐν τῷ στήθει. αὐτό τε γὰρ ἑωυτὸ συνεχές ἐστι τὸ στῆθος καὶ τῇ ῥάχει· ἄγχιστα μὴν ἡ κληὶς πρὸς τὸ τοῦ ὤμου ἄρθρον πλοώδης ἐστίν· ἠνάγκασται γὰρ πυκινοκίνητος εἶναι διὰ τὴν τῆς ἀκρωμίης σύζευξιν. ἄλλως τε ὅταν τρωθῇ, φεύγει ἐς τὸ ἄνω μέρος τὸ πρὸς τῷ στήθει προσεχόμενον, καὶ οὐ μάλα ἐς τὸ κάτω μέρος ἀναγκάζεσθαι ἐθέλει· καὶ γὰρ
100 πέφυκε κοῦφον,[1] καὶ ἡ εὐρυχωρίη αὐτῷ ἄνω πλείων ἢ κάτω. ὁ δὲ ὦμος καὶ ὁ βραχίων καὶ τὰ προσηρτημένα τούτοισιν εὐαπόλυτά ἐστιν ἀπὸ τῶν πλευρέων καὶ τοῦ στήθεος, καὶ διὰ τοῦτο δύναται καὶ ἀνωτέρω πολὺ ἀνάγεσθαι καὶ κατωτέρω· ὅταν οὖν κατεαγῇ ἡ κληίς, τὸ πρὸς τῷ ὤμῳ ὀστέον ἐς τὸ κατωτέρω ἐπιῤῥέπει· ἐς τοῦτο γὰρ ἐπιτροχώτερον αὐτὸ ἅμα τῷ ὤμῳ καὶ τῷ βραχίονι κάτω ῥέψαι μᾶλλον ἢ ἐς τὸ ἄνω. ὁπότε οὖν ταῦτα τοιαῦτά ἐστιν, ἀσυνετέουσιν
110 ὅσοι τὸ ὑπέρεχον τοῦ ὀστέου ἐς τὸ κάτω καταναγκάσαι οἴονται οἷόν τε εἶναι. ἀλλὰ δῆλον ὅτι τὰ κάτω πρὸς τὸ ἄνω προσακτέον ἐστίν· τοῦτο γὰρ ἔχει κίνησιν, τοῦτο γάρ ἐστιν καὶ τὸ ἀποστὰν ἀπὸ τῆς φύσιος. δῆλον οὖν ὅτι ἄλλως μὲν οὐδαμῶς ἔστιν ἀναγκάσαι τοῦτο—αἵ τε γὰρ ἐπιδέσιες οὐδέν τι μᾶλλον προσαναγκάζουσιν ἢ ἀπαναγκάζουσιν—εἰ δέ τις τὸν βραχίονα πρὸς τῇσι πλευρῇσι ἐόντα ἀναγκάζοι ὡς μάλιστα ἄνω, ὡς ὅτι ὀξύτατος ὁ ὦμος φαίνηται[3] εἶναι, δῆλον
120 ὅτι οὕτως ἂν ἁρμοσθείη πρὸς τὸ ὀστέον τὸ ἀπὸ τοῦ στήθεος πεφυκός, ὅθεν ἀπεσπάσθη. εἰ οὖν τις τῇ μὲν ἐπιδέσει χρέοιτο τῇ νομίμῃ τοῦ ταχέως

[1] λορδόν.

the sternum is slight and there is continuous connexion between the breast-bone and the spine, but the clavicle on the side of its connexion with the shoulder is especially[1] loose, for it has to have great freedom of movement owing to the acromial junction. Besides, when it is fractured, the part adherent to the breast-bone flies upwards, and can hardly be pressed down, for it is naturally light and there is a larger vacancy for it above than below. But the shoulder, upper arm and parts annexed are easily separated from the ribs and breast-bone and therefore can be moved through a large space upwards and downwards. Thus, when the collar-bone is broken, the part towards the shoulder sinks downwards, for with the shoulder and arm it is more readily disposed to move down than upwards. So whenever this state of things occurs, they are unintelligent who think it possible to press the projecting part of the bone downwards; while it is obvious that one must bring the lower part up, for this is the moveable part, and this too is the one out of its natural place. It is obvious then that other methods are useless in reducing this fracture —for bandagings are no more likely to bring the parts together than to separate them—but if one presses the arm upwards as much as possible, keeping it to the side, so that the shoulder appears very pointed, it is clear that the fragment will thus be brought into connexion with the bone arising from the sternum from which it was torn. If, then, one should use the ordinary dressing for the sake of

[1] Erotian refers twice to this use of ἄγχιστα = μάλιστα.

[2] φαίνεται, Galen. M.

συναλθεσθῆναι εἵνεκα, ἡγήσαιτο ἂν τἄλλα πάντα μάτην εἶναι παρὰ τὸ σχῆμα τὸ εἰρημένον, ὀρθῶς τε ἂν συνίοι, ἰητρεύοι τε ἂν τάχιστα καὶ κάλλιστα. κατακεῖσθαι μέντοι τὸν ἄνθρωπον μέγα τὸ [1] διάφορόν ἐστιν· καὶ ἡμέραι ἱκαναὶ τεσσαρεσ-
127 καίδεκα, εἰ ἀτρεμέοι, εἴκοσι δὲ πάμπολλαι.

XV. Εἰ μέντοι τινὶ ἐπὶ τἀναντία ἡ κληὶς κατεαγείη, ὃ οὐ μάλα γίνεται, ὥστε τὸ μὲν ἀπὸ τοῦ στήθεος ὀστέον ὑποδεδυκέναι, τὸ δὲ ἀπὸ τῆς ἀκρωμίης ὀστέον ὑπερέχειν καὶ ἐποχεῖσθαι ἐπὶ τοῦ ἑτέρου, οὐδεμιῆς μεγάλης ἰητρείης ταῦτά γ' ἂν δέοιτο· αὐτὸς γὰρ ὁ ὦμος ἀφιέμενος καὶ ὁ βραχίων ἱδρύοι ἂν τὰ ὀστέα πρὸς ἄλληλα, καὶ φαύλη ἄν τις ἐπίδεσις ἀρκέοι, καὶ ὀλίγαι ἡμέραι
9 τῆς πωρώσιος γενοίατ' ἄν.

XVI. Εἰ δὲ μὴ κατεαγείη μὲν οὕτως, παρολισθάνοι δὲ ἐς τὸ πλάγιον ἢ τῇ ἢ τῇ, ἐς τὴν φύσιν μὲν ἀπαγαγεῖν ἂν δέοι, ἀναγαγόντα τὸν ὦμον σὺν τῷ βραχίονι, ὥσπερ καὶ πρόσθεν εἴρηται· ὅταν δὲ ἵζηται ἐς τὴν ἀρχαίην φύσιν, ταχείη ἂν ἡ ἄλλη ἰητρείη εἴη. τὰ μὲν οὖν πλεῖστα τῶν παραλλαγμάτων κατορθοῖ αὐτὸς ὁ βραχίων, ἀναγκαζόμενος πρὸς τὰ ἄνω. ὅσα δὲ ἐκ τῶν ἄνωθεν παρολισθάνοντα ἐς τὸ πλάγιον
10 ἦλθεν, ἢ ἐς τὸ κατωτέρω, συμπορσύνοι ἂν τὴν κατόρθωσιν, εἰ ὁ μὲν ἄνθρωπος ὕπτιος κέοιτο, κατὰ δὲ τὸ μεσηγὺ τῶν ὠμοπλατέων ὑψηλότερόν τι ὀλίγῳ ὑποκέοιτο, ὡς περιῤῥηδὲς ᾖ τὸ στῆθος ὡς μάλιστα· καὶ τὸν βραχίονα εἰ ἀνάγοι τις παρὰ τὰς πλευρὰς παρατεταμένον, ὁ δὲ ἰητρὸς τῇ μὲν ἑτέρῃ χειρὶ ἐς τὴν κεφαλὴν τοῦ βραχίονος ἐμβαλὼν τὸ θέναρ τῆς χειρὸς ἀπωθέοι, τῇ δὲ

getting a quick cure, and should consider everything else of no importance compared with the attitude described, his opinion would be right and his treatment most correct and speedy. Still, it makes a great difference if the patient lies down, and fourteen days suffice if he keeps at rest, while twenty are very many.

XV. If, however, a man has his collar-bone broken in the opposite way, which rarely happens—so that the thoracic fragment is underneath and the acromial part projects and overrides the other—no complicated treatment will be required here, for the shoulder and arm left to themselves will bring the fragments together. Any ordinary dressing will suffice, and callus will form in a few days.

XVI. If the fracture is not of this kind, but the displacement is to one side or the other, one must reduce it to its natural position by elevating the shoulder and arm as described before, and when it is set in its old natural place the rest of the cure will be rapid. Most lateral displacements are corrected by the arm itself when pressed upwards, but in cases where the upper (sternal)[1] fragment is displaced laterally or downwards adjustment will be favoured by the patient lying flat on his back with some slightly elevated support between the shoulders, so that the chest falls away as much as possible at the sides. Let an assistant push the arm, kept stretched along the side, upwards, while the practitioner with one hand on the head of the humerus presses it back with his palm, and with the other adjusts the

[1] So Galen.

ἑτέρῃ τὰ ὀστέα τὰ κατεηγότα εὐθετίζοι, οὕτως ἂν μάλιστα ἐς τὴν φύσιν ἄγοι· ἀτάρ, ὥσπερ ἤδη
20 εἴρηται, εὖ[1] μάλα τὸ ἄνωθεν ὀστέον ἐς τὸ κάτω φιλεῖ ὑποδύνειν. τοῖσι μὲν οὖν πλείστοισιν, ὅταν ἐπιδεθῶσι, τὸ σχῆμα ἀρήγει, παρ' αὐτὰς τὰς πλευρὰς τὸν ἀγκῶνα ἔχοντα οὕτως ἐς τὸ ἄνω τὸν ὦμον ἀναγκάζεσθαι· ἔστι δὲ οἷσι μὲν τὸν ὦμον ἀναγκάζειν δεῖ ἐς τὸ ἄνω, ὡς εἴρηται, τὸν δὲ ἀγκῶνα πρὸς τὸ στῆθος παράγειν, ἄκρην δὲ τὴν χεῖρα παρὰ τὸ ἀκρώμιον τοῦ ὑγιέος ὤμου ἴσχειν. ἢν μὲν οὖν κατακεῖσθαι τολμᾷ, ἀντιστήριγμά τι προστιθέναι χρή, ὡς ἂν ὁ ὦμος
30 ἀνωτάτω ᾖ· ἢν δὲ περιίῃ, σφενδόνην χρὴ ἐκ ταινίης περὶ τὸ ὀξὺ τοῦ ἀγκῶνος ποιήσαντα
32 ἀναλαμβάνειν περὶ τὸν αὐχένα.

XVII. Ἀγκῶνος δὲ ἄρθρον παράλλαξαν μὲν ἢ παραρθρῆσαν πρὸς πλευρὴν ἢ ἔξω, μένοντος τοῦ ὀξέος τοῦ ἐν τῷ κοίλῳ τοῦ βραχίονος, ἐς εὐθὺ κατατείναντα, τὸ ἐξέχον ἀπωθεῖν ὀπίσω καὶ
5 ἐς τὸ πλάγιον.

XVIII. Τὰ δὲ τελέως ἐκβάντα ἢ ἔνθα ἢ ἔνθα, κατάτασις μέν, ἐν ᾗ ὁ βραχίων κατεαγεὶς ἐπιδεῖται· οὕτω γὰρ ἂν τὸ καμπύλον τοῦ ἀγκῶνος οὐ κωλύσει. ἐκπίπτει δὲ μάλιστα ἐς τὸ πρὸς πλευρὰς[2] μέρος. τὰς δὲ κατορθώσιας, ἀπάγοντα ὅτι πλεῖστον, ὡς μὴ ψαύῃ τῆς κορώνης ἡ κεφαλή, μετέωρον περιάγειν καὶ περικάμπτειν,[3] καὶ μὴ ἐς

[1] οὐ Littré, Erm., Kw. [2] πλευρὴν. [3] περικάμψαι.

---

[1] Reading οὐ. εὖ (Galen, Pq, and all MSS.) would accentu-

broken bones; in this way one will best bring them to the natural position; but as was said before the upper (sternal) fragment is not[1] much wont to be displaced downwards.[2] In most cases, the position after bandaging with the elbow to the side suffices to keep the shoulder up, but in some it is necessary to press the shoulder up as described, bring the elbow towards the chest and fix the hand at the point of the sound shoulder. If, then, the patient brings himself to lie down one should supply a prop to keep the shoulder as far up as possible, but if he goes about one should suspend the part by a sling bandage round the neck to include the point of the elbow.

XVII.[3] (Subluxation of the radius.) When there is displacement or subluxation of the elbow-joint towards the side or outwards, the point (olecranon) in the cavity of the humerus retaining its position, make direct extension and push the projecting part obliquely backwards.[4]

XVIII. Complete dislocations of the elbow in either direction require extension in the position in which a fractured humerus is bandaged; for so the curved part of the elbow will not get in the way. The usual dislocation is that towards the ribs.[4] For adjustment separate the bones as much as possible so that the head (of the humerus) may not hit the coronoid process, keep it up and use movements of circumduction and flexion, and do not force it back

ate the statement that the sternal fragment may be displaced downwards.

[2] Or, following Pq and the MSS., "the upper fragment may very well be displaced downwards."

[3] For the sources of XVII—XXIX see Introduction, p. 86.

[4] = our forearm backwards, cf. *Fractures* XLI.

εὐθὺ βιάζεσθαι, ἅμα δὲ ὠθεῖν τἀναντία ἐφ' ἑκάτερα καὶ παρωθεῖν ἐς χώρην· συνωφελοίη
10 δ' ἂν καὶ ἐπίστρεψις ἀγκῶνος ἐν τούτοισιν, ἐν τῷ μὲν ἐς τὸ ὕπτιον, ἐν τῷ δὲ ἐς τὸ πρηνές. ἴησις δέ, σχήματος μέν, ὀλίγῳ ἀνωτέρω ἄκρην τὴν χεῖρα τοῦ ἀγκῶνος ἔχειν, βραχίονα κατὰ πλευράς· οὕτω δὲ καὶ ἀνάληψις καὶ θέσις· καὶ εὔφορον καὶ φύσις, καὶ χρῆσις ἐν τῷ κοινῷ, ἢν ἄρα μὴ κακῶς πωρωθῇ· πωροῦται δὲ ταχέως. ἴησις δὲ ὀθονίοισι κατὰ τὸν νόμον τὸν ἀρθριτικόν,[1]
18 καὶ τὸ ὀξὺ προσεπιδεῖν.

XIX. Παλιγκοτώτατον δὲ ὁ ἀγκὼν πυρετοῖσιν, ὀδύνῃσιν, ἀσώδει, ἀκρητοχόλῳ, ἀγκῶνος δὲ μάλιστα τοὐπίσω διὰ τὸ ναρκῶδες, δεύτερον δὲ τοὔμπροσθεν. ἴησις δὲ ἡ αὐτή· ἐμβολαὶ δέ, τοῦ μὲν ὀπίσω, ἐκτείναντα κατατεῖναι. σημεῖον δέ· οὐ γὰρ δύνανται ἐκτείνειν· τοῦ δὲ ἔμπροσθεν, οὐ δύνανται συγκάμπτειν. τούτῳ δὲ ἐνθέντα τι συνειλιγμένον σκληρόν, περὶ τοῦτο συγκάμψαι
9 ἐξ ἐκτάσιος ἐξαίφνης.

XX. Διαστάσιος δὲ ὀστέων σημεῖον, κατὰ τὴν φλέβα τὴν κατὰ βραχίονα σχιζομένην δια-
3 ψαύοντι.

XXI. Ταῦτα δὲ ταχέως διαπωροῦται· ἐκ γενεῆς δὲ βραχύτερα τὰ κάτω τοῦ σίνεος ὀστέα, πλεῖστον τὰ ἐγγύτατα τοῦ πήχεος· δεύτερον χειρός· τρίτον δακτύλων· βραχίων δὲ καὶ ὦμος,

[1] Cf. *Fract.* XLVIII.

---

[1] "Evidently complete lateral luxation of the forearm," Adams.

[2] Our "external lateral."

[3] Internal lateral, but Adams "forwards or backwards."

in a straight line, but at the same time press on the two bones in opposite directions and bring them round into place. In these cases turning of the elbow sometimes towards supination, sometimes towards pronation will contribute to success. For after treatment, as regards position, keep the hand rather higher than the elbow, and the arm to the side: this applies both to suspension and fixation. The position is easy and natural and serves for ordinary use, if indeed the ankylosis [stiffening of the joint] is not unfavourable; but ankylosis comes on quickly. Treatment with bandages according to what is customary with joints; and include the point of the elbow in the bandaging.[1]

XIX. Elbow injury is very liable to exacerbation with fever, pain, nausea and bilious vomiting, especially the dislocation backwards[2] owing to the numbness [injury of the ulnar nerve], and secondly dislocation forwards.[3] Treatment is the same. Modes of reduction—for backward dislocation, extension and counter-extension: sign—they cannot extend the arm, while in dislocation forward they cannot flex it. In this case, when something rolled up hard has been put in the bend of the elbow, flex the arm suddenly upon it after extension.

XX. Separation of the bones (of the forearm) is recognised by palpation at the point where the blood vessel of the upper arm bifurcates.

XXI. In these cases there is rapid and complete ankylosis, and when it is congenital, the bones below the injury are shortened, those of the forearm nearest the injury most; secondly, those of the hand, third those of the fingers; while the upper arm and shoulder are stronger because they get

ἐγκρατέστερα διὰ τὴν τροφήν· ἡ δὲ ἑτέρη χεὶρ διὰ τὰ ἔργα ἔτι πλείω ἐγκρατεστέρη. μινύθησις δὲ σαρκῶν, εἰ μὲν ἔξω ἐξέπεσεν, ἔσωθεν· εἰ δὲ μή, 8 ἐς τοὐναντίον ᾗ ἐξέπεσεν.

XXII. Ἀγκὼν δὲ ἢν ἔσω ἢ ἔξω ἐκβῇ, κατάτασις μὲν ἐν σχήματι ἐγγωνίῳ τῷ πήχει πρὸς βραχίονα· τὴν μὲν γὰρ μασχάλην ἀναλαβόντα ταινίῃ ἀνακρεμάσαι, ἀγκῶνι δὲ ἄκρῳ ὑποθέντα τι παρὰ τὸ ἄρθρον βάρος, ἐκκρεμάσαι, ἢ χερσὶ καταναγκάζειν· ὑπεραιωρηθέντος δὲ τοῦ ἄρθρου, αἱ παραγωγαὶ τοῖσι θέναρσι ὡς τὰ ἐν χερσίν· ἐπίδεσις ἐν τούτῳ τῷ σχήματι, καὶ ἀνάληψις 9 καὶ θέσις.

XXIII. Τὰ δὲ ὄπισθεν, ἐξαίφνης ἐκτείνοντα διορθοῦν τοῖσι θέναρσι· ἅμα δὲ δεῖ ἐν τῇ διορθώσει καὶ ἐν τοῖσι ἑτέροισιν. ἢν δὲ ἔμπροσθεν ἀμφὶ ὀθόνιον συνειλιγμένον, εὔογκον συγκάμπ- 5 τοντα ἅμα διορθοῦν.

XXIV. Ἢν ἑτεροκλινὲς ᾖ, ἐν τῇ διορθώσει ἀμφότερα ἅμα χρὴ ποιεῖν. τῆς δὲ μελέτης τῆς θεραπείης κοινόν, καὶ τὸ σχῆμα καὶ ἡ ἐπίδεσις. δύναται δὲ καὶ ἐκ τῆς διαστάσιος κοινῇ συμπίπ- 5 τειν ἅπαντα.

XXV. Τῶν δὲ ἐμβολέων, αἱ μὲν ἐξ ὑπεραιωρήσιος ἐμβάλλονται, αἱ δὲ ἐκ κατατάσιος, αἱ δὲ ἐκ περισφάλσιος· αὗται δὲ ἐκ τῶν ὑπερ- 4 βολέων τῶν σχημάτων ᾗ τῇ ᾗ τῇ σὺν τῷ τάχει.

XXVI. Χειρὸς δὲ ἄρθρον ὀλισθάνει ἢ ἔσω ἢ ἔξω, ἔσω δὲ τὰ πλεῖστα. σημεῖα δὲ εὔσημα·

[1] XXII and XXIII are notes partly repeating XVIII and XIX.

more nourishment. The other arm is stronger still because of the work it does. Attenuation of the soft parts is on the inner side if the dislocation is outwards, otherwise on the side opposite to the dislocation.

XXII. When the elbow is dislocated inwards or outwards, extension should be made with the forearm at right angles to the upper arm. Take up and suspend the armpit by a band, and hang a weight from the point of the elbow near the joint, or press it down with the hands. The articular end of the humerus being lifted up, adjustments are made with the palms, as in dislocations of the hand. Bandaging, suspension, and fixation in this attitude.

XXIII. Backward dislocations, sudden extension and adjustment with the palms of the hands; the actions must be combined as in the other cases. If the dislocation is forwards make combined flexion and adjustment round a large rolled bandage.[1]

XXIV. If there is deviation to one side, in the adjustment both movements should be combined. Position and bandaging follow the common rule of treatment. It is also possible to put in all these cases by the common method of double extension.[2]

XXV. Some reductions are brought about by a lifting over, others by extension, others by circumduction; and these are by exaggerations of attitude in one direction or another combined with rapidity.

XXVI. The wrist is dislocated inwards or outwards, but chiefly inwards.[3] The signs are obvious,

[2] Partial lateral dislocations (cf. XVII), probably of radius.
[3] Partial dislocation of wrist, Celsus VIII. 17.

συγκάμπτειν τοὺς δακτύλους οὐ δύνανται· ἢν δὲ ἔξω, μὴ ἐκτείνειν. ἐμβολὴ δέ, ὑπὲρ τραπέζης τοὺς δακτύλους ἔχων, τοὺς μὲν τείνειν, τοὺς δὲ ἀντιτείνειν, τὸ δὲ ἐξέχον ἢ θέναρι ἢ πτέρνῃ ἅμα ἀπωθεῖν καὶ ὠθεῖν πρόσω κάτω, κάτωθεν δὲ κατὰ τὸ ἕτερον ὀστέον, ὄγκον μαλθακὸν ὑποθείς, ἢν μὲν ἄνω, καταστρέψας τὴν χεῖρα, ἢν δὲ κάτω, 10 ὑπτίην. ἴησις δὲ ὀθονίοισιν.

XXVII. Ὅλη δὲ ἡ χεὶρ ὀλισθάνει ἢ ἔσω ἢ ἔξω, ἢ ἔνθα ἢ ἔνθα, μάλιστα δὲ ἔσω· ἔστι δὲ ὅτε καὶ ἡ ἐπίφυσις ἐκινήθη· ἔστι δ' ὅτε τὸ ἕτερον τῶν ὀστέων διέστη. τούτοισι κατάτασις ἰσχυρὴ ποιητέη· καὶ τὸ μὲν ἐξέχον ἀπωθεῖν, τὸ δὲ ἕτερον ἀντωθεῖν, δύο εἴδεα ἅμα καὶ ἐς τοὐπίσω καὶ ἐς τὸ πλάγιον, ἢ χερσὶν ἐπὶ τραπέζης ἢ πτέρνῃ. παλίγκοτα δὲ καὶ ἀσχήμονα· τῷ δὲ χρόνῳ κρατύνεται ἐς χρῆσιν. ἴησις, ὀθονίοισι σὺν τῇ 10 χειρὶ καὶ τῷ πήχει· καὶ νάρθηκας μέχρι δακτύλων τιθέναι· ἐν νάρθηξι δὲ δεθέντα ταῦτα πυκνότερον[1] λύειν ἢ τὰ κατήγματα καὶ καταχύσει 13 πλέονι χρῆσθαι.

XXVIII. Ἐκ γενεῆς δὲ βραχυτέρη ἡ χεὶρ γίνεται καὶ μινύθησις σαρκῶν μάλιστα τἀναντία ᾗ ᾗ τὸ ἔκπτωμα· ηὐξημένῳ δέ, τὰ ὀστέα 4 μένει.

XXIX. Δακτύλου δὲ ἄρθρον, ὀλισθὸν μέν,

[1] πυκνότερα.

---

[1] "In a great measure ideal," Adams. Seems connected with LXIV, but the epitomist may have seen lost chapters.

[2] Complete dislocation of wrist. *Mochl.* XVII; cf. *Fract.* XIII.

if inwards they cannot flex the fingers, if outwards they cannot extend them. Reduction: placing the fingers on a table, assistants should make extension and counter-extension, while the operator with palm or heel presses the projecting part back, with a downward and forward pressure, having put something thick and soft under the other bone. The hand should be prone if the dislocation is upwards and supine if it is downwards. Treatment with bandages.[1]

XXVII. The hand is completely dislocated, inwards, outwards, or to either side, but chiefly inwards, and the epiphysis is sometimes displaced [fracture of lower end of radius], sometimes one of the bones is separated. In these cases one must make strong extension. Press back the projecting part and make counter-pressure on the other side, the two kinds of movement backward and lateral being simultaneous, and performed on a table with the hands or heel. These are serious injuries and cause deformity, but in time the joints get strong enough for use. Treatment with bandages to include the hand and forearm, and apply splints reaching to the fingers. When put up in splints change more frequently than with fractures and use more copious douching.[2]

XXVIII. When the dislocation is congenital the hand becomes relatively shorter, and there is attenuation of the tissues most pronounced on the side opposite the displacement, but in an adult the bones are unaltered.[3]

XXIX. Dislocation of a finger-joint is easily

[3] *Mochl.* XVIII. These obscure accounts of elbow and wrist dislocations are discussed, p. 411.

εὔσημον. ἐμβυλὴ δέ, κατατείναντα ἐς ἰθύ, τὸ μὲν ἐξέχον ἀπωθεῖν, τὸ δὲ ἐναντίον ἀντωθεῖν· ἴησις δέ, ταινίοισιν ὀθονίοισιν. μὴ ἐμπεσὸν δέ, ἐπιπωροῦται ἔξωθεν. ἐκ γενεῆς δὲ ἢ ἐν αὐξήσει ἐξαρθρήσαντα, τὰ ὀστέα βραχύνεται τὰ κάτω τοῦ ὀλισθήματος, καὶ σάρκες μινύθουσι τἀναντία μάλιστα ἢ ὡς[1] τὸ ἔκπτωμα· ηὐξημένῳ δέ, τὰ
9 ὀστέα μένει.

XXX. Γνάθος δὲ ὀλίγοισιν ἤδη τελέως ἐξήρθρησεν· ὀστέον[2] τε γὰρ τὸ ἀπὸ τῆς ἄνω γνάθου πεφυκὸς ὑπεζύγωται πρὸς τῷ ὑπὸ τὸ οὖς ὀστέῳ προσπεφυκότι, ὅπερ ἀποκλείει τὰς κεφαλὰς τῆς κάτω γνάθου, τῆς μὲν ἀνωτέρω ἐόν, τῆς δὲ κατωτέρω τῶν κεφαλέων· τά τε ἄκρεα τῆς κάτω γνάθου, τὸ μὲν διὰ τὸ μῆκος οὐκ εὐπαρείσδυτον,[3] τὸ δὲ αὖ τὸ κορωνόν τε καὶ ὑπερέχον ὑπὲρ τοῦ ζυγώματος· ἅμα τε ἀπ᾽ ἀμφοτέρων τῶν ἄκρων
10 τούτων νευρώδεις τένοντες πεφύκασιν, ἐξ ὧν ἐξήρτηνται οἱ μύες οἱ κροταφῖται καὶ μασητῆρες καλεόμενοι. διὰ τοῦτο δὲ καλέονται καὶ διὰ τοῦτο κινέονται, ὅτι ἐντεῦθεν ἐξήρτηνται· ἐν γὰρ τῇ ἐδωδῇ καὶ ἐν τῇ διαλέκτῳ καὶ ἐν τῇ ἄλλῃ χρήσει τοῦ στόματος, ἡ μὲν ἄνω γνάθος ἀτρεμεῖ· συνήρτηται γὰρ τῇ κεφαλῇ καὶ οὐ διήρθρωται· ἡ δὲ κάτω γνάθος κινεῖται· ἀπήρθρωται γὰρ ὑπὸ τῆς ἄνω γνάθου καὶ ἀπὸ τῆς κεφαλῆς. διότι μὲν οὖν ἐν σπασμοῖσί τε καὶ τετάνοισι πρῶτον
20 τοῦτο τὸ ἄρθρον ἐπισημαίνει συντεταμένον, καὶ διότι πληγαὶ καίριοι καὶ καροῦσαι αἱ κροταφίτιδες γίνονται, ἐν ἄλλῳ λόγῳ εἰρήσεται. περὶ

[1] ᾗ Kw. Mochl. [2] τὸ ὀστέον Erm., K.
[3] εὐπαρέκδυτον Foës in note, Erm., Kw.; εὐπαρείσδυτον MSS.

recognised. Reduction: while extending in a direct line, press back the projecting part, and make counter-pressure on the opposite side. Treatment with tapes and as (narrow bandages). If not reduced, it gets fixed outside. When the dislocation is congenital or during growth, the bones below the laxation are shortened and the tissues waste, especially on the side opposite the displacement; but in an adult the bones are unaltered.

XXX. Complete dislocation of the lower jaw rarely occurs, for the bone which arises from the upper jaw forms a yoke[1] with that which is attached below the ear, and shuts off the heads of the lower jaw, being above the one and below the other. As to these extremities of the lower jaw, one of them is not easily dislocated[2] because of its length, while the other is the coronoid, and projects above the zygoma. And besides, ligamentous tendons arise from both these summits, into which are inserted the muscles called temporals and masseters. They derive their names and functions from being so attached; for in eating, speech, and other uses of the mouth the upper jaw is at rest, being connected with the head directly, not by a joint.[3] But the lower jaw moves, for it is articulated with the upper jaw and the head. Now, the reason why the joint first shows rigidity in spasms and tetanus, and why wounds of the temporal muscles are dangerous and apt to cause coma will be stated in another treatise.[4] The above are the

[1] The "zygoma."

[2] "Accessible," MSS. reading.

[3] Or, "by synarthrosis, not diarthrosis" (Galen). Some read συνήρθρωται.

[4] Pq. thinks this is *Wounds in the head*, but that seems to be the older treatise, and is written in a less finished style: also it hardly gives a full account of the matter.

δὲ τοῦ μὴ κάρτα ἐξαρθρεῖν, τάδε τὰ αἴτια· αἴτιον δὲ καὶ τόδε, ὅτι οὐ μάλα καταλαμβάνουσι τοιαῦται ἀνάγκαι βρωμάτων, ὥστε τὸν ἄνθρωπον χανεῖν μέζον ἢ ὅσον δύναται· ἐκπέσοι δ' ἂν ἀπ' οὐδενὸς ἄλλου σχήματος ἢ ἀπὸ τοῦ μέγα χανόντα παραγαγεῖν τὴν γένυν ἐπὶ θάτερα. προσσυμβάλλεται μέντοι καὶ τόδε πρὸς τὸ ἐκπίπτειν·
30 ὁπόσα γὰρ νεῦρα καὶ ὁπόσοι μύες παρὰ ἄρθρα εἰσίν, ἢ ἀπὸ ἄρθρων ἀφ' ὧν συνδέδενται, τούτων ὅσα ἐν τῇ χρήσει πλειστάκις διακινεῖται, ταῦτα καὶ ἐς τὰς κατατάσιας δυνατώτατα ἐπιδιδόναι, ὥσπερ καὶ τὰ δέρματα τὰ εὐδεψητότατα πλείστην ἐπίδοσιν ἔχει. περὶ οὗ οὖν ὁ λόγος, ἐκπίπτει μὲν γνάθος ὀλιγάκις, σχᾶται μέντοι πολλάκις ἐν χάσμῃσιν, ὥσπερ καὶ ἄλλαι πολλαὶ μυῶν παραλλαγαὶ καὶ νεύρων τοῦτο ποιέουσιν. δῆλον μὲν οὖν ἐκ τῶνδε μάλιστά ἐστιν, ὁπόταν
40 ἐκπεπτώκῃ· προΐσχεται[1] γὰρ ἡ κάτω γνάθος ἐς τοὔμπροσθεν καὶ παρῆκται τἀναντία τοῦ ὀλισθήματος καὶ τοῦ ὀστέου τὸ κορωνὸν ὀγκηρότερον φαίνεται παρὰ τὴν ἄνω γνάθον καὶ χαλεπῶς συμβάλλουσι τὰς [κάτω][2] γνάθους.

Τούτοισι δὲ ἐμβολὴ πρόδηλος, ἥτις γίνοιτ' ἂν ἁρμόζουσα· χρὴ γὰρ τὸν μέν τινα κατέχειν τὴν κεφαλὴν τοῦ τετρωμένου, τὸν δὲ περιλαβόντα τὴν κάτω γνάθον καὶ ἔσωθεν καὶ ἔξωθεν τοῖσι δακτύλοισι κατὰ τὸ γένειον, χάσκοντος τοῦ
50 ἀνθρώπου ὅσον μετρίως δύναται, πρῶτον μὲν διακινεῖν τὴν [κάτω][3] γνάθον χρόνον τινά, τῇ καὶ τῇ παράγοντα τῇ χειρί, καὶ αὐτὸν τὸν ἄνθρωπον κελεύειν χαλαρὴν τὴν γνάθον ἔχειν, καὶ συμπαράγειν καὶ συνδιδόναι ὡς μάλιστα· ἔπειτα ἐξ-

reasons why the dislocation is rare; and one may add this—that the necessities of eating are rarely such as to make a man open his mouth wider than is normally possible, and the dislocation would occur from no other position than that of lateral displacement of the chin while widely gaping. Still, the following circumstance also favours dislocation: among the tendons and muscles which surround joints or arise from them and hold them together, those whose functions involve most frequent movement are most capable of yielding to extension, just as the best tanned skins have the greatest elasticity. To come then to our subject, the jaw is rarely dislocated, but often makes a side-slip[1] in yawning, a thing which changes of position in muscles and tendons also often produce. When dislocation occurs, the following are the most obvious signs: the lower jaw is thrown forward and deviates to the side opposite the dislocation; the coronoid process appears more projecting on the upper jaw, and patients bring the jaws together with difficulty.

The appropriate mode of reduction in these cases is obvious. Someone should hold the patient's head, while the operator grasping the jaw with his fingers inside and out near the chin—the patient keeping it open as wide as he conveniently can—should move the jaw this way and that with his hand, and bid the patient keep it relaxed and assist the movement by yielding to it as far as possible.

[1] *σχᾶται*, a gymnastic term for a sudden lateral movement, Galen (XVIII (1), 438).

---

[1] προΐσχει Kw. [2] Omit Kw.
[3] Omit Galen, Erm., etc.

απίνης σχᾶσαι, τρισὶ σχήμασι ὁμοῦ προσέχοντα τὸν νόον· χρὴ μὲν γὰρ παράγεσθαι ἐκ τῆς διαστροφῆς ἐς τὴν φύσιν, δεῖ δὲ ἐς τοὐπίσω ἀπωσθῆναι τὴν γνάθον τὴν κάτω, δεῖ δὲ ἑπόμενον τούτοισι συμβάλλειν τὰς γνάθους, καὶ μὴ χάσκειν.
60 ἐμβολὴ μὲν οὖν αὕτη, καὶ οὐκ ἂν γένοιτο ἀπ' ἄλλων σχημάτων. ἰητρείη δὲ βραχείη ἀρκέσει·[1] σπλῆνα προστιθέντα κεκηρωμένον χαλαρῷ ἐπιδέσμῳ ἐπιδεῖν. ἀσφαλέστερον δὲ χειρίζειν ἐστὶν ὕπτιον κατακλίναντα τὸν ἄνθρωπον, ἐρείσαντα τὴν κεφαλὴν αὐτοῦ ἐπὶ σκυτίνου ὑποκεφαλαίου ὡς πληρεστάτου, ἵνα ὡς ἥκιστα ὑπείκῃ· προσκατ-
67 έχειν δέ τινα χρὴ τὴν κεφαλὴν τοῦ τετρωμένου.

XXXI. Ἢν δὲ ἀμφότεραι αἱ γνάθοι ἐξαρθρήσωσιν, ἡ μὲν ἴησις ἡ αὐτή. συμβάλλειν δέ τι[2] ἧσσον οὗτοι τὸ στόμα δύνανται· καὶ γὰρ προπετέστεραι αἱ γένυες τούτοισι, ἀστραβέες δέ. τὸ δὲ ἀστραβὲς μάλιστ' ἂν γνοίης τοῖσιν ὁρίοισι τῶν ὀδόντων τῶν τε ἄνω καὶ τῶν κάτω κατ' ἴξιν. τούτοισι συμφέρει ὡς τάχιστα ἐμβάλλειν· ἐμβολῆς δὲ τρόπος πρόσθεν εἴρηται. ἢν δὲ μὴ ἐμπέσῃ, κίνδυνος περὶ τῆς ψυχῆς ὑπὸ πυρετῶν συνεχέων
10 καὶ νωθρῆς καρώσιος—καρώδεες γὰρ οἱ μύες οὗτοι, καὶ ἀλλοιούμενοι καὶ ἐντεινόμενοι παρὰ φύσιν—φιλεῖ δὲ καὶ ἡ γαστὴρ ὑποχωρεῖν τούτοισι χολώδεα ἄκρητα ὀλίγα· καὶ ἢν ἐμέωσιν, ἄκρητα ἐμέουσιν· οὗτοι οὖν καὶ θνήσκουσι
15 δεκαταῖοι μάλιστα.

XXXII. Ἢν δὲ κατεαγῇ ἡ κάτω γνάθος, ἢν μὲν μὴ ἀποκαυλισθῇ παντάπασιν, ἀλλὰ συνέχηται τὸ ὀστέον, ἐγκεκλιμένον δὲ ᾖ, κατορθῶσαι μὲν χρὴ τὸ ὀστέον, παρά γε τὴν γλῶσσαν

Then suddenly do a side-slip, having in mind three positions in the manœuvre. For the deviation must be reduced to the natural direction, the jaw must be pressed backwards, and, following this, the patient must close his jaws and not gape. This, then, is the reduction, and it will not succeed with other manœuvres. A short treatment will suffice. Apply a compress with cerate and a loose bandage over it. The safest way of operating is with the patient recumbent, his head being supported on a well-stuffed leather pillow, that it may yield as little as possible; and someone should also keep the patient's head fixed.

XXXI. If both lower jaws are dislocated [*i.e* both sides of the lower jaw], the treatment is the same. These patients are rather less able to close the mouth, for the chin is more projecting, though without deviation. You will best recognize the absence of deviation by the vertical correspondence of the upper and lower rows of teeth. It is well to reduce these cases as quickly as possible; and the mode of reduction is described above. If not reduced there is risk of death from acute fever and deep coma—for these muscles when displaced or abnormally stretched produce coma—and there are small evacuations of pure bile; if there is vomiting, it is also unmixed. These patients, then, die about the tenth day.

XXXII. In fracture of the lower jaw, if it is not entirely broken across, but the bone preserves its continuity though distorted, one should adjust the bone by making suitable lateral pressure with the

[1] ἀρκεῖ. [2] δ' ἔτι.

πλαγίην ὑπείραντα τοὺς δακτύλους, τὸ δὲ ἔξωθεν ἀντερείδοντα, ὡς ἂν συμφέρῃ· καὶ ἢν μὲν διεστραμμένοι ἔωσιν οἱ ὀδόντες οἱ κατὰ τὸ τρῶμα καὶ κεκινημένοι, ὁπόταν[1] τὸ ὀστέον κατορθωθῇ, ζεῦξαι τοὺς ὀδόντας χρὴ πρὸς ἀλλήλους, μὴ
10 μοῦνον τοὺς δύο, ἀλλὰ καὶ πλέονας,[2] μάλιστα μὲν δὴ χρυσίῳ, ἔστ' ἂν κρατυνθῇ τὸ ὀστέον, εἰ δὲ μή, λίνῳ· ἔπειτα ἐπιδεῖν κηρωτῇ καὶ σπλήνεσιν ὀλίγοισι καὶ ὀθονίοισιν ὀλίγοισι, μὴ ἄγαν ἐρείδοντα, ἀλλὰ χαλαροῖσιν. εὖ γὰρ εἰδέναι χρή, ὅτι ἐπίδεσις ὀθονίων γνάθῳ κατεαγείσῃ[3] σμικρὰ μὲν ἂν ὠφελέοι, εἰ χρηστῶς ἐπιδέοιτο, μεγάλα δ' ἂν βλάπτοι, εἰ κακῶς ἐπιδέοιτο. πυκινὰ δὲ παρὰ τὴν γλῶσσαν ἐσματεῖσθαι χρή, καὶ πολὺν χρόνον ἀντέχειν τοῖσι δακτύλοισι
20 κατορθοῦντα τοῦ ὀστέου τὸ ἐκκλιθέν·[4] ἄριστον
21 δέ. εἰ αἰεὶ δύναιτο· ἀλλ' οὐχ οἷόν τε.

XXXIII. Ἢν δὲ ἀποκαυλισθῇ παντάπασιν τὸ ὀστέον—ὀλιγάκις δὲ τοῦτο γίνεται—κατορθοῦν μὲν χρὴ τὸ ὀστέον οὕτω, καθάπερ εἴρηται. ὅταν δὲ κατορθώσῃς, τοὺς ὀδόντας χρὴ ζευγνύναι, ὡς πρόσθεν εἴρηται· μέγα γὰρ ἂν συλλαμβάνοι ἐς τὴν ἀτρεμίην,[5] προσέτι καὶ εἴ τις ὀρθῶς ζεύξει ὥσπερ χρή, τὰς ἀρχὰς ῥάψας. ἀλλὰ γὰρ οὐ ῥηΐδιον ἐν γραφῇ χειρουργίην πᾶσαν διηγεῖσθαι, ἀλλὰ καὶ αὐτὸν ὑποτοπεῖσθαι[6] χρὴ ἐκ τῶν
10 γεγραμμένων. ἔπειτα χρὴ δέρματος Καρχηδονίου· ἢν μὲν νηπιώτερος[7] ᾖ ὁ τρωθείς, ἀρκεῖ τῷ λοπῷ χρῆσθαι, ἢν δὲ τελειότερος ᾖ, αὐτῷ τῷ δέρματι· ταμόντα δὲ χρὴ εὖρος ὡς τριδάκτυλον, ἢ ὅπως ἂν ἁρμόζῃ, ὑπαλείψαντα

[1] ὅταν.
[2] ἐπὶ πλείονας.
[3] γνάθου κατεαγείσης.

fingers on the tongue side, and counter-pressure from without. If the teeth at the point of injury are displaced or loosened, when the bone is adjusted fasten them to one another, not merely the two, but several, preferably with the gold wire, but failing that, with thread, till consolidation takes place. Afterwards dress with cerate and a few compresses and bandages, also few, and with no great pressure, but lax. For one should bear in mind that bandaging a fractured jaw will do little good when well done, but will do great harm when it is done badly. One should make frequent palpation on the tongue side, and hold the distorted part of the bone adjusted with the fingers for a long time. It would be best if one could do so throughout; but that is impossible.

XXXIII. If the jaw is broken right across, which rarely happens, one should adjust it in the manner described. After adjustment you should fasten the teeth together as was described above, for this will contribute greatly to immobility, especially if one joins them up properly and fastens off the ends as they should be. For the rest, it is not easy to give exact and complete details of an operation in writing; but the reader should form an outline of it from the description. Next, one should take Carthaginian leather; if the patient is more of a child, the outer layer is sufficient, but if he is more adult, use the skin itself. Cut a three-finger breadth, or as much as may be suitable, and, anointing the jaw with

[4] ἐγκλιθέν. [5] ἐς τὸ ἀτρεμεῖν.
[6] ὑποτυπεῖσθαι MSS.: ὑποτοπεῖσθαι Erot., Littré.
[7] νεώτερος.

κόμμι τὴν γνάθον—εὐμενέστερον γάρ κόλλης—[1] προσκολλῆσαι τὴν δέῤῥιν ἄκρον πρὸς τὸ ἀποκεκαυλισμένον τῆς γνάθου, ἀπολείποντα ὡς δάκτυλον ἀπὸ τοῦ τρώματος ἢ ὀλίγῳ πλέον. τοῦτο μὲν ἐς τὸ κάτω μέρος· ἐχέτω δὲ ἐντομὴν
20 κατὰ τὴν ἴξιν τοῦ γενείου ὁ ἱμάς, ὡς ἀμφιβεβήκῃ ἀμφὶ τὸ ὀξὺ τοῦ γενείου. ἕτερον δὲ ἱμάντα τοιοῦτον, ἢ ὀλίγῳ πλατύτερον, προσκολλῆσαι χρὴ πρὸς τὸ ἄνω μέρος τῆς γνάθου, ἀπολείποντα καὶ τοῦτον ἀπὸ τοῦ τρώματος, ὅσονπερ ὁ ἕτερος ἀπέλιπεν· ἐσχίσθω δὲ καὶ οὗτος ὁ ἱμὰς τὴν ἀμφὶ τὸ οὖς περίβασιν. ἀποξέες δὲ ἔστωσαν οἱ ἱμάντες ἀμφὶ τὴν συναφήν· [ἔνθα συνάπτεσθαί τε καὶ συνδεῖσθαι ἐς τὰ πέρατα τῶν ἱμάντων·][2] ἐν δὲ τῇ κολλήσει ἡ σὰρξ τοῦ σκύτεος πρὸς τοῦ
30 χρωτὸς ἔστω, ἐχεκολλότερον γὰρ οὕτως. ἔπειτα κατατείναντα χρὴ καὶ τοῦτον τὸν ἱμάντα, μᾶλλον δέ τι τὸν περὶ τὸ γένειον, ὡς ὅτι μάλιστα μὴ ἀπομυλλαίνῃ[3] ἡ γνάθος, συνάψαι τοὺς ἱμάντας κατὰ τὴν κορυφήν· κἄπειτα περὶ τὸ μέτωπον ὀθονίῳ καταδῆσαι, καὶ κατάβλημα χρὴ εἶναι, ὥσπερ νομίζεται, ὡς ἀτρεμέῃ τὰ δεσμά. τὴν δὲ κατάκλισιν ποιείσθω ἐπὶ τὴν ὑγιέα γνάθον, μὴ τῇ γνάθῳ ἐρηρεισμένος, ἀλλὰ τῇ κεφαλῇ. ἰσχναίνειν δὲ χρὴ τὸ σῶμα ἄχρις ἡμέρων δέκα, ἔπειτα
40 ἀνατρέφειν μὴ βραδέως· ἢν δὲ ἐν τῇσι προτέρῃσι ἡμέρῃσι μὴ φλεγμήνῃ, ἐν εἴκοσιν ἡμέρῃσιν ἡ γνάθος κρατύνεται· ταχέως γὰρ ἐπιπωροῦται, ὥσπερ καὶ τὰ ἄλλα τὰ ἀραιὰ ὀστέα, ἢν μὴ ἐπισφακελίσῃ. ἀλλὰ γαρ περὶ σφακελισμῶν τῶν συμπάντων ὀστέων ἄλλος μακρὸς λόγος

[1] εὐμενέστερον γὰρ κόλλης B. ; κόλλῃ M.V.

gum—for it is more agreeable than glue—fasten the end of the leather to the broken-off part of the jaw at a finger's breadth or rather more from the fracture. This is for the lower part; and let the strap have a slit in the line of the chin, so as to include the chin point. Another strap, similar or a little broader, should be gummed to the upper part of the jaw at the same interval from the fracture as the former one; and let it also be split for going round the ear. Let the straps taper off at their junction, where the ends meet and are tied together. In the gumming, let the fleshy side of the leather be towards the skin; for so it adheres more firmly. One should then make traction on the thong, but rather more on the one that goes round the chin, to avoid so far as possible any distortion [1] of the jaw. Fasten the straps together at the top of the head, and afterwards pass a bandage round the forehead; and there should be the usual outer covering to keep the bands steady. The patient should lie on the side of the sound jaw, the pressure being not on the jaw, but on the head. Keep him on low diet for ten days, and afterwards feed him up without delay; for if there is no inflammation in the first period, the jaw consolidates in twenty days, since callus forms quickly as in other porous bones, unless necrosis supervenes. Now, necrosis of bones generally remains to be treated at length elsewhere.

[1] Erotian *s.v.*: probably "snout-like distortion." "In acutum" (Foës).

---

[2] Omit Kw. and most MSS.

[3] ἀποσμιλαίνει Galen ("draw to a point"); ἀπομυλλήνῃ Erot. ("be distorted").

λείπεται.[1] αὕτη ἡ διάτασις ἡ ἀπὸ τῶν κολλημάτων εὐμενὴς καὶ εὐταμίευτος, καὶ ἐς πολλὰ καὶ πολλαχοῦ διορθώματα εὔχρηστος. τῶν δὲ ἰητρῶν οἱ μὴ σὺν νόῳ εὔχειρες καὶ ἐν ἄλλοισι
50 τρώμασι τοιοῦτοί εἰσι καὶ ἐν γνάθων καθήξεσιν· ἐπιδέουσι γὰρ γνάθον κατεαγεῖσαν ποικίλως, καὶ καλῶς καὶ κακῶς· πᾶσα γὰρ ἐπίδεσις γνάθου οὕτως κατεαγείσης ἐκκλίνει[2] τὰ ὀστέα τὰ ἐς τὸ
54 κάτηγμα ῥέποντα μᾶλλον ἢ ἐς τὴν φύσιν ἄγει.

XXXIV. Ἢν δὲ ἡ κάτω γνάθος κατὰ τὴν σύμφυσιν τὴν κατὰ τὸ γένειον διασπασθῇ—μούνη δὲ αὕτη ἡ σύμφυσις ἐν τῇ κάτω γνάθῳ ἐστίν, ἐν δὲ τῇ ἄνω πολλαί· ἀλλ' οὐ βούλομαι ἀποπλανᾶν τοῦ λόγου, ἐν ἄλλοισι γὰρ εἴδεσι νοσημάτων περὶ τούτων λεκτέον—ἢν οὖν διαστῇ ἡ κατὰ τὸ γένειον σύμφυσις, κατορθῶσαι μὲν παντὸς ἀνδρός ἐστιν. τὸ μὲν γὰρ ἐξεστεὸς ἐσωθεῖν χρὴ ἐς τὸ ἔσω μέρος, προσβαλόντα τοὺς
10 δακτύλους, τὸ δ' ἔσω ῥέπον ἀνάγειν ἐς τὸ ἔξω μέρος, ἐνερείσαντα τοὺς δακτύλους. ἐς διάστασιν μέντοι διατεινάμενον ταῦτα χρὴ ποιεῖν· ῥᾷον γὰρ οὕτως ἐς τὴν φύσιν ἥξει ἢ εἴ τις ἐγχρίμπτοντα ἐς ἄλληλα τὰ ὀστέα παραναγκάζειν πειρᾶται· τοῦτο παρὰ πάντα τὰ τοιαῦτα [ὑπομνήματα][3] χαρίεν εἰδέναι. ὁπόταν δὲ κατορθώσῃς, ζεῦξαι μὲν χρὴ τοὺς ὀδόντας τοὺς ἔνθεν καὶ ἔνθεν πρὸς ἀλλήλους, ὥσπερ καὶ πρόσθεν εἴρηται. ἰῆσθαι

[1] Cf. LXIX. [2] ἐγκλίνει B Kw.
[3] κατήγματα Littré. Erm. omits the whole sentence.

This mode of extension by straps gummed on is convenient, easy to manage, and very useful for a variety of adjustments. Practitioners who have manual skill without intelligence show themselves such in fractures of the jaw above all other injuries. They bandage a fractured jaw in a variety of ways, sometimes well, sometimes badly; but any bandaging of a jaw fractured in this way tends to turn the fragments inwards[1] at the lesion rather than bring them to their natural position.

XXXIV. When the lower jaw is torn apart at the symphysis which is at the chin[2]—this is the only symphysis in the lower jaw, while in the upper there are many, but I do not want to digress, for one must discuss these matters in relation to other maladies. When, therefore, the symphysis at the chin is separated, anyone can make the adjustment. For one should thrust the projecting part inwards, making pressure with the fingers, and force out that which inclines inwards, using the fingers for counter-pressure. This, however, must be done while the parts are separated by tension; for they will thus be reduced more easily than if one tries to force the bones into position while they override one another (this is a thing it is well to bear in mind in all such cases[3]). After adjustment, you should join up the teeth on either side as described above. Treat with

[1] Kw.'s reading; Adams prudently has "derange."

[2] The idea that the lower jaw consists of two bones with a symphysis at the chin is corrected in Celsus VIII 1, but repeated by Galen (perhaps out of respect for Hippocrates), though he admits that it is hard to demonstrate.

[3] Perhaps an insertion, but read by Galen.

δὲ χρὴ κηρωτῇ καὶ σπλήνεσιν ὀλίγοισι καὶ
20 ὀθονίοισιν. ἐπίδεσιν δὲ βραχείην ἢ[1] ποικίλην
μάλιστα τοῦτο τὸ χωρίον ἐπιδέχεται, ἐγγὺς γάρ
τι τοῦ ἰσοῤῥόπου ἐστίν, ὡς δὴ μὴ ἰσόῤῥοπον ἐόν.
τοῦ δὲ ὀθονίου τὴν περιβολὴν ποιεῖσθαι χρή,
ἢν μὲν ἡ δεξιὴ γνάθος ἐξεστήκῃ, ἐπὶ δεξιά (ἐπὶ
δεξιὰ γὰρ νομίζεται εἶναι, ἢν ἡ δεξιὴ χεὶρ προ-
ηγῆται τῆς ἐπιδέσιος)· ἢν δὲ ἡ ἑτέρη γνάθος
ἐξεστήκῃ, ὡς ἑτέρως χρὴ τὴν ἐπίδεσιν ἄγειν. κἢν
μὲν ὀρθῶς τις κατορθώσηται καὶ ἐπατρεμήσῃ
ὡς χρή, ταχείη μὲν ἡ ἄλθεξις, οἱ δὲ ὀδόντες
30 ἀσινέες γίνονται· ἢν δὲ μή, χρονιωτέρη ἡ ἄλθεξις,
διαστροφὴν δὲ ἴσχουσιν οἱ ὀδόντες, καὶ σιναροὶ
32 καὶ ἀχρεῖοι γίνονται.

XXXV. Ἢν δὲ ἡ ῥὶς κατεαγῇ, τρόπος μὲν οὐχ
εἷς ἐστὶ κατήξιος· ἀτὰρ πολλὰ μὲν δὴ καὶ ἄλλα
λωβέονται οἱ χαίροντες τῇσι καλῇσιν ἐπιδέσεσιν
ἄνευ νόου, ἐν δὲ τοῖσι περὶ τὴν ῥῖνα μάλιστα·
ἐπιδεσίων γάρ ἐστιν αὕτη ποικιλωτάτη καὶ
πλείστους μὲν σκεπάρνους ἔχουσα, διαῤῥωγὰς
δὲ καὶ διαλείψιας ποικιλωτάτας τοῦ χρωτὸς
ῥομβοειδέας. ὡς οὖν εἴρηται, οἱ τὴν ἀνόητον
εὐχειρίην ἐπιτηδεύοντες ἄσμενοι ῥινὸς κατεηγυίης
10 ἐπιτυγχάνουσι, ὡς ἐπιδήσωσιν. μίην μὲν οὖν
ἡμέραν ἢ δύο ἀγάλλεται μὲν ὁ ἰητρός, χαίρει δὲ
ὁ ἐπιδεδεμένος· ἔπειτα ταχέως μὲν ὁ ἐπιδεδεμένος
κορίσκεται, ἀσηρὸν γὰρ τὸ φόρημα· ἀρκεῖ δὲ τῷ
ἰητρῷ, ἐπειδὴ ἐπέδειξεν ὅτι ἐπίσταται ποικίλως
ῥῖνα ἐπιδεῖν. ποιεῖ δὲ ἡ ἐπίδεσις ἡ τοιαύτη

[1] "Rather than"; cf. *Surg.* XIV, Luke 17. 2. "Simple rather than complex"; but cf. Galen, who says that the

cerate and a few pads and bandages. A simple dressing rather than a complicated one is specially suited to this part, for it is nearly cylindrical[1] without actually being so. The bandage should be carried round to the right if the right jaw sticks out (it is said to be "to the right" if the right hand precedes in bandaging[2]): while if the other jaw projects, make the bandaging the other way. If the bandaging is well done and the patient keeps at rest, as he should, recovery is rapid, and the teeth are not damaged; if not, recovery is slow, and the teeth remain distorted and become damaged and useless.

XXXV. If the nose is broken, which happens in more than one way, those who delight in fine bandaging without judgment do more damage than usual. For this is the most varied of bandagings, having the most adze-like turns and diverse rhomboid intervals and vacancies.[3] Now, as I said, those who devote themselves to a foolish parade of manual skill are especially delighted to find a fractured nose to bandage. The result is that the practitioner rejoices, and the patient is pleased for one or two days; afterwards the patient soon has enough of it, for the burden is tiresome; and as for the practitioner, he is satisfied with showing that he knows how to apply complicated nasal bandages. But such bandaging

[1] *ἰσόῤῥοπος*= "cylindrical" (Galen). "Semicircular" is perhaps clearer.

[2] *I.e.* to the surgeon's right, but from right to left of the patient's jaw (Galen).

[3] *διαλάμψιας* (Kw., Apollon.).

---

lower jaw is the part on which students exercised their skill in complex forms of bandaging. (XVIII. (1) 462).

πάντα τἀναντία τοῦ δέοντος· τοῦτο μὲν γάρ, ὁπόσοι σιμοῦνται διὰ τὴν κάτηξιν, δηλονότι εἰ ἄνωθέν τις μᾶλλον πιέζοι, σιμώτεροι ἂν ἔτι εἶεν· τοῦτο δέ, ὅσοισι παραστρέφεται ἢ ἔνθα ἢ ἔνθα
20 ἡ ῥίς, ἢ κατὰ τὸν χόνδρον ἢ ἀνωτέρω, δηλονότι οὐδὲν αὐτοὺς ἡ ἄνωθεν ἐπίδεσις ὠφελήσειεν,[1] ἀλλὰ καὶ βλάψειε[2] μᾶλλον· οὐχ οὕτω γὰρ εὖ συναρμόσει σπλήνεσι τὸ ἐπὶ θάτερον τῆς ῥινός·
24 καίτοι οὐδὲ τοῦτο ποιέουσιν οἱ ἐπιδέοντες.

XXXVI. Ἄγχιστα δὲ ἡ ἐπίδεσίς μοι δοκεῖ ἄν τι ποιεῖν, εἰ κατὰ μέσην τὴν ῥῖνα κατὰ τὸ ὀξὺ ἀμφιφλασθείη ἡ σὰρξ κατὰ τὸ ὀστέον, ἢ εἰ κατὰ τὸ ὀστέον σμικρόν τι σίνος εἴη,[3] καὶ μὴ μέγα· τοῖσι γὰρ τοιούτοισιν ἐπιπώρωμα ἴσχει ἡ ῥίς, καὶ ὀκριοειδεστέρη τινὶ γίνεται· ἀλλ᾽ ὅμως οὐδὲ τούτοισι δή που πολλοῦ ὄχλου δεῖται ἡ ἐπίδεσις, εἰ δή τι καὶ δεῖ ἐπιδεῖν. ἀρκεῖ δὲ ἐπὶ μὲν τὸ φλάσμα σπληνίον ἐπιτείναντα κεκηρω-
10 μένον, ἔπειτα ὡς ἀπὸ δύο ἀρχέων ἐπιδεῖται, οὕτως ὀθονίῳ ἐς ἅπαξ περιβάλλειν. ἀρίστη μέντοι ἰητρείη τῷ ἀλήτῳ, τῷ σητανίῳ, τῷ πλυτῷ, γλίσχρῳ, πεφυρμένῳ, ὀλίγῳ, καταπλάσσειν τὰ τοιαῦτα· χρὴ δέ, ἢν μὲν ἐξ ἀγαθῶν ᾖ τῶν πυρῶν τὸ ἄλητον καὶ εὐόλκιμον, τούτῳ χρῆσθαι ἐς πάντα τὰ τοιαῦτα· ἢν δὲ μὴ πάνυ ὅλκιμον ᾖ, ἐς ὀλίγην μάννην ὕδατι ὡς λειοτάτην διέντα. τούτῳ φυρᾶν τὸ ἄλητον, ἢ κόμμι πάνυ ὀλίγον ὡσαύτως
19 μίσγειν.

XXXVII. Ὁπόσοισι μὲν οὖν ῥὶς ἐς τὸ κάτω

[1] ὠφελήσει. [2] βλάψει. [3] ἔχοι.

acts in every way contrary to what is proper; for first, in cases where the nose is rendered concave by the fracture, if more pressure is applied from above, it will obviously be more concave, and again in cases where the nose is distorted to either side, whether in the cartilaginous part or higher up, bandaging will obviously be useless in either case, and will rather do harm; for so one will not arrange the pads well on the other side of the nose, and in fact those who put on bandages omit this.

XXXVI. Bandaging seems to me to be most directly[1] useful where the soft parts are contused against the bone in the middle of the nose at the ridge, or when, without great damage, there is some small injury at the bone; for in such cases the nose gets a superficial callus and a certain jagged outline. But not even in these cases is there need of very troublesome bandaging, even if it is required at all. It suffices to stretch a small compress soaked in cerate over the contusion and then take one turn of bandage round it, as from a two-headed roller. After all, the best treatment is to use a little fresh flour, worked and kneaded into a glutinous mass, as a plaster for such lesions. If one has wheat flour[2] of good quality forming a ductile paste, one should use it in all such cases; but if it is not very ductile, soak a little frankincense powdered as finely as possible in water, and knead the flour with this, or mix a very little gum in the same way.[3]

XXXVII. In cases where the nose is fractured with

[1] ἄγχιστα = μάλιστα (Erotian).

[2] σητάνιος may be either summer wheat or a special kind rich in gluten (Galen).

[3] μάννα = powder of frankincense (Dioscorides 1.68).

καὶ ἐς τὸ σιμὸν ῥέπουσα καταγῇ, ἢν μὲν ἐκ τοῦ ἔμπροσθεν μέρεος κατὰ τὸν χόνδρον ἵζηται, οἷόν τέ ἐστι καὶ ἐντιθέναι τι διόρθωμα ἐς τοὺς μυκτῆρας· ἢν δὲ μή, ἀνορθοῦν μὲν χρὴ πάντα τὰ τοιαῦτα, τοὺς δακτύλους ἐς τοὺς μυκτῆρας ἐντιθέντα, ἢν ἐνδέχηται, ἢν δὲ μή, πάγχυ ὑπάλειπτρον, μὴ ἐς τὸ ἔμπροσθεν τῆς ῥινὸς ἀνάγοντα τοῖσι δακτύλοισι, ἀλλ' ᾗ ἵδρυται· ἔξωθεν δὲ τῆς
10 ῥινὸς ἔνθεν καὶ ἔνθεν ἀμφιλαμβάνοντα τοῖσι δακτύλοισι, συναναγκάζειν τε ἅμα καὶ ἀναφέρειν ἐς τὸ ἄνω. καὶ ἢν μὲν πάνυ ἐν τῷ ἔμπροσθεν τὸ κάτηγμα ᾖ,[1] οἷόν τέ τι καὶ ἔσω τῶν μυκτήρων ἐντιθέναι, ὥσπερ ἤδη εἴρηται, ἢ ἄχνην τὴν ἀφ' ἡμιτυβίου ἢ ἄλλο τι τοιοῦτον, ἐν ὀθονίῳ εἱλίσσοντα, μᾶλλον δὲ ἐν Καρχηδονίῳ δέρματι ἐρράψαντα· σχηματίσαντα τὸ ἅρμοσσον σχῆμα τῷ χωρίῳ, ἵνα ἐγκείσεται. ἢν μέντοι προσωτέρω ᾖ τὸ κάτηγμα, οὐδὲν οἷόν τε ἔσω ἐντιθέναι· καὶ
20 γὰρ εἰ ἐν τῷ ἔμπροσθεν ἀσηρὸν τὸ φόρημα, πῶς γε δὴ οὐκ ἐν τῷ ἐσωτέρω; τὸ μὲν οὖν πρῶτον καὶ ἔξωθεν ἀναπλάσασθαι καὶ ἔσωθεν ἀφειδήσαντα χρὴ ἀναγαγεῖν ἐς τὴν ἀρχαίην φύσιν καὶ διορθώσασθαι. κάρτα γὰρ οἵη τε ῥὶς καταγεῖσα ἀναπλάσσεσθαι, μάλιστα μὲν αὐθημερόν,[2] ἢν δὲ μή, ὀλίγῳ ὕστερον· ἀλλὰ καταβλακεύουσιν οἱ ἰητροί, καὶ ἁπαλωτέρως τὸ πρῶτον ἅπτονται ἢ ὡς χρή· παραβάλλοντα γὰρ τοὺς δακτύλους χρὴ ἔνθεν καὶ ἔνθεν κατὰ τὴν φύσιν τῆς ῥινὸς
30 ὡς κατωτάτω, κάτωθεν συναναγκάζειν, καὶ οὕτω μάλιστα ἀνορθοῦσθαι[3] σὺν τῇ ἔσωθεν διορθώσει

[1] εἰ . . . εἴη. [2] αὐθήμερος. [3] ἀνορθοῦντα Kw.

depression and tends to become snub, if the depression is in the front part of the cartilage, it is possible to insert some rectifying support into the nostrils. Failing this, one should elevate all such cases, if possible by inserting the finger into the nostrils, but if not, a thick spatula should be inserted, directing it with the fingers, not to the front of the nose, but to the depressed part: then getting a grip on each side of the nose outside with the fingers, combine the two movements of compression and lifting. If the fracture is quite in front, it is possible, as was said, to insert something into the nostrils, either lint from linen or something of the kind, rolling it up in a rag, or better, sewing it up in Carthaginian leather, adapting its shape to fit the part where it will lie. But if the fracture be further in, nothing can be inserted; for if it is irksome to endure anything in front, how should it not be more so further in? The first thing, then, is to reshape it from outside, and internally to spare no pains in adjusting it and bringing it to its natural position; for it is quite possible for a broken nose to be reshaped, especially on the day of the accident, or, failing that, a little later. But practitioners act feebly, and treat it at first more gently than they should. For one ought to insert[1] the fingers on each side as far as the conformation of the nose allows, and then force it up from below, thus best combining elevation with the rectification from within. Further, no practi-

[1] Editors discuss the obscurity of this passage at great length. The main point is whether the fingers are inserted or applied to the outside of the nose. I follow Ermerins and Petrequin as against Littré-Adams: though there is much to be said on both sides.

[διορθοῦντα].[1] ἔπειτα δὲ ἐς ταῦτα ἰητρὸς οὐδεὶς ἄλλος ἐστὶ τοιοῦτος, εἰ ἐθέλοι καὶ μελετᾶν καὶ τολμᾶν, ὡς οἱ δάκτυλοι αὐτοῦ οἱ λιχανοί· οὗτοι γὰρ κατὰ φύσιν μάλιστά εἰσιν. παραβάλλοντα γὰρ χρὴ τῶν δακτύλων ἑκάτερον, παρὰ πᾶσαν τὴν ῥῖνα ἐρείδοντα, ἡσύχως οὕτως ἔχειν, μάλιστα μέν, εἰ οἷόν τε εἴη, αἰεί, ἔστ' ἂν κρατυνθῇ· εἰ δὲ μή, ὡς πλεῖστον χρόνον, αὐτόν, ὡς εἴρηται· εἰ
40 δὲ μή, ἢ παῖδα ἢ γυναῖκά τινα· μαλθακὰς γὰρ τὰς χεῖρας δεῖ εἶναι· οὕτω γὰρ ἂν κάλλιστα ἰητρευθείη ὁτέῳ ἡ ῥὶς μὴ ἐς τὸ σκολιόν, ἀλλ' ἐς τὸ κάτω ἱδρυμένη, ἰσόῤῥοπος εἴη. ἐγὼ μὲν οὖν οὐδεμίην που ῥῖνα εἶδον ἥτις οὕτω καταγεῖσα οὐχ οἵη τε διορθωθῆναι αὐτίκα πρὶν πωρωθῆναι συναναγκαζομένη ἐγένετο, εἴ τις ὀρθῶς ἐθέλοι ἰητρεύειν· ἀλλὰ γὰρ οἱ ἄνθρωποι αἰσχροὶ μὲν εἶναι πολλοῦ ἀποτιμῶσι, μελετᾶν δὲ ἅμα μὲν οὐκ ἐπίστανται, ἅμα δὲ οὐ τολμῶσιν, ἢν μὴ ὀδυνῶν-
50 ται, ἢ θάνατον δεδοίκωσιν· καίτοι ὀλιγοχρόνιος ἡ πώρωσις τῆς ῥινός· ἐν γὰρ δέκα ἡμέρῃσι
52 κρατύνεται, ἢν μὴ ἐπισφακελίσῃ.

XXXVIII. Ὁπόσοισι δὲ τὸ ὀστέον ἐς τὸ πλάγιον κατάγνυται, ἡ μὲν ἴησις ἡ αὐτή· τὴν δὲ διόρθωσιν δηλονότι χρὴ ποιεῖσθαι οὐκ ἰσόῤῥοπον ἀμφοτέρωθεν, ἀλλὰ τό τε ἐκκεκλιμένον[2] ὠθεῖν ἐς τὴν φύσιν, ἔκτοσθεν ἀναγκάζοντα καὶ ἐσματευόμενον ἐς τοὺς μυκτῆρας, καὶ τὰ ἔσω ῥέψαντα διορθοῦν ἀόκνως, ἔστ' ἂν κατορθώσῃς, εὖ εἰδότα ὅτι, ἢν μὴ αὐτίκα κατορθώσηται, οὐχ οἷόν τε μὴ οὐχὶ διεστράφθαι τὴν ῥῖνα. ὅταν δὲ ἀγάγῃς ἐς

[1] Galen. Omit most MSS., Littré, etc.

tioner is so suitable for the job as are the index fingers of the patient himself, if he is willing to be careful and courageous, for these fingers are especially conformable to the nose. He should insert the fingers alternately,[1] making pressure along the whole course of the nose, and keeping it steady; especially let him continue it, if he can, till consolidation occurs, failing that, as long as possible. As was said, he should do it himself; but if not, a boy or woman must do it, for the hands should be soft. This is the best treatment when the nose is not distorted laterally, but keeps evenly balanced though depressed. Now, I never saw a nose fractured in this way which could not be adjusted by immediate forcible manipulation before consolidation set in, if one chose to treat it properly. But while men will give much to avoid being ugly, they do not know how to combine care with endurance, unless they suffer pain or fear death. Yet the formation of callus in the nose takes little time, for it is consolidated in ten days, unless necrosis supervenes.

XXXVIII. In cases where the bone is fractured with deviation, the treatment is the same. Adjustment should obviously not be made evenly on both sides, but press the bent-out part into its natural position by force from without, and, introducing the finger into the nostrils, boldly rectify the internal deviation till you get it straight, bearing in mind that, if it is not straightened at once, the nose will infallibly be distorted. And when you bring it to

[1] This seems the surgical implication of ἑκάτερον. Cf. *Surg.* X.

---

[2] ἐγκεκλιμένον.

10 τὴν φύσιν, προσβάλλοντα χρὴ ἐς τὸ χωρίον ἢ τοὺς δακτύλους ἢ τὸν ἕνα δάκτυλον, ἢ ἐξέσχεν ἀνακωχεῖν ἢ αὐτὸν ἢ ἄλλον τινά, ἔστ' ἂν κρατυνθῇ τὸ τρῶμα. ἀτὰρ καὶ ἐς τὸν μυκτῆρα τὸν σμικρὸν δάκτυλον ἀπωθέοντα ἄλλοτε καὶ ἄλλοτε διορθοῦν χρὴ τὰ ἐγκλιθέντα. ὅ τι δ' ἂν φλεγμονῆς ὑπογίνηται τούτοισι, δεῖ τῷ σταιτὶ χρῆσθαι· τοῖσι μέντοι δακτύλοισι προσέχειν ὁμοίως καὶ τοῦ σταιτὸς ἐπικειμένου.

Ἢν δέ που κατὰ τὸν χόνδρον ἐς τὰ πλάγια 20 καταγῇ, ἀνάγκη τὴν ῥῖνα ἄκρην παρεστράφθαι. χρὴ οὖν τοῖσι τοιούτοισιν ἐς τὸν μυκτῆρα ἄκρον διόρθωμά τι τῶν εἰρημένων ἢ ὅ τι τούτοισιν ἔοικεν ἐντιθέναι. πολλὰ δ' ἄν τις εὕροι τὰ ἐπιτήδεια, ὅσα μήτε ὀδμὴν ἴσχει, ἄλλως τε καὶ προσηνέα ἐστίν· ἐγὼ δέ ποτε πλεύμονος προβάτου ἀπότμημα ἐνέθηκα, τοῦτο γάρ πως παρέτυχεν· οἱ γὰρ σπόγγοι ἐντιθέμενοι ὑγράσματα δέχονται. ἔπειτα χρὴ Καρχηδονίου δέρματος λοπόν, πλάτος ὡς τοῦ μεγάλου δακτύλου 30 τετμημένον, ἢ ὅπως ἂν συμφέρῃ, προσκολλῆσαι ἐς τὸ ἔκτοσθεν πρὸς τὸν μυκτῆρα τὸν ἐκκεκλιμένον.[1] κἄπειτα κατατεῖναι τὸν ἱμάντα ὅπως ἂν συμφέρῃ· μᾶλλον δὲ ὀλίγῳ τείνειν χρή, ὥστε[2] ὀρθὴν καὶ ἀπαρτῆ[3] τὴν ῥῖνα εἶναι. ἔπειτα—μακρὸς γὰρ ἔστω ὁ ἱμάς—κάτωθεν[4] τοῦ ὠτὸς ἀγαγόντα αὐτὸν ἀναγαγεῖν περὶ τὴν κεφαλήν· καὶ ἔξεστι μὲν κατὰ τὸ μέτωπον προσκολλῆσαι τὴν τελευτὴν τοῦ ἱμάντος, ἔξεστι δὲ καὶ μακρότερον [ἄγειν, ἔπειτα] περιελίσσοντα[5] περὶ τὴν 40 κεφαλὴν καταδεῖν. τοῦτο ἅμα μὲν δικαίην τὴν

[1] ἐγκεκλιμένον. [2] ἢ ὥστε.

the normal, one or more fingers should be applied at the place where it stuck out, and either the patient or someone else should support it till the lesion is consolidated. One should also insert the little finger from time to time into the nostril and adjust the depressed part. If inflammation arises in these cases, one should use the dough, but keep up the finger application as before, even when the dough is on.

If fracture with deviation occurs in the cartilage, the end of the nose will infallibly be distorted. In such cases, insert one of the internal props mentioned above, or something of the kind, into the nasal opening. One could find many suitable substances without odour and otherwise comfortable. I once inserted a slice from a sheep's lung which happened to be handy; for when sponges are put in, they absorb moisture. Then one should take the outer layer of Carthaginian leather, cut a strip of a thumb's breadth, or what is suitable, and gum it to the outer part of the nostril on the bent side. Next, make suitable tension on the strap —one should pull rather more than suffices to make the nose straight and outstanding.[1] Then—the strap should be a long one—bring it under the ear and up round the head. One may gum the end of the strap on to the forehead. One may also carry it further, and after making a turn round the head, fasten it off. This gives an adjustment which is at

[1] ἀπαρτητὴν Kw. ἀπαρτῆ Galen, Littré, vulg.

---

[3] ἀπαρτητὴν. [4] ἐς τὰ κάτωθεν.
[5] ἐπιπεριελίσσοντα, Littré, Kw., who omit ἄγειν, ἔπειτα.

διόρθωσιν ἔχει, ἅμα δὲ εὐταμίευτον, καὶ μᾶλλον, ἢν ἐθέλῃ, καὶ ἧσσον τὴν ἀντιῤῥοπίην ποιήσεται[1] τῆς ῥινός. ἀτὰρ καὶ ὁπόσοισιν ἐς τὸ πλάγιον ἡ ῥὶς κατάγνυται, τὰ μὲν ἄλλα ἰητρεύειν χρὴ ὡς προείρηται· προσδεῖται δὲ τοῖσι πλείστοισι καὶ τοῦ ἱμάντος πρὸς ἄκρην τὴν ῥῖνα προσκολληθῆναι
47 τῆς ἀντιῤῥοπίης εἵνεκα.

XXXIX. Ὁπόσοισι δὲ σὺν τῇ κατήξει καὶ ἕλκεα προσγίνεται, οὐδὲν δεῖ ταράσσεσθαι διὰ τοῦτο· ἀλλ' ἐπὶ μὲν τὰ ἕλκεα ἐπιτιθέναι ἢ πισσήρην ἢ τῶν ἐναίμων τι· εὐαλθέα γὰρ τῶν τοιούτων τὰ πλεῖστά ἐστιν ὁμοίως, κἢν ὀστέα μέλλῃ ἀπιέναι. τὴν δὲ διόρθωσιν τὴν πρώτην ἀόκνως χρὴ ποιεῖσθαι, μηδὲν ἐπιλείποντα, καὶ τὰς διορθώσιας τοῖσι δακτύλοισι ἐν τῷ ἔπειτα χρόνῳ[2] χαλαρωτέροισι μὲν χρεόμενον, χρεόμενον
10 δέ· εὔπλαστότατον γάρ τι παντὸς τοῦ σώματος ἡ ῥίς ἐστιν. τῶν δὲ ἱμάντων τῇ κολλήσει καὶ τῇ ἀντιῤῥοπίῃ παντάπασιν οὐδὲν κωλύει χρῆσθαι, οὔτ' ἢν ἕλκος ᾖ, οὔτ' ἢν ἐπιφλεγμήνῃ·
14 ἀλυπόταται γάρ εἰσιν.

XL. Ἢν δὲ οὖς καταγῇ, ἐπιδέσιες μὲν πᾶσαι πολέμιαι· οὐ γὰρ οὕτω τις χαλαρὸν περιβάλλοι·[3] ἢν δὲ μᾶλλον πιέζῃ, πλέον κακὸν ἐργάσεται· ἐπεὶ καὶ ὑγιὲς οὖς, ἐπιδέσει πιεχθέν, ὀδυνηρὸν καὶ σφυγματῶδες καὶ πυρετῶδες γίνεται. ἀτὰρ καὶ τὰ ἐπιπλάσματα, κάκιστα μὲν τὰ βαρύτατα τὸ ἐπίπαν· ἀτὰρ καὶ πλεῖστα φλαῦρα καὶ ἀποστατικά, καὶ μύξαν τε ὑποποιεῖ [πλείω],[4]

[1] ποιῆσαι. [2] τοῖσιν . . . χρόνοις.
[3] περιβάλλει. [4] Omit.

once normal and easily arranged; and one can make the counter-deviation of the nose more or less, as one chooses. Again, when the [bone of the] nose is fractured with deviation, besides the other treatment mentioned, it is also necessary in most cases that some of the leather should be gummed on to the tip of the nose to make counter-deviation.[1]

XXXIX. In cases where the fracture is complicated with wounds, there should be no alarm on that account, but one should apply an ointment containing pitch or some other remedy for fresh wounds; for the majority of such cases heal no less readily, even if bones are going to come away. The first adjustment should be made without delay and with completeness; the later rectifications with the fingers are to be done more moderately, yet they are to be done, for of all parts of the body the nose is most easily modelled. There is absolutely no objection to the gumming on of straps and counter-deviation, not even if there is a wound or inflammation supervening, for the manipulations are quite painless.

XL. If the ear is fractured, all bandaging is harmful, for one cannot apply a circular bandage so as to be lax; and if one uses more pressure one will do further damage, for even a sound ear under pressure of a bandage becomes painful, throbbing, and heated. Besides, as to plasters, the heaviest on the whole are the worst; they have also for the most part harmful qualities producing abscess, excessive formation of mucus, and afterwards troublesome dis-

[1] Galen found this gummed leather method very unsatisfactory; "if you pull hard enough to do any good, it comes off" (XVIII (1) 481).

κἄπειτα ἐκπυήσιας ἀσηράς· τούτων δὲ ἥκιστα
10 οὖς καταγὲν προσδεῖται· ἄγχιστα μήν, εἴπερ
χρή, τὸ γλίσχρον ἄλητον, χρὴ δὲ μηδὲ τοῦτο
βάρος ἔχειν. ψαύειν δὲ ὡς ἥκιστα συμφέρει·
ἀγαθὸν γὰρ φάρμακόν ἐστιν ἐνίοτε καὶ τὸ μηδὲν
προσφέρειν, καὶ πρὸς τὸ οὖς καὶ πρὸς ἄλλα
πολλά. χρὴ δὲ καὶ τὴν ἐπικοίμησιν φυλάσ-
σεσθαι· τὸ δὲ σῶμα ἰσχναίνειν, καὶ μᾶλλον ᾧ
ἂν κίνδυνος ᾖ ἔμπυον τὸ οὖς γενέσθαι· ἄμεινον
δὲ καὶ μαλθάξαι τὴν κοιλίην· ἢν δὲ καὶ εὐήμετος[1]
ᾖ, ἐμεῖν ἐκ συρμαισμοῦ. ἢν δὲ ἐς ἐμπύησιν ἔλθῃ,
20 ταχέως μὲν οὐ χρὴ στομοῦν· πολλὰ γὰρ καὶ
τῶν δοκεόντων ἐκπυεῖσθαι ἀναπίνεταί ποτε,
κἢν μηδέν τις καταπλάσσῃ. ἢν δὲ ἀναγκασθῇ
στομῶσαι, τάχιστα μὲν ὑγιὲς γίνεται, ἤν τις
πέρην διακαύσῃ· εἰδέναι μέντοι χρὴ σαφῶς ὅτι
κυλλὸν ἔσται τὸ οὖς, καὶ μεῖον τοῦ ἑτέρου,
ἢν πέρην διακαυθῇ. ἢν δὲ μὴ πέρην καίηται,
τάμνειν χρὴ τὸ μετέωρον, μὴ πάνυ σμικρὴν
τομήν· διὰ παχυτέρου καὶ τὸ πύον εὑρίσκεται
ἢ ὡς ἄν τις δοκέοι· ὡς δ' ἐν κεφαλαίῳ εἰπεῖν,[2]
30 καὶ πάντα τἄλλα τὰ μυξώδεα καὶ μυξοποιά,
ἅτε γλίσχρα ἐόντα, ὑποθιγγανόμενα διολισθάνει
ταχέως ὑπὸ τοὺς δακτύλους καὶ ἔνθα καὶ ἔνθα·
διὰ τοῦτο διὰ παχυτέρου εὑρίσκουσι τὰ τοιαῦτα οἱ
ἰητροὶ ἢ ὡς οἴονται· ἐπεὶ καὶ τῶν γαγγλιωδέων
ἔνια, ὅσα ἂν πλαδαρὰ ᾖ, καὶ μυξώδεα σάρκα
ἔχῃ, πολλοὶ στομοῦσιν, οἰόμενοι ῥεῦμα ἀνευρή-
σειν ἐς τὰ τοιαῦτα· ἡ μὲν οὖν γνώμη τοῦ ἰητροῦ
ἐξαπατᾶται· τῷ δὲ πρήγματι τῷ τοιούτῳ οὐδεμία
βλάβη στομωθέντι. ὅσα δὲ ὑδατώδεα χωρία

[1] εὐεμέτης Kw. [2] εἰρῆσθαι.

charges of pus. A fractured ear is far from needing these as well. If need be, the best application is the glutinous flour plaster; but even this should not be heavy. It is well to touch the part as little as possible, for it is a good remedy sometimes to use nothing, both in the case of the ear and many others. Care must be taken as to the way of lying. Keep the patient on low diet, the more so if there is danger of an abscess in the ear. It is also good to loosen the bowels, and, if he vomits easily, cause emesis by "syrmaism."[1] If it comes to suppuration, do not be in a hurry to open the abscess, for in many cases when there seems to be suppuration, it is absorbed, and that without any application. If one is forced to open an abscess, it will heal most quickly by cauterising right through; but bear well in mind that the ear, if cauterised right through, will be deformed and smaller than the other. If it is not cauterised through, one should make an incision in the swollen part, not very small, for the pus will be found under a thicker covering than one would expect. And, speaking generally, all other parts of a mucous nature, or which secrete mucus, being viscous slip about readily hither and thither when palpated, wherefore practitioners find them thicker to penetrate than they expected. Thus, in the case of some ganglionic tumours which are flabby and have mucoid flesh, many open them, thinking to find a flux of humours to such parts. The practitioner is deceived in his opinion; but in practice no harm is done by such a tumour being opened. Now, as to watery parts,

[1] An emetic of radishes and salt water (Erotian): cf. Herod. II. 88.

40 ἐστὶν ἢ μύξης πεπληρωμένα, καὶ ἐν οἵοισι χωρίοισιν ἕκαστα θάνατον φέρει στομούμενα ἢ καὶ ἀλλοίας βλάβας, περὶ τούτων ἐν ἄλλῳ λόγῳ γεγράψεται. ὅταν οὖν τάμῃ τις τὸ οὖς, πάντων μὲν καταπλασμάτων, πάσης δὲ μοτώσιος ἀπέχεσθαι χρή· ἰητρεύειν δὲ ἢ ἐναίμῳ ἢ ἄλλῳ τῳ ὅ τι μήτε βάρος μήτε πόνον παρασχήσει· ἢν γὰρ ὁ χόνδρος ἄρξηται ψιλοῦσθαι, καὶ ὑποστάσιας ἴσχῃ [πυρώδεας ἢ χολώδεας],[1] ὀχλῶδες[2] [καὶ] μοχθηρόν· γίνεται δὲ τοῦτο δι' ἐκείνας τὰς ἰήσιας.
50 πάντων δὲ τῶν παλιγκοτησάντων ἡ πέρην διά-
51 καυσις αὐταρκέστατον.

XLI. Σπόνδυλοι δὲ οἱ κατὰ ῥάχιν, ὅσοισι μὲν ὑπὸ νοσημάτων ἕλκονται ἐς τὸ κυφόν, τὰ μὲν πλεῖστα ἀδύνατα λύεσθαι, ποτὶ καὶ ὅσα ἀνωτέρω τῶν φρενῶν τῆς προσφύσιος κυφοῦται. τῶν δὲ κατωτέρω μετεξέτερα λύουσι κιρσοὶ γενόμενοι ἐν τοῖς σκέλεσι, μᾶλλον δ' ἔτι ἐγγινόμενοι κιρσοὶ ἐν τῇ κατὰ ἰγνύην φλεβί· οἷσι δ' ἂν τὰ κυφώματα λύηται, ἐγγίνονται δὲ ἐν τῇ κατὰ βουβῶνα· ἤδη δέ τισιν ἔλυσε καὶ δυσεντερίη πολυχρόνιος γενο-
10 μένη. καὶ οἷσι μὲν κυφοῦται ῥάχις παισὶν ἐοῦσι, πρὶν ἢ τὸ σῶμα τελειωθῆναι ἐς αὔξησιν· τούτοισι μὲν οὐδὲ συναύξεσθαι ἐθέλει κατὰ τὴν ῥάχιν τὸ σῶμα, ἀλλὰ σκέλεα μὲν καὶ χεῖρες τελειοῦνται· ταῦτα δὲ ἐνδεέστερα γίνεται. καὶ ὅσοισιν ἂν ᾖ ἀνωτέρω τῶν φρενῶν τὸ κῦφος, τούτοισι μὲν αἵ τε πλευραὶ οὐκ ἐθέλουσιν ἐς τὸ εὐρὺ αὔξεσθαι, ἀλλὰ ἐς τοὔμπροσθεν, τὸ δὲ στῆθος ὀξὺ γίνεται,

or those filled with mucus, and in what parts severally opening brings death or other damage, these matters will be discussed in another treatise.[1] When, then, one incises the ear, all plasters[2] and all plugging should be avoided. Treat with an application for fresh wounds, or something else neither heavy nor painful. For if the cartilage begins to get denuded and has troublesome abscesses,[3] it is bad, and this is the result of that treatment [viz. plasters and plugging with tents]. Perforating cautery is most effective by itself for all supervening aggravations.

XLI. When the spinal vertebrae are drawn into a hump by diseases, most cases are incurable, especially when the hump is formed above the attachment of the diaphragm. Some of those lower down are resolved when varicosities form in the legs, and still more when these are in the vein at the back of the knee. In cases where curvatures resolve, varicosities may also arise in the groin; and, in some, prolonged dysentery causes resolution. When hump-back occurs in children before the body has completed its growth, the legs and arms attain full size, but the body will not grow correspondingly at the spine; these parts are defective. And where the hump is above the diaphragm, the ribs do not enlarge in breadth, but forwards, and the chest becomes pointed

[1] Not extant.
[2] "Plasters bandaged on": cf. *Wounds in the Head* XVII.
[3] Kw.'s reading.

---

[1] Littré, Kw. omit.
[2] ὀχλώδεας, Kw The MSS. are very confused.

ἀλλ' οὐ πλατύ, αὐτοί τε δύσπνοοι γίνονται καὶ κερχνώδεες· ἧσσον γὰρ εὐρυχωρίην ἔχουσιν αἱ κοι-
20 λίαι αἱ τὸ πνεῦμα δεχόμεναι καὶ προπέμπουσαι. καὶ γὰρ δὴ καὶ ἀναγκάζονται κατὰ τὸν μέγαν σπόνδυλον λορδὸν καὶ[1] αὐχένα ἔχειν, ὡς μὴ προπετὴς ᾖ αὐτοῖσι ἡ κεφαλή· στενοχωρίην μὲν οὖν πολλὴν τῇ φάρυγγι παρέχει καὶ τοῦτο ἐς τὸ ἔσω ῥέπον· καὶ γὰρ τοῖσιν ὀρθοῖσι φύσει δύσπνοιαν παρέχει τοῦτο τὸ ὀστέον, ἢν ἔσω ῥέψῃ, ἔστ' ἂν ἀναπιεχθῇ. δι' οὖν τὸ τοιοῦτον σχῆμα ἐξεχέβρογχοι οἱ τοιοῦτοι τῶν ἀνθρώπων μᾶλλον φαίνονται ἢ οἱ ὑγιέες· φυματίαι τε ὡς ἐπὶ τὸ
30 πολὺ κατὰ τὸν πλεύμονά εἰσιν οἱ τοιοῦτοι σκληρῶν φυμάτων καὶ ἀπέπτων· καὶ γὰρ ἡ πρόφασις τοῦ κυφώματος καὶ ἡ σύντασις τοῖσι πλείστοισι διὰ τοιαύτας συστροφὰς γίνεται, ᾗσιν ἂν κοινωνήσωσιν οἱ τόνοι οἱ σύνεγγυς. ὅσοισι δὲ κατωτέρω τῶν φρενῶν τὸ κύφωμά ἐστι, τούτοισι νοσήματα μὲν ἐνίοισι προσγίνεται νεφριτικὰ καὶ κατὰ κύστιν· ἀτὰρ καὶ ἀποστάσιες ἐμπυηματικαὶ κατὰ κενεῶνας καὶ κατὰ βουβῶνας, χρόνιαι καὶ δυσαλθέες, καὶ τούτων οὐδετέρη λύει τὰ κυφώ-
40 ματα· ἰσχία δὲ τοιούτοισιν ἔτι ἀσαρκότερα γίνεται ἢ τοῖσιν ἄνωθεν κυφοῖσιν· ἡ μέντοι σύμπασα ῥάχις μακροτέρη τούτοισιν ἢ τοῖσιν ἄνωθεν κυφοῖσιν. ἥβη δὲ καὶ γένειον βραδύτερα καὶ ἀτελέστερα, καὶ ἀγονώτεροι οὗτοι τῶν ἄνωθεν κυφῶν. οἷσι δ' ἂν ηὐξημένοισι ἤδη τὸ σῶμα ἡ κύφωσις γένηται, τούτοισι ἀπαντικρὺ μὲν τῆς νούσου τῆς τότε παρεούσης κρίσιν ποιεῖ ἡ

[1] τὸν.

instead of broad; the patients also get short of breath and hoarse, for the cavities which receive and send out the breath have smaller capacity. Besides, they are also obliged to hold the neck concave at the great vertebra,[1] that the head may not be thrown forwards. This, then, causes great constriction in the gullet, since it inclines inwards; for this bone, if it inclines inwards, causes difficult breathing even in undeformed persons, until it is pushed back. In consequence of this attitude, such persons seem to have the larynx more projecting than the healthy. They have also, as a rule, hard and unripened[2] tubercles in the lungs; for the origin of the curvature and contraction is in most cases due to such gatherings, in which the neighbouring ligaments take part. Cases where the curvature is below the diaphragm are sometimes complicated with affections of the kidneys and parts about the bladder, and besides there are purulent abscessions in the lumbar region and about the groins, chronic and hard to cure; and neither of these causes resolution of the curvatures. The hips are still more attenuated in such cases than where the hump is high up; yet the spine as a whole is longer in these than in high curvatures. But the hair on the pubes and chin is later and more defective, and they are less capable of generation than those who have the hump higher up. When curvature comes on in persons whose bodily growth is complete, its occurrence produces an apparent[3] crisis

[1] Axis or second cervical, according to Galen, but perhaps the seventh. Cf. XLV.

[2] Unmatured or softened.

[3] Or, "to begin with": most translators, "obviously."

κύφωσις· ἀνὰ χρόνον μέντοι ἐπισημαίνει τι τῶν αὐτῶν, ὥσπερ καὶ τοῖσι νεωτέροισιν,[1] ἢ πλέον ἢ
50 ἔλασσον· ἧσσον δὲ κακοήθως ὡς τὸ ἐπίπαν μὴν τοιαῦτα πάντα ἐστίν. πολλοὶ μέντοι ἤδη καὶ εὐφόρως ἤνεγκαν καὶ ὑγιεινῶς[2] τὴν κύφωσιν ἄχρι γήραος, μάλιστα δὲ οὗτοι, οἷσιν ἂν ἐς τὸ εὔσαρκον καὶ πιμελῶδες προτράπηται τὸ σῶμα· ὀλίγοι μὴν ἤδη καὶ τῶν τοιούτων ὑπὲρ ἑξήκοντα ἔτη ἐβίωσαν· οἱ δὲ πλεῖστοι βραχυβιώτεροί εἰσιν. ἔστι δ' οἷσι καὶ ἐς τὸ πλάγιον σκολιοῦνται σπόνδυλοι ἢ τῇ ἢ τῇ· πάντα μὴν ἢ τὰ πλεῖστα τὰ τοιαῦτα γίνεται διὰ συστροφὰς τὰς ἔσωθεν
60 τῆς ῥάχιος· προσσυμβάλλεται δὲ ἐνίοισι σὺν τῇ νούσῳ καὶ τὰ σχήματα, ἐφ' ὁποῖα ἂν ἐθισθέωσι κεκλίσθαι. ἀλλὰ περὶ μὲν τούτων ἐν τοῖσι χρονίοισι κατὰ πλεύμονα νοσήμασιν εἰρήσεται· ἐκεῖ γάρ εἰσιν αὐτῶν χαριέσταται προγνώσιες
65 περὶ τῶν μελλόντων ἔσεσθαι.

XLII. Ὅσοισι δ' ἐκ καταπτώσιος ῥάχις κυφοῦται, ὀλίγα δὴ τούτων ἐκρατήθη ὥστε ἐξιθυθῆναι. τοῦτο μὲν γάρ, αἱ ἐν τῇ κλίμακι κατασείσιες οὐδένα πω ἐξίθυναν, ὧν γε ἐγὼ οἶδα· χρέονται δὲ οἱ ἰητροὶ μάλιστα αὐτῇ οἱ ἐπιθυμέοντες ἐκχαυνοῦν τὸν πολὺν ὄχλον· τοῖσι γὰρ τοιούτοισι ταῦτα θαυμάσιά ἐστιν, ἢν ἢ κρεμάμενον ἴδωσιν ἢ ῥιπτεόμενον, ἢ ὅσα τοῖσι τοιούτοισιν ἔοικε, καὶ ταῦτα κληΐζουσιν αἰεί,
10 καὶ οὐκέτι αὐτοῖσι μέλει ὁποῖόν τι ἀπέβη ἀπὸ τοῦ χειρίσματος, εἴτε κακὸν εἴτε ἀγαθόν. οἱ μέντοι ἰητροὶ οἱ τὰ τοιαῦτα ἐπιτηδεύοντες σκαιοί εἰσιν, οὕς γε ἐγὼ ἔγνων· τὸ μὲν γὰρ ἐπινόημα ἀρχαῖον, καὶ ἐπαινέω ἔγωγε σφόδρα τὸν πρῶτον ἐπι-

in the disease then present. In time, however, some of the same symptoms found in younger patients show themselves to a greater or lesser degree; but in general they are all less malignant. Many patients, too, have borne curvature well and with good health up to old age, especially those whose bodies tend to be fleshy and plump; but few even of these survive sixty years, and the majority are rather short-lived. There are some in whom the vertebrae are curved laterally to one side or the other. All such affections, or most of them, are due to gatherings on the inner side of the spine, while in some cases the positions the patients are accustomed to take in bed are accessory to the malady. But these will be discussed among chronic diseases of the lung; for the most satisfactory prognoses as to their issue come in that department.

XLII. When the hump-back is due to a fall, attempts at straightening rarely succeed. For, to begin with, succussions on a ladder never straightened any case, so far as I know, and the practitioners who use this method are chiefly those who want to make the vulgar herd gape, for to such it seems marvellous to see a man suspended or shaken or treated in such ways; and they always applaud these performances, never troubling themselves about the result of the operation, whether bad or good. As to the practitioners who devote themselves to this kind of thing, those at least whom I have known are incompetent. Yet the contrivance is an ancient one, and for my part I have great admiration for the

---

[1] νέοισι.
[2] ὑγιηρῶς.

νοήσαντα καὶ τοῦτο καὶ ἄλλο πᾶν ὅ τι μηχάνημα κατὰ φύσιν ἐπενοήθη· οὐδὲν γάρ μοι ἄελπτον, εἴ τις καλῶς σκευάσας καλῶς κατασείσειε, κἂν ἐξιθυνθῆναι ἔνια. αὐτὸς μέντοι κατησχύνθην πάντα τὰ τοιουτότροπα ἰητρεύειν οὕτω, διὰ τοῦτο ὅτι 20 πρὸς ἀπατεώνων μᾶλλον οἱ τοιοῦτοι τρόποι.

XLIII. Ὁπόσοισι μὲν οὖν ἐγγὺς τοῦ αὐχένος ἡ κύφωσις γίνεται, ἧσσον εἰκὸς ὠφελεῖν τὰς κατατάσιας ταύτας τὰς ἐπὶ τὴν κεφαλήν· σμικρὸν γὰρ τὸ βάρος ἡ κεφαλὴ καὶ τὰ ἀκρώμια καταῤῥέποντα· ἀλλὰ τούς γε τοιούτους εἰκὸς ἐπὶ [τοὺς][1] πόδας κατασεισθέντας μᾶλλον ἐξιθυνθῆναι· μέζων γὰρ οὕτως ἡ καταῤῥοπίη ἡ ἐπὶ ταῦτα· ὅσοισι δὲ κατωτέρω τὸ ὕβωμα, τούτοισιν εἰκὸς μᾶλλον ἐπὶ κεφαλὴν κατασείεσθαι. εἰ οὖν 10 τις ἐθέλοι κατασείειν, ὀρθῶς ἂν ὧδε σκευάζοι· τὴν μὲν κλίμακα χρὴ σκυτίνοισιν ὑποκεφαλαίοισι πλαγίοισιν, ἢ ἐρινέοισι, καταστρῶσαι εὖ προσδεδεμένοισιν, ὀλίγῳ πλέον καὶ ἐπὶ μῆκος καὶ ἔνθεν καὶ ἔνθεν, ἢ ὅσον ἂν τὸ σῶμα τοῦ ἀνθρώπου κατάσχοι· ἔπειτα τὸν ἄνθρωπον ὕπτιον κατακλῖναι ἐπὶ τὴν κλίμακα χρή· κἄπειτα προσδῆσαι μὲν τοὺς πόδας παρὰ τὰ σφυρὰ πρὸς τὴν κλίμακα μὴ διαβεβῶτας, δεσμῷ εὐόχῳ μέν, μαλθακῷ δέ· προσδῆσαι δὲ κατωτέρω ἑκάτερον τῶν γουνάτων 20 καὶ ἀνωτέρω· προσδῆσαι δὲ καὶ κατὰ τὰ ἰσχία· κατὰ δὲ τοὺς κενεῶνας καὶ κατὰ τὸ στῆθος χαλαρῇσι ταινίῃσι[2] περιβαλεῖν οὕτως, ὅπως μὴ κωλύωσι[3] τὴν κατάσεισιν· τὰς δὲ χεῖρας παρὰ τὰς πλευρὰς παρατείναντα προσκαταλαβεῖν πρὸς αὐτὸ τὸ σῶμα, καὶ μὴ πρὸς τὴν κλίμακα. ὅταν

[1] Omit Erm., Kw.

man who first invented it, or thought out any other mechanism in accordance with nature; for I think it is not hopeless, if one has proper apparatus and does the succussion properly, that some cases may be straightened out. For myself, however, I felt ashamed to treat all such cases in this way, and that because such methods appertain rather to charlatans.

XLIII. In cases where the curvature is near the neck, extension of this kind with the head downwards is naturally less effective; for the downward-pulling weight of the head and shoulders is small. Such cases are more likely to be straightened out by succussion with the feet downwards; for the downward pull is greater thus than in the former position. Cases where the hump is lower may more appropriately undergo succussion head downwards. If then one desires to do succussion, the following is the proper arrangement. One should cover the ladder with transverse leather or linen pillows, well tied on, to a rather greater length and breadth than the patient's body will occupy. Next, the patient should be laid on his back upon the ladder; and then his feet should be tied at the ankles to the ladder, without being separated, with a strong but soft band. Fasten besides a band above and below each of the knees, and also at the hips; but the flanks and chest should have bandages passed loosely round them, so as not to interfere with the succussion. Tie also the hands, extended along the sides, to the body itself, and not to the ladder. When you have

---

[2] χαλαρῇ ταινίῃ.
[3] κωλύσει.

δὲ ταῦτα κατασκευάσῃς οὕτως, ἀνέλκειν τὴν κλίμακα ἢ πρὸς τύρσιν τινὰ ὑψηλὴν ἢ πρὸς ἀέτωμα οἴκου· τὸ δὲ χωρίον ἵνα κατασείεις[1] ἀντίτυπον ἔστω· τοὺς δὲ ἀντιτείνοντας εὐπαιδεύ-
30 τους χρὴ εἶναι, ὅπως ὁμαλῶς [καὶ καλῶς][2] καὶ ἰσοῤῥόπως καὶ ἐξαπιναίως ἀφήσουσι, καὶ μήτε ἡ κλῖμαξ ἑτερόῤῥοπος ἐπὶ τὴν γῆν ἀφίξεται, μήτε αὐτοὶ προπετέες ἔσονται. ἀπὸ μέντοι τύρσιος ἀφιεὶς ἢ ἀπὸ ἱστοῦ καταπεπηγότος καρχήσιον ἔχοντος ἔτι κάλλιον ἄν τις σκευάσαιτο, ὥστε ἀπὸ τροχιλίης τὰ χαλώμενα εἶναι ὅπλα ἢ ἀπὸ ὄνου. ἀηδὲς μὴν καὶ μακρολογεῖν περὶ τούτων· ὅμως δὲ ἐκ τούτων ἂν τῶν κατασκευῶν
39 κάλλιστ'[3] ἄν τις κατασεισθείη.

XLIV. Εἰ μέντοι κάρτα ἄνω εἴη τὸ ὕβωμα, δέοι δὲ κατασείειν πάντως, ἐπὶ πόδας κατασείειν λυσιτελεῖ, ὥσπερ ἤδη εἴρηται· πλείων γὰρ οὕτω γίνεται ἡ καταῤῥοπίη ἐπὶ ταῦτα. ἑρμάσαι δὲ χρὴ κατὰ μὲν τὸ στῆθος πρὸς τὴν κλίμακα προσδήσαντα ἰσχυρῶς, κατὰ δὲ τὸν αὐχένα ὡς χαλαρωτάτῃ ταινίῃ, ὅσον τοῦ κατορθοῦσθαι εἵνεκα· καὶ αὐτὴν τὴν κεφαλὴν κατὰ τὸ μέτωπον προσδῆσαι πρὸς τὴν κλίμακα· τὰς δὲ χεῖρας
10 παρατανύσαντα πρὸς τὸ σῶμα προσδῆσαι, καὶ μὴ πρὸς τὴν κλίμακα· τὸ μέντοι ἄλλο σῶμα ἄδετον εἶναι χρή, πλήν, ὅσον τοῦ κατορθοῦσθαι εἵνεκα, ἄλλῃ καὶ ἄλλῃ ταινίῃ χαλαρῇ περιβεβλῆσθαι· ὅπως δὲ μὴ κωλύωσιν οὗτοι οἱ δεσμοὶ τὴν κατάσεισιν, σκοπεῖν· τὰ δὲ σκέλεα πρὸς μὲν τὴν κλίμακα μὴ προσδεδέσθω, πρὸς ἄλληλα δέ, ὡς κατὰ τὴν ῥάχιν ἰθύῤῥοπα ᾖ. ταῦτα μέντοι τοιουτοτρόπως ποιητέα, εἰ πάντως

arranged things thus, lift the ladder against some high tower or house-gable. The ground where you do the succussion should be solid, and the assistants who lift well trained, that they may let it down smoothly, neatly, vertically, and at once, so that neither the ladder shall come to the ground unevenly, nor they themselves be pulled forwards. When it is let down from a tower, or from a mast fixed in the ground and provided with a truck, it is a still better arrangement to have lowering tackle from a pulley or wheel and axle. It is truly disagreeable to enlarge on these matters; but all the same, succussion would be best done by aid of this apparatus.[1]

XLIV. If the hump is very high up and succussion absolutely required, it is advantageous to do it towards the feet, as was said before; for in this direction the downward impulsion is greater. One should fix the patient by binding him to the ladder firmly at the chest, but at the neck with the loosest possible band sufficient to keep it straight; bind the head itself also to the ladder at the forehead. Extend the arms along, and fasten them to, the body, not to the ladder. The rest of the body should not be tied, except in so far as is requisite to keep it vertical with a loose band round it here and there. But see that these attachments do not hinder the succussion. Do not fasten the legs to the ladder, but to one another, that they may hang in a straight line with the back. This is the sort of thing that

[1] Surgeons will remember that methods no less violent than these and those described below were practised for a time on high authority at the end of last century.

---

[1] κατασείσεις. [2] Apoll., Galen, but most omit. [3] μάλιστα.

δέοι ἐν κλίμακι κατασεισθῆναι· αἰσχρὸν μέντοι
20 καὶ ἐν πάσῃ τέχνῃ, καὶ οὐχ ἥκιστα ἐν ἰατρικῇ, πολὺν ὄχλον καὶ πολλὴν ὄψιν καὶ πολὺν λόγον
22 παρασχόντα, ἔπειτα μηδὲν ὠφελῆσαι.

XLV. Χρὴ δὲ πρῶτον μὲν γινώσκειν τὴν φύσιν τῆς ῥάχιος, οἵη τίς ἐστιν· ἐς πολλὰ γὰρ νουσήματα προσδέοι ἂν αὐτῆς. τοῦτο μὲν γάρ, τὸ πρὸς τὴν κοιλίην ῥέπον οἱ σπόνδυλοι ἐντὸς ἄρτιοί εἰσιν ἀλλήλοισι, καὶ δέδενται πρὸς ἀλλήλοις δεσμῷ μυξώδει καὶ νευρώδει, ἀπὸ χόνδρων ἀποπεφυκότι ἄχρι πρὸς τὸν νωτιαῖον. ἄλλοι δέ τινες τόνοι νευρώδεες διανταῖοι πρόσφυτοι παρατέτανται ἔνθεν καὶ ἔνθεν αὐτῶν. αἱ δὲ φλεβῶν
10 καὶ ἀρτηρίων κοινωνίαι ἐν ἑτέρῳ λόγῳ δεδηλώσονται, ὅσαι τε καὶ οἷαι, καὶ ὅθεν ὡρμημέναι, καὶ ἐν οἵοισιν[1] οἷα δύνανται, αὐτὸς δὲ ὁ νωτιαῖος οἷσιν ἐλύτρωται ἐλύτροισιν καὶ ὅθεν ὡρμημένοισι, καὶ ὅπῃ κραίνουσι καὶ οἷσιν κοινωνέουσι, καὶ οἷα δυναμένοισιν· ἐν δὲ τῷ ἐπέκεινα ἐν ἄρθροισι γεγιγγλύμωνται πρὸς ἀλλήλους οἱ σπόνδυλοι. τόνοι δὲ κοινοὶ παρὰ πάντας καὶ ἐν τοῖσιν ἔξω μέρεσι καὶ ἐν τοῖσιν ἔσω παρατέτανται· ἀπόφυσίς τέ ἐστιν ὀστέου ἐς τὸ ἔξω μέρος ἀπὸ πάντων τῶν
20 σπονδύλων, μία ἀπὸ ἑνὸς ἑκάστου, ἀπό τε τῶν μεζόνων ἀπό τε τῶν ἐλασσόνων· ἐπὶ δὲ τῇσιν ἀποφύσεσι ταύτῃσι χονδρίων ἐπιφύσιες, καὶ ἀπ' ἐκείνων νεύρων ἀποβλάστησις ἠδελφισμένη τοῖσιν ἐξωτάτω τόνοισιν. πλευραὶ δὲ προσπεφύκασιν, ἐς τὸ ἔσω μέρος τὰς κεφαλὰς ῥέπουσαι μᾶλλον ἢ ἐς τὸ ἔξω· καθ' ἕνα δὲ ἕκαστον τῶν σπονδύλων προσήρθρωνται· καμπυλώταται δὲ πλευραὶ ἀν-

[1] οἷς.

must be done if succussion on a ladder is absolutely required; but it is disgraceful in any art, and especially in medicine, to make parade of much trouble, display, and talk, and then do no good.

XLV. One should first get a knowledge of the structure of the spine; for this is also requisite for many diseases. Now on the side turned towards the body cavity, the vertebrae are fitted evenly to one another and bound together by a mucous and ligamentous connection extending from the cartilages right to the spinal cord.[1] There are also certain ligamentous cords extending all along, attached on either side of them. The communications of the veins and arteries will be described elsewhere as regards their number, nature, origin, and functions; also the spinal cord itself with its coverings, their origin, endings, connections and functions. Posteriorly, the vertebrae are connected with one another by hinge-like joints. Cords common to them all are stretched along both the inner and outer sides.[2] From every vertebra there is an outgrowth (apophysis) of bone posteriorly [lit. "to the outer part"], one from each, both the larger and smaller; upon the apophyses are epiphyses of cartilage, and from these there is an outgrowth of tendons, which are in relation with the outermost cords. The ribs are articulated severally with each of the vertebrae, their heads being disposed rather inwards (forwards) than outwards (backwards). Man's ribs are the most curved,

[1] Intervertebral cartilage: reference to its mucous centre and cartilaginous anterior layer.

[2] Both these and those mentioned above seem to be the anterior and posterior common ligaments. "Inner" and "outer" = our "front" and "back."

θρώπου εἰσὶ ῥαιβοειδέα τρόπον. τὸ δὲ μεσηγὺ τῶν πλευρέων καὶ τῶν ὀστέων τῶν ἀποπεφυκότων
30 ἀπὸ τῶν σπονδύλων ἀποπληρέουσιν ἑκατέρωθεν οἱ μύες ἀπὸ τοῦ αὐχένος ἀρξάμενοι, ἄχρι τῆς προσφύσιος. αὐτὴ δὲ ἡ ῥάχις κατὰ μῆκος ἰθυσκόλιός ἐστιν· ἀπὸ μὲν τοῦ ἱεροῦ ὀστέου ἄχρι τοῦ μεγάλου σπονδύλου, παρ' ὃν προσήρτηται τῶν σκελέων ἡ πρόσφυσις, ἄχρι μὲν τούτου κυφή· κύστις τε γὰρ καὶ γοναὶ καὶ ἀρχοῦ τὸ χαλαρὸν ἐν τούτῳ ἔκτισται. ἀπὸ δὲ τούτου ἄχρι φρενῶν προσαρτήσιος, ἰθυλόρδη· καὶ παραφύσιας ἔχει μυῶν τοῦτο μοῦνον τὸ χωρίον ἐκ τῶν ἔσωθεν μερῶν, ἃς
40 δὴ καλοῦσιν ψόας. ἀπὸ δὲ τούτου ἄχρι τοῦ μεγάλου σπονδύλου τοῦ ὑπὲρ τῶν ἐπωμίδων, ἰθυκύφη· ἔτι δὲ μᾶλλον δοκεῖ ἢ ἐστιν· ἡ γὰρ ἄκανθα κατὰ μέσον ὑψηλοτάτας τὰς ἐκφύσιας τῶν ὀστέων ἔχει, ἔνθεν δὲ καὶ ἔνθεν ἐλάσσους. αὐτὸ δὲ τὸ
45 ἄρθρον τὸ τοῦ αὐχένος λορδόν ἐστιν.

XLVI. Ὁπόσοισι μὲν οὖν κυφώματα γίνεται κατὰ τοὺς σπονδύλους, ἔξωσις μὲν μεγάλη ἀποῤῥαγεῖσα ἀπὸ τῆς συμφύσιος ἢ ἑνὸς σπονδύλου ἢ καὶ πλεόνων οὐ μάλα πολλοῖσι γίνεται, ἀλλ' ὀλίγοισι. οὐδὲ γὰρ τὰ τρώματα τὰ τοιαῦτα ῥηΐδιον γίνεσθαι· οὔτε γὰρ ἐς τὸ ἔξω ἐξωσθῆναι ῥηΐδιόν ἐστιν, εἰ μὴ ἐκ τοῦ ἔμπροσθεν ἰσχυρῷ τινὶ τρωθείη διὰ τῆς κοιλίης (οὕτω δ' ἂν ἀπόλοιτο), ἢ εἴ τις ἀφ' ὑψηλοῦ του χωρίου πεσὼν ἐρείσειε τοῖσιν ἰσχίοι-
10 σιν ἢ τοῖσιν ὤμοισιν (ἀλλὰ καὶ οὕτως ἂν ἀποθάνοι, παραχρῆμα δὲ οὐκ ἂν ἀποθάνοι)· ἐκ δὲ τοῦ ὄπισθεν οὐ ῥηΐδιον τοιαύτην ἔξαλσιν γενέσθαι ἐς τὸ ἔσω, εἰ μὴ ὑπερβαρύ τι ἄχθος ἐμπέσοι· τῶν τε γὰρ ὀστέων τῶν ἐκπεφυκότων ἔξω ἓν

and they are bandy-shaped. As to the part between the ribs and the bony outgrowths (apophyses) of the vertebrae, it is filled on each side by the muscles which begin at the neck and extend to the attachment[1] [of the diaphragm]. The spine itself is curved vertically through its length. From the sacrum to the great vertebra,[2] near which the origin of the legs is inserted, all this is curved outwards; for the bladder, generative organs, and loose part of the rectum are lodged there. From this point to the attachment of the diaphragm it curves inwards; and this part only of the inside has attachments of muscles, which they call "psoai." From this to the great vertebra[3] over the shoulder-blades it is curved outwards, and seems to be more so than it is; for the ridge has the outgrowths of bone highest here, while above and below they are smaller. The articulation of the neck itself is curved inwards.

XLVI. In cases then of outward curvature at the vertebrae, a great thrusting-out and rupture of the articulation of one or more of them does not very often occur, but is rare. Such injuries, indeed, are hard to produce; nor is it easy for outward thrusting to be brought about, unless a man were violently wounded from the front through the body cavity—and then he would perish—or if a man falling from a height came down on his buttocks or shoulders—but then he would die also, though he might not die at once. And from behind it would not be easy for such sudden luxation to take place inwards, unless some very heavy weight fell on the spine; for each of the external bony epiphyses is of

[1] "To their attachment" (Petrequin).
[2] Fifth lumbar. [3] Seventh cervical.

ἕκαστον τοιοῦτόν ἐστιν, ὥστε πρόσθεν ἂν αὐτὸ καταγῆναι πρὶν ἢ μεγάλην ῥοπὴν ἔσω ποιῆσαι, τούς τε συνδέσμους βιησάμενον καὶ τὰ ἄρθρα τὰ ἐνηλλαγμένα. ὅ τε αὖ νωτιαῖος πονοίη ἄν, εἰ ἐξ ὀλίγου χωρίου τὴν περικαμπὴν ἔχοι, τοιαύτην
20 ἔξαλσιν ἐξαλλομένου σπονδύλου· ὅ τε ἐκπηδήσας σπόνδυλος πιέζοι ἂν τὸν νωτιαῖον, εἰ μὴ καὶ ἀποῤῥήξειεν. πιεχθεὶς δ' ἂν καὶ ἀπολελαμμένος πολλῶν ἂν καὶ μεγάλων καὶ ἐπικαίρων ἀπονάρκωσιν ποιήσειεν· ὥστε οὐκ ἂν μέλοι τῷ ἰητρῷ ὅπως χρὴ τὸν σπόνδυλον κατορθῶσαι, πολλῶν καὶ βιαίων ἄλλων κακῶν παρεόντων. ὥστε δὴ οὐδ' ἐμβαλεῖν οἷόν τε πρόδηλον τὸν τοιοῦτον οὔτε κατασείσει οὔτε ἄλλῳ τρόπῳ οὐδενί, εἰ μή τις διαταμὼν τὸν ἄνθρωπον, ἔπειτα ἐσμασάμενος
30 ἐς τὴν κοιλίην, ἐκ τοῦ ἔσωθεν τῇ χειρὶ ἐς τὸ ἔξω ἀντωθέοι· καὶ τοῦτο νεκρῷ μὲν οἷόν τε ποιεῖν, ζῶντι δὲ οὐ πάνυ. διὰ τί οὖν ταῦτα γράφω; ὅτι οἴονταί τινες ἰητρευκέναι ἀνθρώπους οἷσιν ἔσωθεν ἐνέπεσον σπόνδυλοι, τελέως ὑπερβάντες τὰ ἄρθρα· καίτοι γε ῥηΐστην ἐς τὸ περιγενέσθαι τῶν διαστροφέων ταύτην ἔνιοι νομίζουσι καὶ οὐδὲν δεῖσθαι ἐμβολῆς, ἀλλὰ αὐτόματα ὑγιέα γίνεσθαι τὰ τοιαῦτα. ἀγνοέουσι δὴ πολλοί, καὶ κερδαίνουσιν ὅτι ἀγνοέουσι· πείθουσι γὰρ τοὺς πέλας.
40 ἐξαπατῶνται δὲ διὰ τόδε· οἴονται γὰρ τὴν ἄκανθαν τὴν ἐξέχουσαν κατὰ τὴν ῥάχιν ταύτην τοὺς σπονδύλους αὐτοὺς εἶναι, ὅτι στρογγύλον αὐτῶν ἕκαστον φαίνεται ψαυόμενον, ἀγνοεῦντες ὅτι τὰ ὀστέα ταῦτά ἐστι τὰ ἀπὸ τῶν σπονδύλων πεφυκότα, περὶ ὧν ὁ λόγος ὀλίγῳ πρόσθεν εἴρηται· οἱ δὲ σπόνδυλοι πολὺ προσω-

such a nature as to be fractured itself before overcoming the ligaments and interconnecting joints and making a great deviation inwards. The spinal cord, too, would suffer, if the luxation due to jerking out of a vertebra had made so sharp a curve; and the vertebra in springing out would press on the cord, even if it did not break it. The cord, then, being compressed and intercepted, would produce complete narcosis of many large and important parts, so that the physician would not have to trouble about how to adjust the vertebra, in the presence of many other urgent complications. So, then, the impossibility of reducing such a dislocation either by succussion or any other method is obvious, unless after cutting open the patient, one inserted the hand into the body cavity and made pressure from within outwards. One might do this with a corpse, but hardly with a living patient. Why then am I writing this? Because some think they have cured patients whose vertebrae had fallen inwards with complete disarticulation; and there are even some also who think this is the easiest distortion to recover from, not even requiring reduction, but that such injuries get well of themselves. There are many ignorant practitioners; and they profit by their ignorance, for they get credit with their neighbours. Now this is how they are deceived. They think that the projecting ridge along the spine represents the vertebrae themselves, because each of the processes feels rounded on palpation; not knowing that these bones are the natural outgrowths from the vertebrae which were discussed a little above. But

τέρω ἄπεισιν· στενοτάτην γὰρ πάντων τῶν ζώων ὥνθρωπος κοιλίην ἔχει, ὡς ἐπὶ τῷ μεγέθει, ἀπὸ τοῦ ὄπισθεν ἐς τὸ ἔμπροσθεν, ποτὶ καὶ κατὰ τὸ
50 στῆθος. ὅταν οὖν τι τούτων τῶν ὀστέων τῶν ὑπερεχόντων ἰσχυρῶς καταγῇ, ἤν τε ἓν ἤν τε πλείω, ταύτῃ ταπεινότερον τὸ χωρίον γίνεται ἢ τὸ ἔνθεν καὶ ἔνθεν, καὶ διὰ τοῦτο ἐξαπατῶνται, οἰόμενοι τοὺς σπονδύλους ἔσω οἴχεσθαι. προσεξαπατᾷ δὲ ἔτι αὐτοὺς καὶ τὰ σχήματα τῶν τετρωμένων· ἢν μὲν γὰρ πειρῶνται καμπύλλεσθαι, ὀδυνῶνται, περιτενέος γινομένου ταύτῃ τοῦ δέρματος ᾗ τέτρωνται, καὶ ἅμα τὰ ὀστέα τὰ κατεηγότα ἐνθράσσει οὕτω μᾶλλον τὸν χρῶτα. ἢν δὲ
60 λορδαίνωσι, ῥᾴους εἰσίν· χαλαρώτερον γὰρ τὸ δέρμα κατὰ τὸ τρῶμα ταύτῃ γίνεται, καὶ τὰ ὀστέα ἧσσον ἐνθράσσει· ἀτὰρ καὶ ἤν τις ψαύῃ αὐτῶν, κατὰ τοῦτο ὑπείκουσι λορδοῦντες, καὶ τὸ χωρίον κενεὸν καὶ μαλθακὸν ψαυόμενον ταύτῃ φαίνεται. ταῦτα πάντα τὰ εἰρημένα προσεξαπατᾷ τοὺς ἰητρούς. ὑγιέες δὲ ταχέως καὶ ἀσινέες αὐτόματοι οἱ τοιοῦτοι γίνονται· ταχέως γὰρ πάντα τὰ τοιαῦτα ὀστέα ἐπιπωροῦται, ὅσα
69 χαῦνά ἐστιν.

XLVII. Σκολιαίνεται μὲν οὖν ῥάχις καὶ ὑγιαίνουσι κατὰ πολλοὺς τρόπους· καὶ γὰρ ἐν τῇ φύσει καὶ ἐν τῇ χρήσει οὕτως ἔχει· ἀτὰρ καὶ ὑπὸ γήραος καὶ ὑπὸ ὀδυνημάτων[1] συνδοτική ἐστιν. αἱ δὲ δὴ κυφώσιες αἱ ἐν τοῖσι πτώμασιν ὡς ἐπὶ τὸ πολὺ γίνονται, ἢν ἢ τοῖσιν ἰσχίοισιν ἐρείσῃ ἢ ἐπὶ τοὺς ὤμους πέσῃ. ἀνάγκη γὰρ ἔξω φαίνεσθαι ἐν τῷ κυφώματι ἕνα μέν τινα ὑψηλότερον τῶν σπονδύλων, τοὺς δὲ ἔνθεν καὶ ἔνθεν ἐπὶ ἧσσον·

the vertebrae are much farther in front; for man has the narrowest body cavity of all animals relatively to his size and measured from behind forwards, especially in the thoracic region. Whenever, therefore, there is a violent fracture of these projecting processes, either one or more, the part is more depressed there than on either side; and therefore they are deceived, and think the vertebrae have gone inwards. And the attitudes of the patients help to deceive them still more; for if they try to bend forwards, they suffer pain, the skin being stretched at the level of the injury, while at the same time the fractured bones disturb the flesh more; but if they hollow their backs, they are easier, for thereby the skin gets more relaxed at the wound, and the bones cause less disturbance. Again, if one feels them, they shrink at the part, and bend inwards; and the region appears hollow and soft on palpation. All these things contribute to deceive the physicians, while such patients recover of themselves quickly and without damage; for callus forms rapidly on all bones of this kind, by reason of their being porous.

XLVII. Curvature of the spine occurs even in healthy persons in many ways, for such a condition is connected with its nature and use; and besides, there is a giving way in old age, and on account of pain. But the outward curvatures due to falls usually occur when the patient comes down on his buttocks or falls on his shoulders; and, in the curvature, one of the vertebrae necessarily appears to stand out more prominently, and those on either

---

[1] ὀδύνης Kw.

10 οὔκουν εἷς ἐπὶ πολὺ ἀποπεπηδηκὼς ἀπὸ τῶν ἄλλων ἐστίν, ἀλλὰ σμικρὸν ἕκαστος συνδιδοῖ, ἀθρόον δὲ πολὺ φαίνεται. διὰ οὖν τοῦτο καὶ ὁ νωτιαῖος μυελὸς εὐφόρως φέρει τὰς τοιαύτας διαστροφάς, ὅτι κυκλώδης αὐτῷ ἡ διαστροφὴ γίνεται, ἀλλ' οὐ γωνιώδης.

Χρὴ δὲ τὴν κατασκευὴν τοῦ διαναγκασμοῦ τοιήνδε κατασκευάσαι. ἔξεστι μὲν ξύλον ἰσχυρὸν καὶ πλατύ, ἐντομὴν παραμηκέα ἔχον, κατορύξαι· 20 ἔξεστι δὲ ἀντὶ τοῦ ξύλου ἐν τοίχῳ ἐντομὴν παραμηκέα ἐνταμεῖν, ἢ πήχει ἀνωτέρω τοῦ ἐδάφεος, ἢ ὅπως ἂν μετρίως ἔχῃ· ἔπειτα οἷον στύλον δρύϊνον τετράγωνον πλάγιον παραβάλλειν, ἀπολείποντα ἀπὸ τοῦ τοίχου ὅσον παρελθεῖν τινά, ἢν δέῃ· καὶ ἐπὶ μὲν τὸν στύλον ἐπιστορέσαι ἢ χλαίνας ἢ ἄλλο τι, ὃ μαλθακὸν μὲν ἔσται, ὑπείξει δὲ μὴ μέγα· τὸν δὲ ἄνθρωπον πυριῆσαι, ἢν ἐνδέχηται, ἢ πολλῷ θερμῷ λοῦσαι· κἄπειτα πρηνέα κατακλῖναι κατατεταμένον, καὶ τὰς μὲν χεῖρας αὐτοῦ 30 παρατείναντα κατὰ φύσιν προσδῆσαι πρὸς τὸ σῶμα, ἱμάντι δὲ μαλθακῷ, ἱκανῶς πλατεῖ τε καὶ μακρῷ, ἐκ δύο διανταίων συμβεβλημένῳ μέσῳ, κατὰ μέσον δὲ τὸ στῆθος δὶς περιβεβλῆσθαι χρὴ ὡς ἐγγυτάτω τῶν μασχαλέων· ἔπειτα τὸ περισσεῦον τῶν ἱμάντων κατὰ τὴν μασχάλην ἑκάτερον περὶ τοὺς ὤμους περιβεβλήσθω· ἔπειτα αἱ ἀρχαὶ πρὸς ξύλον ὑπεροειδές τι προσδεδέσθωσαν, ἁρμόζουσαι τὸ μῆκος τῷ ξύλῳ τῷ ὑποτεταμένῳ, πρὸς ὅ τι πρόσβαλλον τὸ ὑπεροειδὲς ἀντιστηρίζοντα 40 κατατείνειν. τοιούτῳ δέ τινι ἑτέρῳ δεσμῷ χρὴ ἄνωθεν τῶν γουνάτων δήσαντα καὶ ἄνωθεν τῶν πτερνέων τὰς ἀρχὰς τῶν ἱμάντων πρὸς τοιοῦτόν

side less so. It is not that one has sprung out to a distance from the rest; but each gives way a little, and the displacement taken altogether seems great. This is why the spinal marrow does not suffer from such distortion, because the distortion affecting it is curved and not angular.[1]

The apparatus for forcible reduction should be arranged as follows. One may fix in the ground a strong broad plank having in it a transverse groove. Or, instead of the plank, one may cut a transverse groove in a wall, a cubit above the ground, or as may be convenient. Then place a sort of quadrangular oak board parallel with the wall and far enough from it that one may pass between if necessary; and spread cloaks on the board, or something that shall be soft, but not very yielding. Give the patient a vapour bath if possible, or one with plenty of hot water; then make him lie stretched out in a prone position, and fasten his arms, extending them naturally, to the body. A soft band, sufficiently broad and long, composed of two strands, should be applied at its middle to the middle of the chest, and passed twice round it as near as possible to the armpits; then let what remains of the (two) bands be passed round the shoulders at each side, and the ends be attached to a pestle-shaped pole, adjusting their length to that of the underlying board against which the pestle-shaped pole is put, using it as a fulcrum to make extension. A second similar band should be attached above the knees and above the heels, and the ends of the straps fastened to

[1] In spite of this, the strange contradiction "angular curvature" has come to be the technical term for hump-back.

τι ξύλον προσδῆσαι· ἄλλῳ δὲ ἱμάντι πλατεῖ καὶ μαλθακῷ καὶ δυνατῷ, ταινιοειδεῖ, πλάτος ἔχοντι καὶ μῆκος ἱκανόν, ἰσχυρῶς περὶ τὰς ἰξύας κύκλῳ περιδεδέσθαι ὡς ἐγγύτατα τῶν ἰσχίων· ἔπειτα τὸ περισσεῦον τῆς ταινιοειδέος, ἅμα ἀμφοτέρας τὰς ἀρχὰς τῶν ἱμάντων, πρὸς τὸ ξύλον προσδῆσαι τὸ πρὸς τῶν ποδῶν· κἄπειτα κατατείνειν ἐν τούτῳ
50 τῷ σχήματι ἔνθα καὶ ἔνθα, ἅμα μὲν ἰσοῤῥόπως, ἅμα δὲ ἐς ἰθύ. οὐδὲν γὰρ ἂν μέγα κακὸν ἡ τοιαύτη κατάτασις ποιήσειεν, εἰ χρηστῶς σκευασθείη,[1] εἰ μὴ ἄρα ἐξεπίτηδές τις βούλοιτο σίνεσθαι. τὸν δὲ ἰητρὸν χρὴ ἢ ἄλλον, ὅστις ἰσχυρὸς καὶ μὴ ἀμαθής, ἐπιθέντα τὸ θέναρ τῆς χειρὸς ἐπὶ τὸ ὕβωμα, καὶ τὴν ἑτέρην χεῖρα προσεπιθέντα ἐπὶ τὴν ἑτέρην, καταναγκάζειν, προσσυνιέντα ἤν τε ἐς ἰθὺ ἐς τὸ κάτω πεφύκῃ καταναγκάζεσθαι, ἤν τε πρὸς τῆς κεφαλῆς, ἤν τε πρὸς τῶν ἰσχίων. καὶ
60 ἀσινεστάτη μὲν αὕτη ἡ ἀνάγκη· ἀσινὲς δὴ καὶ ἐπικαθέζεσθαί τινα ἐπὶ τὸ κύφωμα, αὐτοῦ ἅμα κατατεινομένου, καὶ ἐνσεῖσαι μετεωρισθέντα. ἀτὰρ καὶ ἐπιβῆναι τῷ ποδὶ καὶ ὀχηθῆναι ἐπὶ τὸ κύφωμα· ἡσύχως τε ἐπενσεῖσαι οὐδὲν κωλύει· τὸ τοιοῦτον δὲ ποιῆσαι μετρίως ἐπιτήδειος ἄν τις εἴη τῶν ἀμφὶ παλαίστρην εἰθισμένων. δυνατωτάτη μέντοι τῶν ἀναγκέων ἐστίν, εἰ ὁ μὲν τοῖχος ἐντετμημένος ἢ τὸ δὲ ξύλον τὸ κατωρυγμένον, ᾗ ἐντέτμηται, κατωτέρω εἴη τῆς ῥάχιος τοῦ ἀνθρώ-
70 που, ὁπόσῳ ἂν δοκῇ μετρίως ἔχειν, σανὶς δὲ φιλυρίνη, μὴ λεπτή, ἐνείη, ἢ καὶ ἄλλου τινὸς ξύλου· ἔπειτα ἐπὶ μὲν τὸ ὕβωμα ἐπιτεθείη ἢ τρύχιόν τι πολύπτυχον ἢ σμικρόν τι σκύτινον ὑποκεφάλαιον· ὡς ἐλάχιστα μὴν ἐπικεῖσθαι

a similar pole. With another soft, strong strap, like a head-band, of sufficient breadth and length, the patient should be bound strongly round the loins, as near as possible to the hips. Then fasten what is over of this band, as well as the ends of both the other straps, to the pole at the foot end; next, make extension in this position towards either end simultaneously, equally and in a straight line. Such extension would do no great harm, if well arranged, unless indeed one deliberately wanted to do harm. The physician, or an assistant who is strong and not untrained, should put the palm of his hand on the hump, and the palm of the other on that, to reduce it forcibly, taking into consideration whether the reduction should naturally be made straight downwards, or towards the head, or towards the hips. This reduction method also is very harmless; indeed, it will do no harm even if one sits on the hump while extension is applied, and makes succussion by raising himself; nay, there is nothing against putting one's foot on the hump and making gentle succussion by bringing one's weight upon it. A suitable person to perform such an operation properly would be one of those habituated to the palaestra. But the most powerful method of reduction is to have the incision in the wall, or that in the post embedded in the ground, at an appropriate level, rather below that of the patient's spine, and a not too thin plank of lime or other wood inserted in it. Then let many thicknesses of cloth or a small leather pillow be put on the hump. It is well that

[1] σκευασθῇ.

συμφέρει, μόνον προμηθεόμενον ὡς μὴ ἡ σανὶς ὑπὸ σκληρότητος ὀδύνην παρὰ καιρὸν προσπαρέχῃ· κατ' ἴξιν δὲ ἔστω ὡς μάλιστα τῇ ἐντομῇ τῇ ἐς τὸν τοῖχον τὸ ὕβωμα, ὡς ἂν ἡ σανίς, ᾗ μάλιστα ἐξέστηκε, ταύτῃ μάλιστα πιέζῃ ἐπιτε-
80 θεῖσα. ὅταν δὲ ἐπιτεθῇ, τὸν μέν τινα καταναγκάζειν χρὴ τὸ ἄκρον τῆς σανίδος, ἤν τε ἕνα δέῃ ἤν τε δύο, τοὺς δὲ κατατείνειν [1] τὸ σῶμα κατὰ μῆκος, ὡς πρόσθεν εἴρηται, τοὺς μὲν τῇ, τοὺς δὲ τῇ. ἔξεστι δὲ καὶ ὀνίσκοισι τὴν κατάτασιν ποιεῖσθαι, ἢ παρακατορύξαντα παρὰ τὸ ξύλον, ἢ ἐν αὐτῷ τῷ ξύλῳ τὰς φλιὰς τῶν ὀνίσκων ἐντεκτηνάμενον, ἤν τε ὀρθὰς ἐθέλῃς, ἑκατέρωθεν σμικρὸν ὑπερεχούσας, ἤ τε κατὰ κορυφὴν τοῦ ξύλου ἔνθεν καὶ ἔνθεν. αὗται αἱ ἀνάγκαι εὐταμίευτοί εἰσι
90 καὶ ἐς τὸ ἰσχυρότερον καὶ ἐς τὸ ἧσσον, καὶ ἰσχὺν ἔχουσι τοιαύτην, ὥστε καὶ εἴ τις ἐπὶ λύμῃ βούλοιτο, ἀλλὰ μὴ ἐπὶ ἰητρείῃ, ἐς τοιαύτας ἀνάγκας ἀγαγεῖν κἂν [2] τούτῳ ἰσχυρῶς δύνασθαι· καὶ γὰρ ἂν κατατείνων κατὰ μῆκος μοῦνον ἔνθεν καὶ ἔνθεν οὕτω καὶ ἄλλην ἀνάγκην οὐδεμίην προστιθείς, ὅμως κατατείνειεν ἄν τις· ἀλλὰ μὴν καὶ ἢν μὴ κατατείνων, αὐτῇ δὲ μοῦνον τῇ σανίδι οὕτως ἰποίη τις, καὶ οὕτως ἂν [ἱκανῶς] [3] καταναγκάσειεν. καλαὶ οὖν αἱ τοιαῦται ἰσχύες εἰσίν, ᾗσιν ἔξεστι
100 καὶ ἀσθενεστέρῃσι καὶ ἰσχυροτέρῃσι χρῆσθαι αὐτὸν ταμιεύοντα. καὶ μὲν δὴ καὶ κατὰ φύσιν γε ἀναγκάζουσι· τὰ μὲν γὰρ ἐξεστεῶτα ἐς τὴν χώρην ἀναγκάζει ἡ ἴπωσις ἰέναι, τὰ δὲ συνεληλυθότα κατὰ φύσιν κατατείνουσι αἱ κατὰ φύσιν κατατάσιες. οὔκουν [ἐγὼ] [4] ἔχω τούτων ἀνάγκας

[1] κατατανύειν. [2] καὶ ἐν.
[3] Kw. omits. [4] Kw. omits.

it should be as small as possible, only sufficient to prevent the plank from causing needless additional pain by its hardness. Let the hump come as nearly as possible in line with the groove in the wall, so that the plank, when in place, makes most pressure on the most projecting part. When it is put in place, an assistant, or two if necessary, should press down the extremity of the plank, while others extend the body lengthwise, some at one end, some at the other, as was described above. But it is possible to make extension by wheel and axle, either embedded in the earth by the board, or with the supports of the axle carpentered on to the board itself; either projecting upwards a little, if you like, or on the top of the board at each end.[1] This reduction apparatus is easy to regulate as regards greater or less force, and has such power that, if one wanted to use such forcible manœuvres for harm and not for healing, it is able to act strongly in this way also. For even by making traction lengthwise, only at both ends and without any other additional force, one would produce extension. On the other hand, if, without making traction, one only pressed downwards with the plank in this way, one would get reduction thus also. Such forces, then, are good where it is possible for the operator to regulate their use as to weaker or stronger, and, what is more, they are exerted in accordance with nature; for the pressure forces the protruding parts into place, and the extensions according to nature draw asunder naturally the parts which have come together. For my part, then, I know no better or more correct modes of

[1] (?) Projecting horizontally.

καλλίους οὐδὲ δικαιοτέρας· ἡ γὰρ κατ᾽ αὐτὴν τὴν ἄκανθαν ἰθυωρίη τῆς κατατάσιος κάτωθέν τε καὶ κατὰ τὸ ἱερὸν ὀστέον καλεόμενον οὐκ ἔχει ἐπιλαβὴν οὐδεμίην· ἄνωθεν δὲ κατὰ τὸν αὐχένα καὶ
110 κατὰ τὴν κεφαλὴν ἐπιλαβὴν μὲν ἔχει, ἀλλ᾽ ἐσιδέειν γε ἀπρεπὴς ταύτῃ τοι γινομένη ἡ κατάτασις καὶ ἄλλας βλάβας ἂν προσπαρέχοι πλεονασθεῖσα. ἐπειρήθην δὲ δή ποτε ὕπτιον τὸν ἄνθρωπον κατατείνειν, ἀσκὸν ἀφύσητον ὑποθεὶς ὑπὸ τὸ ὕβωμα· κἄπειτα αὐλῷ ἐκ χαλκείου ἐς τὸν ἀσκὸν τὸν ὑποκείμενον ἐνιέναι φυσᾶν· ἀλλά μοι οὐκ εὐπορεῖτο· ὅτε μὲν γὰρ εὖ κατατείνοιμι τὸν ἄνθρωπον, ἡσσᾶτο ὁ ἀσκός, καὶ οὐκ ἠδύνατο ἡ φῦσα ἐσαναγκάζεσθαι· καὶ ἄλλως ἕτοιμον περιο-
120 λισθάνειν ἦν, ἅτε ἐς τὸ αὐτὸ ἀναγκαζόμενον τό τε τοῦ ἀνθρώπου ὕβωμα καὶ τὸ τοῦ ἀσκοῦ πληρουμένου κύρτωμα. ὅτε δ᾽ αὖ μὴ κάρτα κατατείνοιμι τὸν ἄνθρωπον, ὁ μὲν ἀσκὸς ὑπὸ τῆς φύσης ἐκυρτοῦτο· ὁ δὲ ἄνθρωπος πάντῃ μᾶλλον ἐλορδαίνετο ἢ ᾗ συνέφερεν. ἔγραψα δὲ ἐπίτηδες τοῦτο· καλὰ γὰρ καὶ ταῦτα τὰ μαθήματά ἐστιν, ἃ πειρηθέντα ἀπορηθέντα ἐφάνη, καὶ δι᾽ ἅσσα
128 ἠπορήθη.

XLVIII. Ὁπόσοισι δὲ ἐς τὸ ἔσω σκολιαίνονται οἱ σπόνδυλοι ὑπὸ πτώματος, ἢ καὶ ἐμπεσόντος τινὸς βαρέος, εἷς μὲν οὐδεὶς τῶν σπονδύλων μέγα ἐξίσταται κάρτα ὡς ἐπὶ τὸ πολὺ ἐκ τῶν ἄλλων, ἢν δὲ ἐκστῇ μέγα ἢ εἷς ἢ πλείονες, θάνατον φέρουσι· ὥσπερ δὴ καὶ πρόσθεν εἴρηται, κυκλώδης καὶ αὕτη καὶ οὐ γωνιώδης γίνεται ἡ παραλλαγή. οὖρα μὲν οὖν τοῖσι τοιούτοισι καὶ ἀπόπατος μᾶλλον ἵσταται ἢ τοῖσιν ἔξω κυφοῖσι,

reduction than these. For straight-line extension on the spine itself, from below, at the so-called sacred bone (sacrum), gets no grip; from above, at the neck and head, it gets a grip indeed, but extension made here looks unseemly, and would also cause harm if carried to excess. I once tried to make extension with the patient on his back, and, after putting an unblown-up bag under the hump, then tried to blow air into the bag with a bronze tube. But my attempt was not a success, for when I got the man well stretched, the bag collapsed, and air could not be forced into it; it also kept slipping round at any attempt to bring the patient's hump and the convexity of the blown-up bag forcibly together; while when I made no great extension of the patient, but got the bag well blown up, the man's back was hollowed as a whole rather than where it should have been. I relate this on purpose; for those things also give good instruction which after trial show themselves failures,[1] and show why they failed.

XLVIII. In cases where the vertebrae are curved inwards from a fall or the impact of some heavy weight, no single vertebra is much displaced from the others as a rule; and if there is great displacement of one or more, it brings death. But, as was said before, this dislocation also is in the form of a curve and not angular. In such cases, then, retention of urine and faeces is more frequent than in outward curvatures;

[1] "On essay show there's no way" might indicate the play on words.

10 καὶ πόδες καὶ ὅλα τὰ σκέλεα ψύχεται μᾶλλον, καὶ θανατηφόρα ταῦτα μᾶλλον ἐκείνων, καὶ ἢν περιγένωνται δέ, ῥυώδεες τὰ οὖρα μᾶλλον οὗτοι, καὶ τῶν σκελέων ἀκρατέστεροι καὶ ναρκωδέστεροι· ἢν δὲ καὶ ἐν τῷ ἄνω μέρει μᾶλλον τὸ λόρδωμα γένηται, παντὸς τοῦ σώματος ἀκρατέες καὶ νεναρκωμένοι γίνονται. μηχανὴν δὲ οὐκ ἔχω οὐδεμίην ἔγωγε, ὅπως χρὴ τὸν τοιοῦτον ἐς τὸ αὐτὸ καταστῆσαι, εἰ μή τινα ἡ κατὰ[1] τῆς κλίμακος κατάσεισις ὠφελεῖν οἵη τε εἴη, ἢ καὶ 20 ἄλλη τις τοιαύτη ἴησις ἢ κατάτασις, οἵηπερ ὀλίγῳ πρόσθεν εἴρηται. κατανάγκασιν δὲ σὺν τῇ κατατάσει οὐδεμίην ἔχω, ἥτις ἂν γίνοιτο ὥσπερ τῷ κυφώματι τὴν κατανάγκασιν ἡ σανὶς ἐποιεῖτο. πῶς γὰρ ἄν τις ἐκ τοῦ ἔμπροσθεν διὰ τῆς κοιλίης ἀναγκάσαι δύναιτο; οὐ γὰρ οἷόν τε. ἀλλὰ μὴν οὔτε βῆχες οὔτε πταρμοὶ οὐδεμίην δύναμιν ἔχουσιν, ὥστε τῇ κατατάσει συντιμωρεῖν· οὐ μὴν οὐδ' ἔνεσις φύσης ἐνιεμένης ἐς τὴν κοιλίην οὐδὲν ἂν δυνηθείη. καὶ μὴν αἱ 30 μεγάλαι σικύαι προσβαλλόμεναι ἀνασπάσιος εἵνεκα δῆθεν τῶν ἔσω ῥεπόντων σπονδύλων μεγάλη ἁμαρτὰς γνώμης ἐστίν· ἀπωθέουσι γὰρ μᾶλλον ἢ ἀνασπῶσιν· καὶ οὐδ' αὐτὸ τοῦτο γιγνώσκουσι οἱ προσβάλλοντες· ὅσῳ γὰρ ἄν τις μέζω προσβάλλῃ, τοσούτῳ μᾶλλον λορδοῦνται οἱ προσβληθέντες, συναναγκαζομένου ἄνω τοῦ δέρματος. τρόπους τε ἄλλους κατατασίων,[2] ἢ οἷοι πρόσθεν εἴρηνται, ἔχοιμι ἂν εἰπεῖν ἁρμόσαι[3] οὓς ἄν τις δοκέοι[4] τῷ παθήματι μᾶλλον· 40 ἀλλ' οὐ κάρτα πιστεύω αὐτοῖσι· διὰ τοῦτο οὐ γράφω. ἀθρόον δὲ συνιέναι χρὴ περὶ τῶν τοιού-

the feet and lower limbs as a whole more usually lose heat, and these injuries are more generally fatal. Even if they survive, they are more liable to incontinence of urine, and have more weakness and torpor of the legs; while if the incurvation occurs higher up, they have loss of power and complete torpor of the whole body. For my part, I know of no method for reducing such an injury, unless succussion on the ladder may possibly be of use, or other such extension treatment as was described a little above. I have no pressure apparatus combined with extension, which might make pressure reduction, as did the plank in the case of humpback. For how could one use force from the front through the body cavity? It is impossible. Certainly neither coughs nor sneezings have any power to assist extension, nor indeed would inflation of air into the body cavity be able to do anything. Nay more, the application of large cupping instruments, with the idea of drawing out the depressed vertebrae, is a great error of judgment, for they push in rather than draw out; and it is just this which those who apply them fail to see. For the larger the instrument applied, the more the patients hollow their backs, as the skin is drawn together and upwards. I might mention other modes of extension, besides those related above, which would appear more suitable to the lesion; but I have no great faith in them, and therefore do not describe them. As to cases like those summarily mentioned, one

---

[1] διὰ. [2] So Erm., Kw. *κατασεισίων* Littré, Pq.
[3] ἁρμόζειν. [4] ἂν δοκέοντας.

των, ὧν[1] ἐν κεφαλαίῳ εἴρηται, ὅτι τὰ μὲν ἐς τὸ λορδὸν ῥέψαντα ὀλέθριά ἐστιν καὶ σινάμωρα, τὰ δὲ ἐς τὸ κυφὸν ἀσινέα θανάτου, καὶ οὔρων σχεσίων καὶ ἀποναρκωσίων τὸ ἐπίπαν· οὐ γὰρ ἐντείνει τοὺς ὀχετοὺς τοὺς κατὰ τὴν κοιλίην, οὐδὲ κωλύει εὐρόους εἶναι ἡ ἐς τὸ ἔξω κύφωσις· ἡ δὲ λόρδωσις ταῦτά τε ἀμφότερα ποιεῖ καὶ ἐς τὰ ἄλλα πολλὰ προσγίνεται. ἐπεί τοι πολὺ
50 πλέονες σκελέων τε καὶ χειρῶν ἀκρατέες γίνονται, καὶ καταναρκοῦνται τὸ σῶμα, καὶ οὖρα ἴσχεται αὐτοῖσιν οἷσιν ἂν μὴ ἐκστῇ μὲν τὸ ὕβωμα μήτε ἔσω μήτε ἔξω, σεισθέωσι δὲ ἰσχυρῶς ἐς τὴν ἰθυωρίην τῆς ῥάχιος· οἷσι δ' ἂν ἐκστῇ τὸ ὕβωμα,
55 ἧσσον τοιαῦτα πάσχουσι.

XLIX. Πολλὰ δὲ καὶ ἄλλα ἐν ἰητρικῇ ἄν τις τοιαῦτα κατίδοι, ὧν τὰ μὲν ἰσχυρὰ ἀσινέα ἐστὶ καὶ καθ' ἑωυτὰ τὴν κρίσιν ὅλην λαμβάνοντα τοῦ νοσήματος, τὰ δὲ ἀσθενέστερα σινάμωρα, καὶ ἀποτόκους νοσημάτων χρονίους ποιέοντα καὶ κοινωνέοντα τῷ ἄλλῳ σώματι ἐπὶ πλέον. ἐπεὶ καὶ πλευρέων κάτηξις τοιοῦτόν τι πέπονθεν· οἷσι μὲν γὰρ ἂν καταγῇ πλευρή, ἢ μίη ἢ πλέονες, ὡς τοῖσι πλείστοισι κατάγνυται, μὴ διασχόντα τὰ
10 ὀστέα ἐς τὸ ἔσω μέρος μηδὲ ψιλωθέντα, ὀλίγοι μὲν ἤδη ἐπυρέτηναν· ἀτὰρ οὐδὲ αἷμα πολλοὶ ἤδη ἔπτυσαν, οὐδὲ ἔμπυοι πολλοὶ γίνονται, οὐδὲ ἔμμοτοι οὐδὲ ἐπισφακελίσιες τῶν ὀστέων· δίαιτά τε φαύλη ἀρκεῖ· ἢν γὰρ μὴ πυρετὸς συνεχὴς ἐπιλαμβάνηται αὐτούς, κενεαγγεῖν κάκιον τοῖσι τοιούτοισιν ἢ μὴ κενεαγγεῖν, καὶ ἐπωδυνέστερον καὶ πυρετωδέστερον καὶ βηχωδέστερον· τὸ γὰρ πλήρωμα

[1] ὡς.

must bear in mind generally that inward deviations cause death or grievous injury, while those in the form of a hump are not as a rule injuries which cause death, retention of urine, or loss of sensation; for external curvature does not stretch the ducts which pass down the body cavity, nor does it hinder free flow, while inward curvature does both these things, and has many other complications. In fact, many more patients get paralysis of legs and arms, loss of sensation in the body, and retention of urine when there is no displacement either inwards or outwards, but a severe concussion in the line of the backbone; while those who have a hump displacement are less liable to such affections.

XLIX. One may observe in medicine many similar examples of violent lesions which are without harm, and contain in themselves the whole crisis of the malady,[1] while slighter injuries are malignant, producing a chronic progeny of diseases and spreading widely into the rest of the body. Fracture of the ribs is such an affection; for in cases of fractured ribs, whether one or more, as the fracture usually occurs, the bones not being separated and driven inwards or laid bare, we rarely find fever; neither does it come to spitting of blood in many cases, nor do they get empyema or wounds requiring plugs, neither is there necrosis of the bones. An ordinary regimen suffices; for if the patients are not attacked by chronic fever, it is worse to use abstinence in such cases than to avoid it; and it involves greater liability to pain, fever, and coughing; for a moderate fullness

[1] *I.e.* it is confined to the injury itself, and steady recovery ensues.

τὸ μέτριον τῆς κοιλίης, διόρθωμα τῶν πλευρέων γίνεται· ἡ δὲ κένωσις κρεμασμὸν μὲν τῇσι πλευ-
20 ρῇσι ποιεῖ· ὁ δὲ κρεμασμός, ὀδύνην. ἔξωθέν τε αὖ φαύλη ἐπίδεσις τοῖσι τοιούτοισιν ἀρκεῖ· κηρωτῇ καὶ σπλήνεσι καὶ ὀθονίοισιν ἡσύχως ἐρείδοντα, ὁμαλὴν τὴν ἐπίδεσιν ποιεῖσθαι καὶ ἐριῶδές τι προσεπιθέντα. κρατύνεται δὲ πλευρὴ ἐν εἴκοσιν ἡμέρῃσιν· ταχεῖαι γὰρ αἱ ἐπιπωρώσιες
26 τῶν τοιούτων ὀστέων.

L. Ἀμφιφλασθείσης μέντοι τῆς σαρκὸς ἀμφὶ τῇσι πλευρῇσιν ἢ ὑπὸ πληγῆς ἢ ὑπὸ πτώματος ἢ ὑπὸ ἀντερείσιος ἢ ἄλλου τινὸς τοιουτοτρόπου, πολλοὶ ἤδη πολὺ αἷμα ἔπτυσαν· οἱ γὰρ ὀχετοὶ οἱ κατὰ τὸ λαπαρὸν τῆς πλευρῆς ἑκάστης παρατεταμένοι, καὶ οἱ τόνοι ἀπὸ τῶν ἐπικαιροτάτων τῶν ἐν τῷ σώματι τὰς ἀφορμὰς ἔχουσιν· πολλοὶ οὖν ἤδη βηχώδεες καὶ φυματίαι καὶ ἔμπυοι ἐγένοντο καὶ ἔμμοτοι, καὶ ἡ πλευρὴ ἐπεσφακέλισεν αὐτοῖσιν.
10 ἀτὰρ καὶ οἷσιν μηδὲν τοιοῦτον προσεγένετο, ἀμφιφλασθείσης τῆς σαρκὸς ἀμφὶ τῇσι πλευρῇσιν, ὅμως δὲ βραδύτερον ὀδυνώμενοι παύονται οὗτοι ἢ οἷσιν ἂν πλευρὴ καταγῇ, καὶ ὑποστροφὰς μᾶλλον ἴσχει ὀδυνημάτων τὸ χωρίον ἐν τοῖσι τοιούτοισι τρώμασιν ἢ τοῖσι ἑτέροισιν. μάλα μὲν οὖν μετεξέτεροι καταμελέουσιν τῶν τοιούτων σινέων, μᾶλλον ἢ ἢν πλευρὴ καταγῇ αὐτοῖσιν· ἀτὰρ καὶ ἰήσιος σκεθροτέρης οἱ τοιοῦτοι δέονται, εἰ σωφρονοῖεν· τῇ τε γὰρ διαίτῃ συμφέρει συνε-
20 στάλθαι, ἀτρεμεῖν τε τῷ σώματι ὡς μάλιστα, ἀφροδισίων τε ἀπέχεσθαι βρωμάτων τε λιπαρῶν καὶ κερχνωδέων, καὶ ἰσχυρῶν πάντων, φλέβα τε κατ᾽ ἀγκῶνα τέμνεσθαι, σιγᾶν τε ὡς μάλιστα,

of the body cavity tends to adjust the ribs, while emptiness leaves them suspended, and the suspension causes pain. Externally, a simple dressing suffices in such cases, with cerate, compresses and bandages, applying them smoothly with gentle pressure, adding also a little wool. A rib consolidates in twenty days, for callus forms rapidly in bones of this kind.

L. When, however, the flesh is contused about the ribs, either by a blow, fall, encounter, or something else of the sort, we find that many have considerable haemoptysis. For the canals extending along the yielding part of each rib, and the cords,[1] have their origin in the most important parts of the body. Thus we find that many get coughs, tubercles, and internal abscesses, and require plugging with lint; also necrosis of the rib is found in these patients. Besides, when nothing of this kind occurs after contusion of the flesh about the ribs, still these patients get rid of the pain more slowly than in cases where a rib is broken; and the part is more liable to recurrences of pain after such injuries than in the other cases. It is true that many neglect such injuries, as compared with a broken rib; yet such need the more careful treatment, if they would be prudent. It is well to reduce the diet, keep the body at rest as far as possible, avoid sexual intercourse, rich foods and those which excite coughing, and all strong nourishment; to open a vein at the elbow, observe silence as much as possible, dress

[1] Nerves.

ἐπιδεῖσθαί τε τὸ χωρίον τὸ φλασθὲν σπλήνεσι μὴ πολυπτύχοισι, συχνοῖσι δὲ καὶ πολὺ πλατυτέροισι πάντη τοῦ φλάσματος, κηρωτῇ τε ὑποχρίειν,[1] ὀθονίοισί τε πλατέσι σὺν ταινίῃσι πλατείῃσι καὶ μαλθακῇσι ἐπιδεῖν, ἐρείδειν τε μετρίως, ὥστε μὴ κάρτα πεπιέχθαι φάναι τὸν
30 ἐπιδεδεμένον, μηδ' αὖ χαλαρόν· ἄρχεσθαι δὲ τὸν ἐπιδέοντα κατὰ τὸ φλάσμα, καὶ ἐρηρεῖσθαι ταύτῃ μάλιστα, τὴν δὲ ἐπίδεσιν ποιεῖσθαι ὡς ἀπὸ δύο ἀρχέων, ἐπιδεῖν τε, ἵνα μὴ περιῤῥεπὲς τὸ δέρμα τὸ περὶ τὰς πλευρὰς ᾖ, ἀλλ' ἰσόῤῥοπον· ἐπιδεῖν δὲ ἢ καθ' ἑκάστην ἡμέρην ἢ παρ' ἑτέρην. ἄμεινον δὲ καὶ κοιλίην μαλθάξαι κούφῳ τινὶ ὅσον κενώσιος εἵνεκεν τοῦ σίτου, καὶ ἐπὶ μὲν δέκα ἡμέρας ἰσχναίνειν, ἔπειτα ἀναθρέψαι τὸ σῶμα καὶ ἁπαλῦναι· τῇ δὲ ἐπιδέσει, ἔστ' ἂν μὲν
40 ἰσχναίνῃς, ἐρηρεισμένῃ μᾶλλον χρῆσθαι, ὁπόταν δὲ ἐς τὸν ἁπαλυσμὸν ἄγῃς, ἐπιχαλαρωτέρῃ. καὶ ἢν μὲν αἷμα ἀποπτύσῃ καταρχάς, τεσσαρακονθήμερον τὴν μελέτην καὶ τὴν ἐπίδεσιν ποιεῖσθαι χρή· ἢν δὲ μὴ πτύσῃ τὸ αἷμα, ἀρκεῖ ἐν εἴκοσιν ἡμέρῃσιν ἡ μελέτη ὡς ἐπὶ τὸ πολύ· τῇ ἰσχύϊ δὲ τοῦ τρώματος τοὺς χρόνους προτεκμαίρεσθαι χρή. ὅσοι δ' ἂν ἀμελήσωσι τῶν τοιούτων ἀμφιφλασμάτων, ἢν καὶ ἄλλο μηδὲν αὐτοῖσι φλαῦρον μέζον γένηται, ὅμως τό γε χωρίον
50 ἀμφιφλασθὲν μυξωδεστέρην τὴν σάρκα ἴσχει ἢ πρόσθεν εἶχεν. ὅπου δέ τι τοιοῦτον ἐγκαταλείπεται, καὶ μὴ εὖ ἐξιποῦται τῇ γε ἀλθέξει, φαυλότερον μέν, ἢν παρ' αὐτὸ τὸ ὀστέον ἐγκαταλειφθῇ τὸ μυξῶδες· οὔτε γὰρ ἔτι ἡ σὰρξ ὁμοίως ἅπτεται τοῦ ὀστέου, τό τε ὀστέον νοση-

the contused part with pads not much folded, but numerous, and extending in every direction a good way beyond the contusion. Anoint first[1] with cerate, and bandage with broad, soft linen bands, making them suitably firm, so that the patient says there is no great pressure, nor on the other hand is it slack. The dresser should begin at the contusion, and make most pressure there; and the bandaging should be done as with a two-headed roller, in such a way that the skin may not get in folds at the ribs, but lie evenly. Change the dressing every day or every other day. It is rather a good thing to relax the bowels with something mild, sufficiently to clear out the food, and give low diet for ten days. Then nourish the body and plump it up. During the attenuation period, use rather tighter bandaging, but more relaxed when you come to the plumping up. If there is haemoptysis to begin with, the treatment and bandaging should be kept up for forty days; if there is no haemoptysis a twenty-day course of treatment usually suffices. The forecast as to time should be made from the gravity of the wound. In cases where such contusions are neglected, even if nothing worse happens to them, still the tissues in the contused part contain more mucus than they did before. When anything of this kind is left behind and not well squeezed out by the curative process, it is worse if the mucoid substance is left in the region of the bone itself; for the flesh no longer adheres so closely to the bone, and the

[1] Cf. *Fract.* XXI for ὑποχρίω.

---

[1] ὑπαλείφειν.

ρότερον γίνεται, σφακελισμοί τε χρόνιοι ὀστέου πολλοῖσιν ἤδη ἀπὸ τῶν τοιούτων προφασίων ἐγένοντο. ἀτὰρ καὶ ἢν μὴ παρὰ τὸ ὀστέον, ἀλλ' αὐτὴ ἡ σὰρξ μυξώδης ᾖ, ὅμως ὑποστροφαὶ
30 γίνονται καὶ ὀδύναι ἄλλοτε καὶ ἄλλοτε, ἤν τις τῷ σώματι τύχῃ πονήσας· καὶ διὰ τοῦτο τῇ ἐπιδέσει χρῆσθαι χρή, ἅμα μὲν ἀγαθῇ, ἅμα δὲ ἐπὶ πολὺ προηκούσῃ, ἕως ἂν ξηρανθῇ μὲν καὶ ἀναποθῇ τὸ ἐκχύμωμα τὸ ἐν τῇ φλάσει ἐγγενόμενον, αὐξηθῇ δὲ σαρκὶ ὑγιέϊ τὸ χωρίον, ἅψηται δὲ τοῦ ὀστέου ἡ σάρξ. οἷσι δ' ἂν ἀμεληθεῖσι χρονιωθῇ καὶ ὀδυνῶδες τὸ χωρίον γένηται, καὶ ἡ σὰρξ ὑπόμυξος [ᾖ],[1] τούτοισι καῦσις ἴησις ἀρίστη. καὶ ἢν μὲν αὐτὴ ἡ σὰρξ μυξώδης ᾖ,
70 ἄχρι τοῦ ὀστέου καίειν χρή, μὴ μὴν διαθερμανθῆναι τὸ ὀστέον· ἢν δὲ μεσηγὺ τῶν πλευρῶν ᾖ, ἐπιπολῆς μὲν οὐδὲ οὕτω χρὴ καίειν, φυλάσσεσθαι μέντοι μὴ διακαύσῃς πέρην. ἢν δὲ πρὸς τῷ ὀστέῳ δοκῇ εἶναι τὸ φλάσμα, καὶ ἔτι νεαρὸν ᾖ, καὶ μήπω σφακελίσῃ τὸ ὀστέον, ἢν μὲν κάρτα ὀλίγον ᾖ, οὕτω καίειν χρὴ ὥσπερ εἴρηται· ἢν μέντοι παραμηκὴς ᾖ ὁ μετεωρισμὸς ὁ κατὰ τὸ ὀστέον, πλέονας ἐσχάρας ἐμβάλλειν χρή· περὶ δὲ σφακελισμοῦ
79 πλευρῆς ἅμα τῇ τῶν ἐμμότων ἰητρείῃ εἰρήσεται.

LI. Ἢν δὲ μηροῦ ἄρθρον ἐξ ἰσχίου ἐκπέσῃ, ἐκπίπτει δὲ κατὰ τέσσαρας τρόπους, ἐς μὲν τὸ ἔσω πολὺ πλειστάκις, ἐς δὲ τὸ ἔξω τῶν ἄλλων πλειστάκις· ἐς δὲ τὸ ὄπισθεν καὶ τὸ ἔμπροσθεν ἐκπίπτει μέν, ὀλιγάκις δέ. ὁπόσοισι μὲν οὖν ἂν ἐκβῇ ἐς τὸ ἔσω, μακρότερον τὸ σκέλος φαίνεται, παραβαλλόμενον πρὸς τὸ ἕτερον, διὰ δισσὰς προ-

[1] B Kw. and most MSS. omit

latter becomes more subject to disease. Chronic necroses of bone are found to arise in many cases from causes like these. Besides, even if the mucoid part is not along the bone, but involves the flesh itself, still relapses occur, and periodical pains, whenever one happens to have bodily trouble; and therefore one should use bandaging, both careful and prolonged, for some time, till the exudation formed in the bruise is dried up and consumed, the part filled with healthy flesh, and the flesh firmly attached to the bone. In neglected cases which have become chronic, when the part is painful and the flesh rather mucous, the best treatment is cauterising. If the flesh itself is mucous, one should cauterise down to the bone, but avoid greatly heating the latter. If it is intercostal, the cauterisation should, even so, not be superficial; yet one should take care not to burn right through. If the contusion appears to have reached the bone, and is still fresh, and the bone not yet necrosed, if it be quite small, one should cauterise as directed; but if there is an elongated tumefaction over the bone, one should make several eschars. Necrosis of a rib will be considered along with the treatment of patients with discharging abscesses.

LI. When the head of the thigh-bone is dislocated from the hip, it is dislocated in four ways, far most frequently inwards; and of the others the most frequent is outwards. Dislocation backwards and forwards occurs, but is rare. In cases where it is displaced inwards, the leg appears longer when placed beside the other, naturally so, for a double

φάσιας εἰκότως· ἐπί τε γὰρ τὸ ἀπὸ τοῦ ἰσχίου πεφυκὸς ὀστέον, τὸ ἄνω φερόμενον πρὸς τὸν
10 κτένα, ἐπὶ τοῦτο ἡ ἐπίβασις τῆς κεφαλῆς τοῦ μηροῦ γίνεται, καὶ ὁ αὐχὴν τοῦ ἄρθρου ἐπὶ τῆς κοτύλης ὀχεῖται· ἔξωθέν τε αὖ γλουτὸς κοῖλος φαίνεται, ἅτε ἔσω ῥεψάσης τῆς κεφαλῆς τοῦ μηροῦ, τό τε αὖ κατὰ τὸ γόνυ τοῦ μηροῦ ἄκρον ἀναγκάζεται ἔξω ῥέπειν, καὶ ἡ κνήμη καὶ ὁ πους ὡσαύτως. ἅτε οὖν ἔξω ῥέποντος τοῦ ποδός, οἱ ἰητροὶ δι' ἀπειρίην τὸν ὑγιέα πόδα πρὸς τοῦτον προσίσχουσιν, ἀλλ' οὐ τοῦτον πρὸς τὸν ὑγιέα· διὰ τοῦτο πολὺ μακρότερον φαίνεται τὸ σιναρὸν
20 τοῦ ὑγιέος· πολλαχῇ δὲ καὶ ἄλλῃ τὰ τοιαῦτα παρασύνεσιν ἔχει. οὐ μὴν οὐδὲ συγκάμπτειν δύνανται κατὰ τὸν βουβῶνα ὁμοίως τῷ ὑγιέϊ· ἀτὰρ καὶ ψαυομένη ἡ κεφαλὴ τοῦ μηροῦ κατὰ τὸν περίναιον ὑπερογκέουσα εὔδηλός ἐστιν. τὰ μὲν οὖν σημεῖα ταῦτά ἐστιν, οἷσιν ἂν ἔσω ἐκπεπτώκῃ
26 ὁ μηρός.

LII. Οἷσι μὲν οὖν ἂν ἐκπεσὼν μὴ ἐμπέσῃ, ἀλλὰ καταπορηθῇ καὶ[1] ἀμεληθῇ, ἥ τε ὁδοιπορίη περιφοράδην τοῦ σκέλεος ὥσπερ τοῖσι βουσὶ γίνεται, καὶ ἡ ὄχησις πλείστη αὐτοῖσιν ἐπὶ τοῦ ὑγιέος σκέλεός ἐστιν. καὶ ἀναγκάζονται κατὰ τὸν κενεῶνα καὶ κατὰ τὸ ἄρθρον τὸ ἐκπεπτωκὸς κοῖλοι καὶ σκολιοὶ εἶναι· κατὰ δὲ τὸ ὑγιὲς ἐς τὸ ἔξω ὁ γλουτὸς ἀναγκάζεται περιφερὴς εἶναι· εἰ γάρ τις ἔξω τῷ ποδὶ τοῦ ὑγιέος σκέλεος βαίνοι,
10 ἀπωθέοι ἂν τὸ σῶμα τὸ ἄλλο ἐς τὸ σιναρὸν σκέλος τὴν ὄχησιν ποιεῖσθαι· τὸ δὲ σιναρὸν οὐκ

[1] καὶ = ἤ. Cf. Thucyd. II. 35.

reason ; for the dislocation of the head of the femur takes place on to the bone arising from the ischium and passing up to the pubes, and its neck is supported against the cotyloid cavity.[1] Besides, the buttock looks hollow on the outer side, because the head of the femur is turned inwards ; again, the end of the femur at the knee is compelled to turn outwards, and the leg and the foot likewise. Thus, as the foot inclines outwards, practitioners through inexperience bring the foot of the sound limb to it, instead of bringing it to the sound one. This makes the damaged limb appear much longer than the sound one ; and this sort of thing causes misapprehension in a variety of other ways. The patients, moreover, cannot bend at the groin so well as one with a sound limb ; and for the rest, on palpating the head of the femur, it is manifest as an abnormal prominence at the perineum.[2] These then are the signs in cases of internal dislocation of the thigh.

LII. In cases where the dislocation is not reduced, but is given up or neglected, progression is accomplished, as in oxen, by bringing the leg round ; and they throw most of their weight on the sound leg. They are also of necessity curved in and distorted in the region of the loin and the dislocated joint, while on the sound side the buttock is necessarily rounded outwards. For if one were to walk with the foot of the sound leg turned out, he would thrust the body over, and put its weight on the injured leg ;

[1] *I.e.* lower rim of the acetabulum ; so Littré, Pq. Adams suggests the perforation below the pubic bone (thyroid). As already remarked the frequency and nature of this dislocation are hard to understand.

[2] Evidently understood in a wide sense, to include inner part of groin.

ἂν δύναιτο ὀχεῖν· πῶς γάρ; ἀναγκάζεται οὖν οὕτω κατὰ τοῦ ὑγιέος σκέλεος τῷ ποδὶ ἔσω βαίνειν, ἀλλὰ μὴ ἔξω· οὕτω γὰρ ὀχεῖ μάλιστα τὸ σκέλος τὸ ὑγιὲς καὶ τὸ ἑωυτοῦ μέρος τοῦ σώματος καὶ τὸ τοῦ σιναροῦ σκέλεος μέρος. κοιλαινόμενοι δὲ κατὰ τὸν κενεῶνα καὶ κατὰ τὰ ἄρθρα, σμικροὶ φαίνονται καὶ[1] ἀντερείδεσθαι ἀναγκάζονται πλάγιοι κατὰ τὸ ὑγιὲς σκέλος·
20 δέονται γὰρ ἀντικοντώσιος ταύτῃ· ἐπὶ τοῦτο γὰρ οἱ γλουτοὶ ῥέπουσι, καὶ τὸ ἄχθος τοῦ σώματος ὀχεῖται[2] ἐπὶ τοῦτο. ἀναγκάζονται δὲ καὶ ἐπικύπτειν· τὴν γὰρ χεῖρα τὴν κατὰ τὸ σκέλος τὸ σιναρὸν ἀναγκάζονται κατὰ πλάγιον τὸν μηρὸν ἐρείδειν· οὐ γὰρ δύναται τὸ σιναρὸν σκέλος ὀχεῖν τὸ σῶμα ἐν τῇ μεταλλαγῇ τῶν σκελέων, ἢν μὴ κατέχηται πρὸς τὴν γῆν πιεζόμενον. ἐν τοιούτοισι[3] οὖν τοῖσι σχήμασιν ἀναγκάζονται ἐσχηματίσθαι, οἷσιν ἂν ἔσω ἐκβὰν
30 τὸ ἄρθρον μὴ ἐμπέσῃ, οὐ προβουλεύσαντος τοῦ ἀνθρώπου ὅπως ἂν ῥήϊστα ἐσχηματισμένον[4] ᾖ, ἀλλ' αὐτὴ ἡ συμφορὴ διδάσκει ἐκ τῶν παρεόντων τὰ ῥήϊστα αἱρεῖσθαι. ἐπεὶ καὶ ὁπόσοι[5] ἕλκος ἔχοντες ἐν ποδὶ ἢ κνήμῃ οὐ κάρτα δύνανται ἐπιβαίνειν τῷ σκέλει, πάντες, καὶ οἱ νήπιοι, οὕτως ὁδοιποροῦσιν· ἔξω γὰρ βαίνουσι τῷ σιναρῷ σκέλει· καὶ δισσὰ κερδαίνουσι, δισσῶν γὰρ δέονται· τό τε γὰρ σῶμα οὐκ ὀχεῖται ὁμοίως ἐπὶ τοῦ ἔξω ἀποβαινομένου ὥσπερ ἐπὶ τοῦ ἔσω·
40 οὐδὲ γὰρ κατ' ἰθυωρίην αὐτῷ γίνεται τὸ ἄχθος, ἀλλὰ πολὺ μᾶλλον ἐπὶ τοῦ ὑποβαινομένου· κατ' ἰθυωρίην γὰρ αὐτῷ γίνεται τὸ ἄχθος, ἔν τε αὐτῇ τῇ ὁδοιπορίῃ καὶ τῇ μεταλλαγῇ τῶν σκελέων.

and the injured limb could not carry it. How should it? He is thus obliged to walk with the foot of the sound leg turned in and not out; for in this way the sound limb is best able to carry both its own share of the body and that of the injured one. But, owing to the inward curvature at the loin and at the joints, they appear short, and patients have to support themselves laterally on the side of the sound leg with a crutch. They want a prop there, because the buttocks incline that way, and the weight of the body lies in that direction. They are also obliged to stoop; for they have to press the hand on the side of the injured leg laterally against the thigh, since the injured limb cannot support the body during the change of legs, unless it is kept down on the ground by pressure. Such then are the attitudes which patients are obliged to assume in unreduced internal dislocation of the hip—not as a result of previous deliberation by the patient as to what will be the easiest attitude; but the lesion itself teaches him to choose the easiest available. So too those who, when they have a wound on the foot or leg, can hardly use the limbs—all of them, even young children, walk in this way. They turn the injured leg out in walking, and get a double boon to match a double need; for the body is not borne equally on the limb brought outwards and on that brought in, since the weight is not perpendicular to it, but comes much more on the limb that is brought under; the weight is perpendicular to the latter both in actual walking and in the

[1] ξύλῳ τῳ K. τῷ ξύλῳ Littré. Pq. omits.
[2] ἐγκεῖται.
[3] τούτοισιν.
[4] ἐσχηματισμένος.
[5] ὅσοι.

ἐν τούτῳ τῷ σχήματι τάχιστα ἂν δύναιτο ὑποτιθέναι τὸ ὑγιὲς σκέλος, ἣν[1] τῷ μὲν σιναρῷ ἐξωτέρω βαίνοι, τῷ δὲ ὑγιέϊ ἐσωτέρω. περὶ οὗ οὖν ὁ λόγος, ἀγαθὸν εὑρίσκεσθαι αὐτὸ ἑωυτῷ τὸ σῶμα ἐς τὰ ῥήϊστα τῶν σχημάτων. ὅσοισι μὲν οὖν μήπω τετελειωμένοισιν ἐς αὔξησιν ἐκπεσὼν
50 μὴ ἐμπέσῃ, γυιοῦται ὁ μηρὸς καὶ ἡ κνήμη καὶ ὁ πούς· οὔτε γὰρ τὰ ὀστέα ἐς τὸ μῆκος ὁμοίως αὔξεται, ἀλλὰ βραχύτερα γίνεται, μάλιστα δὲ τὸ τοῦ μηροῦ, ἄσαρκόν τε ἅπαν τὸ σκέλος καὶ ἄμυον καὶ ἐκτεθηλυσμένον καὶ λεπτότερον γίνεται, ἅμα μὲν διὰ τὴν στέρησιν τῆς χώρης τοῦ ἄρθρου, ἅμα δὲ ὅτι ἀδύνατον χρῆσθαί ἐστιν, ὅτι οὐ κατὰ φύσιν κεῖται· χρῆσις γὰρ μετεξετέρη ῥύεται τῆς ἄγαν ἐκθηλύνσιος· ῥύεται δέ τι καὶ τῆς ἐπὶ μῆκος ἀναυξήσιος. κακοῦται μὲν
60 οὖν μάλιστα οἷσιν ἂν ἐν γαστρὶ ἐοῦσιν ἐξαρθρήσῃ τοῦτο τὸ ἄρθρον, δεύτερον δὲ οἷσιν ἂν ὡς νηπιωτάτοισιν ἐοῦσιν, ἥκιστα δὲ τοῖσι τετελειωμένοισιν. τοῖσι μὲν οὖν τετελειωμένοισιν εἴρηται οἵη τις ἡ ὁδοιπορίη γίνεται· οἷσι δ' ἂν νηπίοισιν ἐοῦσιν ἡ συμφορὴ αὕτη γένηται, οἱ μὲν πλεῖστοι καταβλακεύουσι[2] τὴν διόρθωσιν τοῦ σώματος, ἀλλὰ [κακῶς][3] εἰλέονται ἐπὶ τὸ ὑγιὲς σκέλος, τῇ χειρὶ πρὸς τὴν γῆν ἀπερειδόμενοι τῇ κατὰ τὸ ὑγιὲς σκέλος. καταβλακεύουσι δὲ ἔνιοι τὴν ἐς
70 ὀρθὸν ὁδοιπορίην καὶ οἷσιν ἂν τετελειωμένοισι αὕτη ἡ συμφορὴ γένηται. ὁπόσοι δ' ἂν νήπιοι ἐόντες ταύτῃ τῇ συμφορῇ χρησάμενοι ὀρθῶς παιδαγωγηθέωσι,[4] τῷ μὲν ὑγιέϊ σκέλει χρέονται[5] ἐς ὀρθόν, ὑπὸ δὲ τὴν μασχάλην τὴν κατὰ τὸ

[1] εἰ. [2] καταμβλακεύουσι *bis*.

change of legs. It is in this attitude, with the injured leg rather outwards and the sound one rather inwards, that one can most rapidly put the sound limb under. As regards our subject, then, it is good that the body finds out for itself the easiest posture. When it is in persons who have not yet completed their growth that the hip remains unreduced after dislocation, the thigh is maimed, and the leg and foot also. The bones do not grow to their normal length, but are shorter, especially that of the thigh; while the whole leg is deficient in flesh and muscle, and becomes flaccid and attenuated. This is due at once to the head of the bone being out of place and to the impossibility of using it in its abnormal position; for a certain amount of exercise saves it from excessive flaccidity, and in some degree prevents the defective growth in length. Thus the greatest damage is done to those in whom this joint is dislocated *in utero*; next, to those who are very young; and least to adults. In the case of adults, their mode of walking has been described; but when this accident occurs in those who are very young, for the most part they lack energy to keep the body up, but they crawl about [miserably] on the sound leg, supporting themselves with the hand on the sound side on the ground. Some even among those to whom this accident happens when adult lack the energy to walk standing up; but when persons are afflicted by this accident in early childhood and are properly trained, they use the sound leg to stand up

---

[3] Kw. omits; also B and the best MSS.
[4] Kw.'s correction for παιδαγωγηθῶσι codd.
[5] χρέωνται Kw.

ὑγιὲς σκέλος σκίπωνα περιφέρουσι, μετεξέτεροι δὲ καὶ ὑπ' ἀμφοτέρας τὰς χεῖρας· τὸ δὲ σιναρὸν σκέλος μετέωρον ἔχουσι, καὶ τοσούτῳ ῥηΐους εἰσίν, ὅσῳ ἂν αὐτοῖσιν ἔλασσον τὸ σκέλος τὸ σιναρὸν ᾖ· τὸ δὲ ὑγιὲς ἰσχύει αὐτοῖσιν οὐδὲν
80 ἧσσον ἢ εἰ καὶ ἀμφότερα ὑγιέα ἦν. θηλύνονται δὲ πᾶσι τοῖσι τοιούτοισι αἱ σάρκες τοῦ σκέλεος, μᾶλλον δέ τι θηλύνονται αἱ ἐκ τοῦ ἔξω μέρεος ἢ
83 αἱ ἐκ τοῦ ἔσω ὡς ἐπὶ πολύ.

LIII. Μυθολογοῦσι[1] δέ τινες, ὅτι αἱ Ἀμαζωνίδες τὸ ἄρσεν γένος τὸ ἑωυτῶν αὐτίκα νήπιον ἐὸν ἐξαρθρέουσιν, αἱ μὲν κατὰ [τὰ][2] γούνατα, αἱ δὲ κατὰ τὰ ἰσχία, ὡς δῆθεν χωλὰ γίνοιτο, καὶ μὴ ἐπιβουλεύοι τὸ ἄρσεν γένος τῷ θήλει· χειρώναξιν ἄρα τούτοισι χρέονται,[3] ὁπόσα ἢ σκυτείης ἔργα ἢ χαλκείης, ἢ ἄλλο τι ἑδραῖον ἔργον. εἰ μὲν οὖν ἀληθέα ταῦτά ἐστιν, ἐγὼ μὲν οὐκ οἶδα· ὅτι δὲ γίνοιτο ἂν τοιαῦτα οἶδα, εἴ τις ἐξαρθρέοι αὐτίκα
10 νήπια ἐόντα. κατὰ μὲν οὖν τὰ ἰσχία μέζον τὸ διάφορόν ἐστιν ἐς τὸ ἔσω ἢ ἐς τὸ ἔξω ἐξαρθρῆσαι· κατὰ δὲ τὰ γούνατα διαφέρει μέν τι, ἔλασσον δέ τι διαφέρει. τρόπος δὲ ἑκατέρου τοῦ χωλώματος ἴδιός ἐστιν· κυλλοῦνται[4] μὲν γὰρ μᾶλλον οἷσιν ἂν ἐς τὸ ἔξω ἐξαρθρήσῃ· ὀρθοὶ δὲ ἧσσον ἵστανται οἷσιν ἂν ἐς τὸ ἔσω ἐξαρθρήσῃ. ὡσαύτως δὲ καὶ ἢν παρὰ τὸ σφυρὸν ἐξαρθρήσῃ, ἢν μὲν ἐς τὸ ἔξω μέρος, κυλλοὶ μὲν γίνονται, ἑστάναι δὲ δύνανται· ἢν δὲ ἐς τὸ ἔσω μέρος, βλαισοὶ μὲν γίνονται,
20 ἧσσον δὲ ἑστάναι δύνανται. ἥ γε μὴν συναύξησις τῶν ὀστέων τοιήδε γίνεται· οἷσι μὲν ἂν τὸ κατὰ τὸ

[1] Μυθολογέουσι Kw.
[2] Littré's insertion, but Galen also has it.
[3] χρέωνται Kw.
[4] Erm. Pq. for γυιοῦνται vulg.

on, but carry a crutch under the armpit on that side, and some of them under both arms. As for the injured leg, they keep it off the ground, and do so the more easily, because in them the injured leg is smaller; but their sound leg is as strong as if both were sound. In all such cases the fleshy parts of the leg are flaccid; and, as a general rule, they are more flaccid on the outer than on the inner side.

LIII. Some tell a tale how the Amazons dislocate the joints of their male offspring in early infancy (some at the knees and some at the hips), that they may, so it is said, become lame, and the males be incapable of plotting against the females. They are supposed to use them as artisans in all kinds of leather or copper work, or some other sedentary occupation. For my part, I am ignorant whether this is true; but I know that such would be the result of dislocating the joints of young infants. At the hips there is a marked difference between inward and outward dislocation; but at the knees, though there is a certain difference, it is less. In each case there is a special kind of lameness. Those in whom the dislocation [at the knee] is outwards are more bandy-legged, while those in whom it is inwards[1] are less able to stand erect. Similarly, when the dislocation is at the ankle, if it is outwards, they become club-footed,[2] but are able to stand; while if it is inwards, they become splay-footed, and are less able to stand. As regards growth of the bones, the following is what happens: when the bone of the

[1] *I.e.* the knock-kneed.

[2] *I.e.* leg outwards and foot inwards, and vice versa. The knock-kneed and splay-footed are worse off than the bandy-legged and club-footed.

σφυρὸν ὀστέον τὸ τῆς κνήμης[1] ἐκστῇ, τούτοισι μὲν τὰ τοῦ ποδὸς ὀστέα ἥκιστα συναύξεται, ταῦτα γὰρ ἐγγυτάτω τοῦ τρώματός ἐστιν, τὰ δὲ τῆς κνήμης ὀστέα αὔξεται μέν, οὐ πολὺ δὲ ἐνδεεστέρως, αἱ μέντοι σάρκες μινύθουσι. οἷσι δ' ἂν κατὰ μὲν τὸ σφυρὸν μένῃ τὸ ἄρθρον κατὰ φύσιν, κατὰ δὲ τὸ γόνυ ἐξεστήκῃ, τούτοισι τὸ τῆς κνήμης ὀστέον οὐκ ἐθέλει συναυξάνεσθαι ὁμοίως, ἀλλὰ
30 βραχύτερον γίνεται, τοῦτο γὰρ ἐγγυτάτω τοῦ τρώματός ἐστιν, τοῦ μέντοι ποδὸς τὰ ὀστέα μινύθει μέν, ἀτὰρ οὐχ ὁμοίως, ὥσπερ ὀλίγον τι πρόσθεν εἴρηται, ὅτι τὸ ἄρθρον τὸ παρὰ τὸν πόδα σῶόν ἐστι. εἰ δέ οἱ χρῆσθαι ἠδύναντο, ὥσπερ καὶ τῷ κυλλῷ, ἔτι ἂν ἧσσον ἐμινύθει τὰ τοῦ ποδὸς ὀστέα τούτοισιν. οἷσι δ' ἂν κατὰ τὸ ἰσχίον ἡ ἐξάρθρησις γένηται, τούτοισι τοῦ μηροῦ τὸ ὀστέον οὐκ ἐθέλει συναυξάνεσθαι ὁμοίως, τοῦτο γὰρ ἐγγυτάτω τοῦ τρώματός ἐστιν, ἀλλὰ βραχύτερον
40 τοῦ ὑγιέος γίνεται· τὰ μέντοι τῆς κνήμης ὀστέα οὐχ ὁμοίως τούτοισιν ἀναυξέα γίνεται, οὐδὲ τὰ τοῦ ποδός, διὰ τοῦτο δέ, ὅτι τὸ τοῦ μηροῦ ἄρθρον τὸ παρὰ τὴν κνήμην ἐν τῇ ἑωυτοῦ φύσει μένει, καὶ τὸ τῆς κνήμης τὸ παρὰ τὸν πόδα· σάρκες μέντοι μινύθουσι παντὸς τοῦ σκέλεος τούτοισιν. εἰ μέντοι χρῆσθαι τῷ σκέλει ἠδύναντο, ἔτι ἂν μᾶλλον τὰ ὀστέα συνηύξανετο, ὡς καὶ πρόσθεν εἴρηται, πλὴν τοῦ μηροῦ, κἂν ἧσσον ἄσαρκα εἴη, ἀσαρκότερα δὲ πολλῷ ἢ εἰ ὑγιέα ἦν. σημεῖον δὲ
50 ὅτι ταῦτα τοιαῦτά ἐστιν· ὁπόσοι γάρ, τοῦ βραχίονος ἐκπεσόντος, γαλιάγκωνες ἐγένοντο ἐκ γενεῆς, ἢ καὶ ἐν αὐξήσει πρὶν[2] τελειωθῆναι, οὗτοι τὸ μὲν ὀστέον τοῦ βραχίονος βραχὺ ἴσχουσι, τὸν

leg at the ankle is dislocated, the bones of the foot show least growth, for they are nearest the injury, but growth of the leg-bones is not very deficient; the tissues however are atrophied. In cases where the ankle-joint keeps its natural position while there is dislocation at the knee, the bone of the leg will not grow like the other, but is shortened; for this is nearest the injury. The bones of the foot are atrophied, but not to the same extent as was noticed a little above, because the joint at the foot is intact; and should they be able to use the part, as is the case even in club-foot, the bones of the foot in their case would be still less atrophied. When the dislocation occurs at the hip, the thigh-bone will not grow like the other, for it is nearest the injury; but it gets shorter than the sound one; the bones of the leg, however, do not stop growing in the same way, nor do those of the foot, because the end of the thigh-bone at the knee keeps its natural place, also that of the leg at the foot; but the tissues of the whole leg are atrophied in these cases. But if they were able to use the leg, the bones would correspond in growth to a still greater extent, the thigh excepted, as was said before; and they would be less deficient in flesh, though much more so than if the limb were sound. Here is a proof that these things are so: those who become weasel-armed owing to dislocation of the shoulder either congenitally or during adolescence, and before they become adults, have the bone of the upper arm short, but the forearm and

[1] This is curious phrasing. Cf. remarks on the astragalus in Introduction and notes on ankle dislocation, *Mochl.* XXX.

[2] καὶ πρὶν Kw.

δὲ πῆχυν καὶ ἄκρην τὴν χεῖρα ὀλίγῳ ἐνδεεστέρην τοῦ ὑγιέος, διὰ ταύτας τὰς προφάσιας τὰς εἰρημένας, ὅτι ὁ μὲν βραχίων ἐγγυτάτω [τοῦ ἄρθρου] τοῦ τρώματός ἐστιν, ὥστε διὰ τοῦτο βραχύτερος ἐγένετο· ὁ δὲ αὖ πῆχυς διὰ τοῦτο οὐχ ὁμοίως ἐνακούει τῆς συμφορῆς, ὅτι τὸ τοῦ βραχίονος
60 ἄρθρον τὸ πρὸς τοῦ πήχεος ἐν τῇ ἀρχαίῃ φύσει μένει, ἥ τε αὖ χεὶρ ἄκρη ἔτι τηλοτέρω ἄπεστιν ἢ ὁ πῆχυς ἀπὸ τῆς συμφορῆς. διὰ ταύτας οὖν τὰς εἰρημένας προφάσιας, τῶν ὀστέων τά τε μὴ συναυξανόμενα οὐ συναυξάνεται, τά τε συναυξανόμενα συναυξάνεται. ἐς δὲ τὸ εὔσαρκον τῇ χειρὶ καὶ τῷ βραχίονι ἡ ταλαιπωρίη τῆς χειρὸς μέγα προσωφελεῖ· ὅσα γὰρ χειρῶν ἔργα ἐστί, τὰ πλεῖστα προθυμέονται οἱ γαλιάγκωνες ἐργάζεσθαι τῇ χειρὶ ταύτῃ, ὅσα περ καὶ τῇ ἑτέρῃ δύνανται
70 οὐδὲν ἐνδεεστέρως τῆς ἀσινέος· οὐ γὰρ δεῖ ὀχεῖσθαι τὸ σῶμα ἐπὶ τῶν χειρῶν ὡς ἐπὶ τῶν σκελέων, ἀλλὰ κοῦφα αὐτοῖσι τὰ ἔργα ἐστίν. διὰ δὲ τὴν χρῆσιν οὐ μινύθουσιν αἱ σάρκες αἱ κατὰ τὴν χεῖρα καὶ κατὰ τὸν πῆχυν τοῖσι γαλιάγκωσιν· ἀλλὰ καὶ ὁ βραχίων τι προσωφελεῖται ἐς εὐσαρκίην διὰ ταῦτα.[1] ὅταν δὲ ἰσχίον ἐκπαλὲς γένηται ἐς τὸ ἔσω μέρος ἐκ γενεῆς, ἢ καὶ ἔτι νηπίῳ ἐόντι, μινύθουσιν αἱ σάρκες διὰ τοῦτο μᾶλλον ἢ τῆς χειρός, ὅτι οὐ δύνανται χρῆσθαι τῷ σκέλει.
80 μαρτύριον ἓν[2] δέ τι ἐνέσται καὶ ἐν τοῖσιν ὀλίγον
81 ὕστερον εἰρησομένοισι, ὅτι ταῦτα τοιαῦτά ἐστιν.

LIV. Ὁπόσοισι[3] δ' ἂν ἐς τὸ ἔξω ἡ τοῦ μηροῦ κεφαλὴ ἐκβῇ, τούτοισι βραχύτερον μὲν τὸ σκέλος

[1] ταύτην. [2] Kw. omits.
[3] Οἷσι.

hand little inferior to those on the sound side, for the reasons that have been given, viz., that the upper arm is nearest the injury, and on that account is shorter.[1] The forearm, on the contrary, is not equally influenced by the lesion, because the end of the humerus which articulates with the ulna retains its old position. And the hand, again, is still further away from the lesion than is the forearm. For the aforesaid reasons, then, the bones which do not grow normally are defective in growth, and those which do grow maintain their growth. Manual exercise contributes greatly to the good flesh-development in hand and arm. In fact, taking all sorts of handiwork, the weasel-armed are ready to do with this one most of what they can do with the other arm, and do the work no less efficiently than with the sound limb; for it is not necessary for the body weight to be supported on the arms as on the legs, and the work done by them [*i.e.* the weasel-armed] [2] is light. Owing to use, the flesh of the hand and forearm is not atrophied in the weasel-armed; and even the upper arm gains some further development from this. But when the hip is dislocated inwards, either congenitally or in one still a child, there is more atrophy of flesh than in the arm, just because they cannot use the leg. A special piece of evidence that this is the case will be found in what is about to be said a little below.

LIV. In cases where the head of the thigh-bone is dislocated outwards, the leg is seen to be shorter,

[1] Kw. puts τοῦ ἄρθρου in brackets. It appears a needless gloss.

[2] Littré, Adams, Erm. read αὐτῇσι and refer it to the hands. But hands and arms may do hard work.

φαίνεται παρατεινόμενον παρὰ τὸ ἕτερον, εἰκότως· οὐ γὰρ ἐπ᾽ ὀστέον ἡ ἐπίβασις τῆς κεφαλῆς τοῦ μηροῦ ἐστίν, ὡς ὅτε ἔσω ἐκπέπτωκεν, ἀλλὰ παρ᾽ ὀστέον παρεγκεκλιμένην τὴν φύσιν ἔχον, ἐν σαρκὶ δὲ στηρίζεται ὑγρῇ καὶ ὑπεικούσῃ· διὰ τοῦτο μὲν βραχύτερον φαίνεται. ἔσωθεν δὲ ὁ μηρὸς παρὰ τὴν πλιχάδα καλεομένην κοιλότερος καὶ ἀσαρ-
10 κότερος φαίνεται·[1] ἔξωθεν δὲ ὁ γλουτὸς κυρτότερος, ἅτε ἐς τὸ ἔξω τῆς κεφαλῆς τοῦ μηροῦ ὠλισθηκυίης· ἀτὰρ καὶ ἀνωτέρω φαίνεται ὁ γλουτὸς ἅτε ὑπειξάσης τῆς σαρκὸς τῆς ἐνταῦθα τῇ τοῦ μηροῦ κεφαλῇ· τὸ δὲ παρὰ τὸ γόνυ τοῦ μηροῦ ἄκρον ἔσω ῥέπον φαίνεται, καὶ ἡ κνήμη καὶ ὁ πούς· ἀτὰρ οὐδὲ συγκάμπτειν ὥσπερ τὸ ὑγιὲς σκέλος δύνανται. τὰ μὲν οὖν σημεῖα ταῦτα τοῦ ἔξω
18 ἐκπεπτωκότος μηροῦ εἰσίν.

LV. Οἷσι μὲν οὖν ἂν τετελειωμένοισιν ἤδη ἐκπεσὸν τὸ ἄρθρον μὴ ἐμπέσῃ, τούτοισι βραχύτερον μὲν φαίνεται τὸ σύμπαν σκέλος, ἐν δὲ τῇ ὁδοιπορίῃ τῇ μὲν πτέρνῃ οὐ δύνανται καθικνεῖσθαι [ἐπὶ][2] τῆς γῆς, τῷ δὲ στήθει τοῦ ποδὸς βαίνουσι ἐπὶ τὴν γῆν· ὀλίγον δὲ ἐς τὸ ἔσω μέρος ῥέπουσι τοῖσι δακτύλοισι ἄκροισιν. ὀχεῖν δὲ δύναται τὸ σῶμα τὸ σιναρὸν σκέλος τούτοισι πολλῷ μᾶλλον ἢ οἷσιν ἂν ἐς τὸ ἔσω μέρος ἐκπε-
10 πτώκῃ, ἅμα μὲν ὅτι ἡ κεφαλὴ τοῦ μηροῦ καὶ ὁ αὐχὴν τοῦ ἄρθρου πλάγιος φύσει πεφυκὼς ὑπὸ συχνῷ μέρει τοῦ ἰσχίου τὴν ὑπόστασιν πεποίηται, ἅμα δὲ ὅτι ἄκρος ὁ πους οὐκ ἐς τὸ ἔξω μέρος ἀναγκάζεται ἐκκεκλίσθαι, ἀλλ᾽ ἐγγὺς τῆς ἰθυωρίης τῆς κατὰ τὸ σῶμα καὶ τείνει καὶ ἐσωτέρω. ὅταν οὖν τρίβον μὲν λάβῃ τὸ ἄρθρον ἐν τῇ σαρκὶ ἐς ἣν

when put beside the other. Naturally so, for it is no longer on bone that the head of the thigh-bone has its support, as when it was displaced inwards; but it lies along the natural slope of the hip-bone, and is sustained by soft and yielding flesh; wherefore it is seen to be shorter. The thigh on the inside at what is called the fork appears more hollow and less fleshy, while the buttock is rather more rounded on the outside, since the head of the bone is displaced outwards; besides this, the buttock is seen to be higher, since the flesh at that part gives way before the head of the thigh-bone. But the end of the bone at the knee is seen to turn inwards, and with it the leg and foot; for the rest, they cannot bend it in the same way as the sound leg. These then are the signs of dislocation of the thigh outwards.

LV. In cases of adults, when the joint is not reduced after dislocation, the whole leg is seen to be shorter; and in walking they cannot reach the ground with the heel, but go on the ball of the foot, and turn the toes a little inwards. But the injured leg can bear the weight of the body much better in these cases than where there has been dislocation inwards, partly because the head and neck of the thigh-bone, being naturally oblique, have got a lodging under a large part of the hip, and partly because the foot is not obliged to incline outwards, but is near the vertical line of the body, and even tends rather inwards. As soon, then, as the articular part forms a friction-cavity in the flesh where it is

[1] γίνεται. [2] Omit B Kw.

ἐξεκλίθη, ἡ δὲ σὰρξ γλισχρανθῇ, ἀνώδυνον τῷ χρόνῳ γίνεται· ὅταν δὲ ἀνώδυνον γένηται, δύνανται μὲν ὁδοιπορεῖν ἄνευ ξύλου, ἢν ἄλλως βούλωνται·
20 δύνανται δὲ ὀχεῖν τὸ σῶμα ἐπὶ τὸ σιναρὸν σκέλος. διὰ οὖν τὴν χρῆσιν ἧσσον τοῖσι τοιούτοισι ἐκθηλύνονται αἱ σάρκες ἢ οἷσιν ὀλίγον πρόσθεν εἴρηται· ἐκθηλύνονται δὲ ἢ πλεῖον ἢ ἔλασσον· μᾶλλον δέ τι ἐκθηλύνονται κατὰ τὸ ἔσω μέρος ἢ κατὰ τὸ ἔξω ὡς ἐπὶ τὸ πολύ. τὸ μέντοι ὑπόδημα μετεξέτεροι τούτων ὑποδεῖσθαι οὐ δύνανται, διὰ τὴν ἀκαμπίην τοῦ σκέλεος, οἱ δέ τινες καὶ δύνανται. οἷσιν δ' ἂν ἐν γαστρὶ ἐοῦσιν ἐξαρθρήσῃ τοῦτο τὸ ἄρθρον, ἢ ἔτι ἐν αὐξήσει
30 ἐοῦσι βίῃ ἐκπεσὸν μὴ ἐμπέσῃ, ἢ καὶ ὑπὸ νούσου ἐξαρθρήσῃ τοῦτο τὸ ἄρθρον καὶ ἐκπαλήσῃ—πολλὰ γὰρ τοιαῦτα γίνεται—καὶ ἐνίων μὲν τῶν τοιούτων ἢν ἐπισφακελίσῃ ὁ μηρός, ἐμπυήματα χρόνια καὶ ἔμμοτα γίνεται, καὶ ὀστέων ψιλώσιες ἐνίοισιν· ὁμοίως δὲ καὶ οἷσιν ἐπισφακελίζει καὶ οἷσι μὴ ἐπισφακελίζει, τοῦ μηροῦ τὸ ὀστέον πολλῷ βραχύτερον γίνεται, καὶ οὐκ ἐθέλει συναύξεσθαι ὥσπερ τοῦ ὑγιέος· τὰ μέντοι τῆς κνήμης βραχύτερα μὲν γίνεται ἢ τὰ τῆς ἑτέρης,
40 ὀλίγῳ δέ, διὰ τὰς αὐτὰς προφάσιας αἳ καὶ πρόσθεν εἴρηνται· ὁδοιπορεῖν τε δύνανται οι τοιοῦτοι, οἱ μέν τινες αὐτῶν τοῦτον τὸν τρόπον ὥσπερ οἷσι τετελειωμένοισιν ἐξέπεσε καὶ μὴ ἐνέπεσεν, οἱ δὲ καὶ βαίνουσι μὲν παντὶ τῷ ποδί, διαῤῥέπουσι δὲ ἐν τῇσι ὁδοιπορίῃσιν, ἀναγκαζόμενοι διὰ τὴν βραχύτητα τοῦ σκέλεος. ταῦτα δὲ [1] τοιαῦτα γίνεται, ἢν ἐπιμελέως μὲν παιδαγωγηθέωσιν [2] ἐν τοῖσι σχήμασι καὶ ὀρθῶς ἐν οἷσι

dislocated, and the flesh gets lubricated, it in time becomes painless; and when it becomes painless, they can walk without a crutch, at least should they wish to do so, and can put the weight of the body on the injured leg. Owing to the exercise, the flesh becomes less flaccid in such cases than in those mentioned just above; yet it does get more or less flaccid; and as a rule there is rather greater flaccidity on the inner than on the outer side. Some of these patients are unable to put on a shoe, owing to the stiffness of the leg; but some manage it. In cases where this joint is dislocated before birth, or is forcibly put out and not reduced during adolescence, or when the joint is dislocated and started from its socket by disease—such things often happen—if necrosis of the thigh-bone occurs in some of these cases, chronic abscesses are formed, requiring tents;[1] and in some there is denudation of bone. Likewise, both where there is and where there is not necrosis of the bone, it becomes much shorter, and will not grow correspondingly with the sound one. The bones of the lower leg, however, though shorter than those of the other, are but slightly so, for the same reasons as those given above. These patients can walk, some of them in the aforesaid fashion, like adults who have an unreduced dislocation; while others use the whole foot, but sway from side to side in their gait, being compelled to do so through the shortness of the leg. But such results are only attained if they are carefully instructed in the correct

[1] *I.e.* drainage apparatus.

---

[1] μέντοι Kw. [2] Kw.'s correction.

δεῖ, πρὶν κρατυνθῆναι ἐς τὴν ὁδοιπορίην, ἐπι-
50 μελέως δὲ καὶ ὀρθῶς, ἐπὴν κρατυνθῶσιν. πλείστης δὲ ἐπιμελείης δέονται οἷσιν ἂν νηπιωτάτοισιν ἐοῦσιν αὕτη ἡ συμφορὴ γένηται· ἢν γὰρ ἀμεληθῶσι νήπιοι ἐόντες, ἀχρήϊον παντάπασι καὶ ἀναυξὲς ὅλον τὸ σκέλος γίνεται. αἱ δὲ σάρκες τοῦ σύμπαντος σκέλεος μινύθουσι μᾶλλον ἢ τοῦ ὑγιέος· πάνυ μὲν πολλῷ ἧσσον τούτοισι μινύθουσι ἢ οἷσιν ἂν ἔσω ἐκπεπτώκῃ, διὰ τὴν χρῆσιν καὶ τὴν ταλαιπωρίην, οἷον εὐθέως δύνασθαι χρῆσθαι τῷ σκέλει, ὡς καὶ πρόσθεν ὀλίγῳ περὶ τῶν γαλιαγ-
60 κώνων εἴρηται.

LVI. Εἰσὶ δέ τινες, ὧν τοῖσι μὲν ἐκ γενεῆς αὐτίκα, τοῖσι δὲ καὶ ὑπὸ νούσου ἀμφοτέρων τῶν σκελέων ἐξέστη τὰ ἄρθρα ἐς τὸ ἔξω μέρος. τούτοισιν οὖν τὰ μὲν ὀστέα ταὐτὰ παθήματα πάσχει· αἱ μέντοι σάρκες ἥκιστα ἐκθηλύνονται τοῖσι τοιούτοισιν· εὔσαρκα [1] δὲ καὶ τὰ σκέλεα γίνεται, πλὴν εἴ τι ἄρα κατὰ τὸ ἔσω μέρος ἐλλείποι [2] ὀλίγον. διὰ τοῦτο δὲ εὔσαρκά ἐστιν, ὅτι ἀμφοτέροισι τοῖσι σκέλεσι ὁμοίως ἡ χρῆσις
10 γίνεται· ὁμοίως γὰρ σαλεύουσιν ἐν τῇ ὁδοιπορίῃ ἔνθα καὶ ἔνθα· ἐξεχέγλουτοι δὲ οὗτοι ἰσχυρῶς φαίνονται [3] διὰ τὴν ἔκστασιν τῶν ἄρθρων. ἢν δὲ μὴ ἐπισφακελίσῃ αὐτοῖσι τὰ ὀστέα, μηδὲ κυφοὶ ἀνωτέρω τῶν ἰσχίων γένωνται—ἐνίους γὰρ καὶ τοιαῦτα καταλαμβάνει—ἢν οὖν μὴ τοιοῦτόν τι γένηται, ἱκανῶς ὑγιηροὶ τἄλλα διαφέρονται· ἀναυξέστεροι μέντοι τὸ πᾶν σῶμα οὗτοι γίνον-
18 ται, πλὴν τῆς κεφαλῆς.

LVII. Ὅσοισι δ' ἂν ἐς τοὔπισθεν ἡ κεφαλὴ τοῦ μηροῦ ἐκπέσῃ—ὀλίγοισι δὲ ἐκπίπτει—οὗτοι

attitudes before they have acquired strength for walking, and carefully and rightly guided when they are strong. The greatest care is required in cases where this lesion occurs when they are very young; for if they are neglected when infants, the whole leg gets altogether useless and atrophied. The flesh is attenuated throughout the leg, compared with the sound one; but the attenuation is much less in these cases than where the dislocation is inwards, owing to use and exercise, since they can use the leg at once, as was said a little before concerning the weasel-armed.

LVI. There are some cases in which the hip-joints of both legs are dislocated outwards, either immediately at birth or from disease. Here the bones are affected in the same way as was described, but there is very little flaccidity of the tissues in such cases; for the legs keep plump, except for some little deficiency on the inner side. The plumpness is due to the fact that both legs get exercised alike; for they have an even swaying gait to this side and that. These patients show very prominent haunches, because of the displacement of the hip-joints; but if no necrosis of the bones supervenes, and they do not become humped above the hips—for this is an affection which attacks some—if nothing of this sort occurs, they are distinguished by very fair health in other respects. Still, these patients have defective growth of the whole body, except the head.

LVII. In cases where the head of the thigh-bone is dislocated backwards—this is a rare dislocation—

---

[1] ἅμα γὰρ εὔσαρκα. [2] ἐλλείπει.
[3] καὶ ῥαιβοὶ οἱ μηροί.

ἐκτανύειν οὐ δύνανται τὸ σκέλος, οὔτε κατὰ τὸ ἄρθρον τὸ ἐκπεσὸν οὔτε τι κάρτα κατὰ τὴν ἰγνύην· ἀλλ' ἥκιστα τῶν ἐκπαλησίων οὗτοι [μᾶλλον] [1] ἐκτανύουσι καὶ τὸ κατὰ τὸν βουβῶνα καὶ τὸ κατὰ τὴν ἰγνύην ἄρθρον. προσσυνιέναι μὲν οὖν καὶ τόδε χρή—εὔχρηστον γὰρ καὶ πολλοῦ ἄξιόν ἐστι καὶ τοὺς πλείστους λήθει—ὅτι οὐδ'
10 ὑγιαίνοντες δύνανται κατὰ τὴν ἰγνύην ἐκτανύειν τὸ ἄρθρον, ἢν μὴ συνεκτανύσωσι καὶ τὸ κατὰ τὸν βουβῶνα ἄρθρον, πλὴν ἢν μὴ πάνυ ἄνω ἀείρωσι τὸν πόδα, οὕτω δ' ἂν δύναιντο· οὐ τοίνυν οὐδὲ συγκάμπτειν δύνανται τὸ κατὰ τὴν ἰγνύην ἄρθρον ὁμοίως, ἀλλὰ πολὺ χαλεπώτερον, ἢν μὴ συγκάμψωσι καὶ τὸ κατὰ τὸν βουβῶνα ἄρθρον. πολλὰ δὲ καὶ ἄλλα κατὰ τὸ σῶμα τοιαύτας ἀδελφίξιας ἔχει, καὶ κατὰ νεύρων συντάσιας καὶ κατὰ μυῶν σχήματα, καὶ πλεῖστά τε καὶ
20 πλείστου ἄξια γινώσκεσθαι ἢ ὥς τις οἴεται, καὶ κατὰ τὴν τοῦ ἐντέρου φύσιν καὶ τὴν τῆς συμπάσης κοιλίης, καὶ κατὰ τὰς τῶν ὑστέρων πλάνας καὶ συντάσιας· ἀλλὰ περὶ μὲν τούτων ἑτέρωθι λόγος ἔσται ἠδελφισμένος τοῖσι νῦν λεγομένοισι. περὶ οὗ δὲ ὁ λόγος ἐστίν, οὔτε ἐκτανύειν δύνανται, ὥσπερ ἤδη εἴρηται, βραχύτερόν τε τὸ σκέλος φαίνεται, διὰ δισσὰς προφάσιας· ὅτι τε οὐκ ἐκτανύεται, ὅτι τε πρὸς τὴν σάρκα ὠλίσθηκε τὴν τοῦ πυγαίου· ἡ γὰρ φύσις
30 τοῦ ἰσχίου τοῦ ὀστέου ταύτῃ, ᾗ καὶ ἡ κεφαλὴ καὶ ὁ αὐχὴν τοῦ μηροῦ γίνεται, ὅταν δὲ ἐξαρθρήσῃ, καταφερής τι πέφυκεν ἐπὶ τοῦ πυγαίου τὸ ἔξω μέρος. συγκάμπτειν μέντοι δύνανται, ὅταν μὴ ἡ ὀδύνη κωλύῃ· καὶ ἡ κνήμη τε καὶ ὁ ποὺς ὀρθὰ

the patients cannot extend the leg at the dislocated joint, nor indeed at the ham; in fact, of all displacements, those who suffer this one make least extension, both at the groin and at the ham. One should also bear the following in mind—it is a useful and important matter, of which most are ignorant—that not even sound individuals can extend the joint at the ham, if they do not extend that at the groin as well, unless they lift the foot very high; then they could do it. Nor can they as readily flex the joint at the ham, unless they flex that at the groin as well, but only with much greater difficulty. Many parts of the body have affinities of this kind, both as regards contraction of cords and attitudes of muscles; and they are very numerous, and more important to recognise than one would think, both as regards the nature of the intestine and the whole body cavity, also the irregular movements and contractions of the uterus. But these matters will be discussed elsewhere in connection with the present remarks. To return to our subject—as already observed, the patients cannot extend the leg, also it appears shorter, for a double reason; both because it is not extended, and because it has slipped into the flesh of the buttock; for the hip-bone, at the part where the head and neck of the femur lie when dislocated, has a natural slope towards the outer side of the buttock. They can however flex the limb, when pain does not prevent it; and the lower leg and foot appear fairly straight,

[1] Omit Galen, Littré, Erm.

ἐπιεικῶς φαίνεται, καὶ οὔτε τῇ οὔτε τῇ πολὺ ἐκκεκλιμένα· κατὰ δὲ τὸν βουβῶνα δοκεῖ τι ἡ σὰρξ λαπαρωτέρη εἶναι ποτὶ καὶ ψαυομένη, ἅτε τοῦ ἄρθρου ἐς τὰ ἐπὶ θάτερα μέρη ὠλισθηκότος· κατὰ δὲ αὐτὸ τὸ πυγαῖον διαψαυομένη ἡ κεφαλὴ 40 τοῦ μηροῦ δοκεῖ τι ἐξογκεῖν καὶ μᾶλλον. τὰ μὲν οὖν σημεῖα ταῦτά ἐστιν, ᾧ ἂν ἐς τὸ ὄπισθεν 42 ἐκπεπτώκῃ ὁ μηρός.

LVIII. Ὅτεῳ μὲν οὖν ἂν τετελειωμένῳ ἤδη ἐκπεσὸν μὴ ἐμπέσῃ, ὁδοιπορεῖν μὲν δύναται, ὅταν ὁ χρόνος ἐγγένηται καὶ ἡ ὀδύνη παύσηται, καὶ ἐθισθῇ τὸ ἄρθρον ἐν τῇ σαρκὶ ἐνστρωφᾶσθαι. ἀναγκάζεται μέντοι ἰσχυρῶς συγκάμπτειν[1] κατὰ τοὺς βουβῶνας ὁδοιπορέων,[2] διὰ δισσὰς προφάσιας, ἅμα μὲν ὅτι πολλῷ βραχύτερον τὸ σκέλος γίνεται διὰ τὰ προειρημένα, καὶ τῇ μὲν πτέρνῃ καὶ πάνυ πολλοῦ δεῖται ψαύειν τῆς γῆς·[3] 10 εἰ γὰρ πειρήσαιτο καὶ ἐπ' ὀλίγον τοῦ ποδὸς ὀχηθῆναι, μηδενὶ ἄλλῳ ἀντιστηριζόμενος, ἐς τοὐπίσω ἂν πέσοι· ἡ γὰρ ῥοπὴ πολλὴ ἂν εἴη, τῶν ἰσχίων ἐπὶ πολὺ ἐς τοὐπίσω ὑπερεχόντων ὑπὲρ τοῦ ποδὸς τῆς βάσιος καὶ τῆς ῥάχιος ἐς τὰ ἰσχία ῥεπούσης. μόλις δὲ τῷ στήθει τοῦ ποδὸς καθικνεῖται, καὶ οὐδὲ οὕτως, ἢν μὴ κάμψῃ αὐτὸς ἑωυτὸν κατὰ τοὺς βουβῶνας, καὶ τῷ ἑτέρῳ σκέλει κατὰ τὴν ἰγνύην ἐπισυγκάμψῃ. ἐπὶ δὲ τούτοισιν ἀναγκάζεται ὥστε τῇ χειρὶ τῇ κατὰ τὸ σιναρὸν 20 σκέλος ἐρείδεσθαι ἐς τὸ ἄνω τοῦ μηροῦ ἐφ' ἑκάστῃ συμβάσει. ἀναγκάζει οὖν τι καὶ τοῦτο αὐτὸ ὥστε κάμπτεσθαι κατὰ τοὺς βουβῶνας· ἐν γὰρ τῇ μεταλλαγῇ τῶν σκελέων ἐν τῇ ὁδοιπορίῃ

[1] συγκάμπτων. [2] ὁδοιπορεῖν.

without much inclination to either side. At the groin the flesh seems rather relaxed, especially on palpation, since the joint[1] has slipped to the other side; while at the buttock itself the head of the bone seems, on deep palpation, to stick out abnormally. These then are the signs in a case of dislocation of the thigh backwards.

LVIII. When the dislocation occurs in an adult, and is not reduced, the patient can walk, indeed, after an interval, when the pain subsides, and the head of the bone has become accustomed to rotate in the tissues; but he is obliged in walking to flex his body strongly at the groin, for a double reason, both because the leg is much shorter, owing to the causes above mentioned, and is very far from touching the ground with the heel; for if he should try even for a moment to have his weight on the foot with no opposite support, he would fall backwards, as there would be a great inclination that way, the hips coming far beyond the sole of the foot behind, and the spine inclining towards the hips.[2] He hardly reaches the ground with the ball of the foot, and cannot do this without a simultaneous flexure of the other leg at the ham. Besides, he is forced at every step to make pressure with the hand at the side of the injured leg on the upper part of the thigh. This of itself would compel him to bend the body somewhat at the groin; for at the change of

[1] "Joint" here means "articular head."

[2] L. and Erm. put the above from "for if he should try" after "displaced backwards at the hip." It gives better sense, but has no authority.

---

[3] Littré, followed by Ermerins, rearranges the text in an arbitrary manner.

οὐ δύναται τὸ σῶμα ὀχεῖσθαι ἐπὶ τοῦ σιναροῦ σκέλεος, ἢν μὴ προσκατερείδηται τὸ σιναρὸν πρὸς τὴν γῆν ὑπὸ τῆς χειρός, οὐχ[1] ὑφεστεῶτος τοῦ ἄρθρου ὑπὸ τῷ σώματι, ἀλλ' ἐς τὸ ὄπισθεν ἐξεστεῶτος κατὰ τὸ ἰσχίον. ἄνευ μὲν οὖν ξύλου δύνανται ὁδοιπορεῖν οἱ τοιοῦτοι, ἢν ἄλλως 30 ἐθισθέωσιν, διὰ τοῦτο, ὅτι ἡ βάσις τοῦ ποδὸς κατὰ τὴν ἀρχαίην ἰθυωρίην ἐστίν, ἀλλ' οὐκ ἐς τὸ ἔξω ἐκκεκλιμένη· διὰ τοῦτο οὖν οὐδὲν δέονται τῆς ἀντικοντώσιος. ὅσοι μέντοι βούλονται ἀντὶ τῆς τοῦ μηροῦ ἐπιλαβῆς ὑπὸ τὴν μασχάλην τὴν κατὰ τὸ σιναρὸν σκέλος ὑποτιθέμενοι σκίπωνα ἀντερείδειν, ἐκεῖνοι, ἢν[2] μὲν μακρότερον τὸν σκίπωνα ὑποτιθέοιντο, ὀρθότερον μὲν ὁδοιποροῦσι, τῷ δὲ ποδὶ πρὸς τὴν γῆν οὐκ ἐρείδονται· εἰ δ' αὖ βούλονται ἐρείδεσθαι τῷ ποδί, βραχύτερον μὲν 40 τὸ ξύλον φορητέον, κατὰ δὲ τοὺς βουβῶνας ἐπισυγκάμπτεσθαι ἂν δέοι αὐτούς. τῶν δὲ σαρκῶν αἱ μινυθήσιες κατὰ λόγον γίγνονται καὶ τούτοισιν, ὥσπερ καὶ πρόσθεν εἴρηται· τοῖσι μὲν γὰρ μετέωρον ἔχουσι τὸ σκέλος καὶ μηδὲν ταλαιπωρέουσι, τούτοισι καὶ μάλιστα μινύθουσιν· οἳ δ' ἂν πλεῖστα χρέωνται τῇ ἐπιβάσει, τούτοισιν ἥκιστα μινύθουσι. τὸ μέντοι ὑγιὲς σκέλος οὐκ ὠφελεῖται, ἀλλὰ μᾶλλον[3] καὶ ἀσχημονέστερον γίνεται, ἢν χρέωνται τῷ σιναρῷ σκέλει ἐπὶ τὴν 50 γῆν· συνυπουργέον γὰρ ἐκείνῳ ἐξίσχιόν τε ἀπαναγκάζεται εἶναι, καὶ κατὰ τὴν ἰγνύην συγκάμπτειν, ἤν γε[4] μὴ προσχρέηται τῷ σιναρῷ ἐπὶ τὴν γῆν, ἀλλὰ μετέωρον ἔχων σκίπωνι ἀντερείδηται, οὕτω δὲ καρτερὸν γίνεται τὸ ὑγιὲς σκέλος· ἔν τε γὰρ τῇ φύσει διαιτᾶται, καὶ τὰ

legs in walking, the body weight cannot be carried by the injured leg unless it be further pressed to the ground by the hand, the articular head not being in line under the body, but displaced backwards at the hip.[1] Still, such patients can walk without a crutch, at any rate after practice, for this reason, viz., that the sole of the foot keeps its old straight line, and is not inclined outwards; wherefore they have no need for counter-propping. Those who prefer, instead of the grasp on the thigh, to have the support of a crutch under the arm on the side of the injured leg, if they have a rather long crutch, walk more erect; but they do not press with the foot on the ground. But if they want to make pressure with the foot, a shorter crutch must be carried; and they must also flex the body at the groin. Wasting of the flesh takes place in these cases also according to rule, as was said before; in those who keep the leg off the ground and give it no exercise the wasting is greatest, while in those who use it most in walking it is least. Still, the sound leg gets no benefit, but rather becomes also somewhat deformed, if patients use the injured leg on the ground; for in giving assistance to the latter, it is forced outwards at the hip, and bends at the ham; but if one does not use the injured leg on the ground as well, but, keeping it suspended, gets support from a crutch, the sound limb thus becomes strong; for it is employed in the natural way, and

[1] See previous note.

---

[1] ἅτε οὐχ. [2] εἰ.
[3] Omit. [4] ἢν δὲ.

γυμνάσια προσκρατύνει αὐτό. φαίη μὲν οὖν ἄν τις, ἔξω ἰητρικῆς τὰ τοιαῦτα εἶναι· τί γὰρ δῆθεν δεῖ περὶ τῶν ἤδη ἀνηκέστων γεγονότων ἔτι προσσυνιέναι; πολλοῦ δὲ δεῖ οὕτως ἔχειν· τῆς 60 γὰρ αὐτῆς γνώμης καὶ ταῦτα συνιέναι· οὐ γὰρ οἷόν τε ἀπαλλοτριωθῆναι ἀπ' ἀλλήλων. δεῖ μὲν γὰρ ἐς τὰ ἀκεστὰ μηχανάασθαι, ὅπως μὴ ἀνήκεστα ἔσται, συνιέντα ὅπῃ ἂν μάλιστα κωλυτέα ἐς τὸ ἀνήκεστον ἐλθεῖν· δεῖ δὲ τὰ ἀνήκεστα συνιέναι, ὡς μὴ μάτην λυμαίνηται· τὰ δὲ προῤῥήματα λαμπρὰ καὶ ἀγωνιστικὰ ἀπὸ τοῦ διαγινώσκειν ὅπῃ ἕκαστον καὶ οἵως καὶ ὁπότε τελευτήσει, ἤν τε ἐς τὸ ἀκεστὸν τράπηται, ἤν τε ἐς τὸ ἀνήκεστον. ὁπόσοισι δ' ἂν ἐκ γενεῆς 70 ἢ καὶ ἄλλως πως ἐν αὐξήσει ἐοῦσιν οὕτως ὀλίσθῃ τὸ ἄρθρον ὀπίσω καὶ μὴ ἐμπέσῃ, ἤν τε βίῃ ὀλίσθῃ, ἤν τε καὶ ὑπὸ νούσου—πολλὰ γὰρ τοιαῦτα ἐξαρθρήματα γίνεται ἐν νούσοισιν· οἷαι δέ τινές εἰσιν αἱ νοῦσοι, ἐν ᾗσιν ἐξαρθρεῖται τὰ τοιαῦτα, ὕστερον γεγράψεται—ἢν οὖν ἐκστὰν μὴ ἐμπέσῃ, τοῦ μὲν μηροῦ τὸ ὀστέον βραχὺ γίνεται, κακοῦται δὲ καὶ πᾶν τὸ σκέλος, καὶ ἀναυξέστερον γίνεται καὶ ἀσαρκότερον πολλῷ διὰ τὸ μηδὲν προσχρῆσθαι αὐτῷ· κακοῦται γὰρ 80 τούτοισι καὶ τὸ κατὰ τὴν ἰγνύην ἄρθρον· τὰ γὰρ νεῦρα ἐντεταμένα γίνεται διὰ τὰ πρόσθεν εἰρημένα. διὸ οὐ δύνανται τὸ κατὰ τὴν ἰγνύην ἄρθρον ἐκτανύειν, οἷσιν ἂν οὕτως ἰσχίον ἐκπέσῃ. ὡς γὰρ ἐν κεφαλαίῳ εἰρῆσθαι, πάντα τὰ ἐν τῷ σώματι, ὁπόσα ἐπὶ χρήσει γέγονε, χρεομένοισι μὲν μέτρια καὶ γυμναζομένοισιν ἐν τῇσι ταλαιπωρίῃσιν, ἐν ᾗσιν ἕκαστα εἴθισται, οὕτω μὲν

the exercises strengthen it more. One might say that such matters are outside the healing art. Why, forsooth, trouble one's mind further about cases which have become incurable? This is far from the right attitude. The investigation of these matters too belongs to the same science; it is impossible to separate them from one another. In curable cases we must contrive ways to prevent their becoming incurable, studying the best means for hindering their advance to incurability; while one must study incurable cases so as to avoid doing harm by useless efforts. Brilliant and effective forecasts are made by distinguishing the way, manner and time in which each case will end, whether it takes the turn to recovery or to incurability. In cases where such a dislocation backwards occurs and is not reduced, whether congenitally or during the period of growth, and whether the displacement is due to violence or disease—many such dislocations occur in diseases, and the diseases which cause such dislocations will be described later—if, then, the displacement is unreduced, the thigh-bone gets short, and the whole leg deteriorates, and becomes much more undeveloped and devoid of flesh, because it gets no exercise. For in these cases, the joint at the ham is also maimed, since the ligaments get contracted, for the reasons given above; and therefore patients in whom the leg is thus dislocated cannot extend the joint at the ham. Speaking generally, all parts of the body which have a function, if used in moderation and exercised in labours to which each is accustomed, become thereby healthy and well-

ὑγιηρὰ καὶ αὔξιμα καὶ εὔγηρα γίνεται· μὴ χρεομένοισι δέ, ἀλλ' ἐλινύουσι, νοσηρότερα γίνε-
90 ται καὶ ἀναυξέα καὶ ταχύγηρα. ἐν δὲ τούτοισιν οὐχ ἥκιστα τὰ ἄρθρα τοῦτο πέπονθε καὶ τὰ νεῦρα, ἢν μή τις αὐτοῖσι χρέηται· κακοῦνται μὲν οὖν διὰ ταύτας τὰς προφάσιας μᾶλλόν τι ἐν τούτῳ τῷ τρόπῳ τοῦ ὀλισθήματος ἢ ἐν τοῖσι ἄλλοισιν· ὅλον γὰρ τὸ σκέλος ἀναυξὲς γίνεται, καὶ τῇ ἀπὸ τῶν ὀστέων φύσει καὶ τῇ ἀπὸ τῶν σαρκῶν. οἱ οὖν τοιοῦτοι ὁπόταν ἀνδρωθῶσι, μετέωρον καὶ συγκεκαμμένον τὸ σκέλος ἴσχουσιν, ἐπὶ δὲ τοῦ ἑτέρου ὀχέονται, καὶ τῷ ξύλῳ
100 ἀντιστηριζόμενοι, οἱ μὲν ἑνί, οἱ δὲ δυσίν.

LIX. Οἷσι δ' ἂν ἐς τοὔμπροσθεν ἡ κεφαλη τοῦ μηροῦ ἐκπέσῃ—ὀλίγοισι δὲ τοῦτο γίνεται—οὗτοι ἐκτανύειν μὲν τὸ σκέλος δύνανται τελέως, συγκάμπτειν δὲ ἥκιστα οὗτοι δύνανται τὰ κατὰ τὸν βουβῶνα· πονέουσι δέ, καὶ ἢν κατὰ τὴν ἰγνύην ἀναγκάζωνται συγκάμπτειν. μῆκος δὲ τοῦ σκέλεος παραπλήσιον φαίνεται, κατὰ μὲν τὴν πτέρνην καὶ πάνυ· ἄκρος δὲ ὁ ποὺς ἧσσόν τι προκύπτειν ἐθέλει·[1] ὅλον δὲ τὸ σκέλος ἔχει
10 τὴν ἰθυωρίην τὴν κατὰ φύσιν, καὶ οὔτε τῇ οὔτε τῇ ῥέπει. ὀδυνῶνται δὲ αὐτίκα οὗτοι μάλιστα, καὶ οὖρον ἴσχεται τὸ πρῶτον τούτοισι μᾶλλόν τι ἢ τοῖσιν ἄλλοισιν ἐξαρθρήμασιν· ἔγκειται γὰρ ἡ κεφαλὴ τοῦ μηροῦ ἐγγυτάτω τούτοισι τῶν τόνων τῶν ἐπικαίρων. καὶ κατὰ μὲν τὸν βουβῶνα ἐξόγκεόν τε καὶ κατατεταμένον τὸ χωρίον φαίνεται, κατὰ δὲ τὸ πυγαῖον στολιδωδέστερον καὶ ἀσαρκότερον. ταῦτα μὲν οὖν σημεῖά ἐστι
19 τὰ εἰρημένα, ὧν ἂν οὕτως ἐκπεπτώκῃ ὁ μηρός.

developed, and age slowly; but if unused and left idle, they become liable to disease, defective in growth, and age quickly. This is especially the case with joints and ligaments, if one does not use them. For these reasons, patients are more troubled by this sort of dislocation than by the other; for the whole leg is atrophied in the natural growth both of bone and flesh. Such patients, then, when they become adults, keep the leg raised and contracted, and walk on the other, supporting themselves, some with one and some with two crutches.

LIX. Those in whom the head of the thigh-bone is dislocated forwards—a rare occurrence—can extend the leg completely, but are least able to flex it at the groin; and they suffer pain even if they are compelled to bend it at the ham. The length of the leg seems about equal, and quite so at the heel; but there is less power of pointing the foot. The whole leg preserves its natural straight line, inclining neither to one side nor the other. It is in these cases that the immediate pain is greatest, and retention of urine occurs from the first more than in other dislocations; for the head of the femur in these cases lies very close to important cords. The region of the groin appears prominent and tense; but at the buttock it is rather wrinkled and fleshless. The above-mentioned signs, then, occur in patients whose thigh is put out in this way.

---

[1] ἐθέλει = δύναται, says Galen, comparing *Iliad* XXI. 366.

LX. Ὁπόσοισι μὲν οὖν ἂν ἤδη ἠνδρωμένοισι τοῦτο τὸ ἄρθρον ἐκπεσὸν μὴ ἐμπέσῃ, οὗτοι, ὁπόταν αὐτοῖσιν ἡ ὀδύνη παύσηται καὶ τὸ ἄρθρον ἐθισθῇ ἐν τῷ χωρίῳ τούτῳ στρωφᾶσθαι, ἵνα ἐξέπεσεν, οὗτοι δύνανται σχεδὸν εὐθὺς[1] ὀρθοὶ ὁδοιπορεῖν ἄνευ ξύλου, καὶ πάνυ μέντοι εὐθέες, ἐπὶ δὲ[2] τὸ σιναρόν, ἅτε οὔτε κατὰ τὸν βουβῶνα εὔκαμπτοι ἐόντες, οὔτε κατὰ τὴν ἰγνύην· διὰ οὖν τοῦ βουβῶνος τὴν ἀκαμπίην εὐθυτέρῳ ὅλῳ
10 τῷ σκέλει ἐν τῇ ὁδοιπορίῃ χρέονται[3] ἢ ὅτε ὑγίαινον. καὶ σύρουσι δὲ ἐνίοτε πρὸς τὴν γῆν τὸν πόδα, ἅτε οὐ ῥηϊδίως συγκάμπτοντες τὰ ἄνω ἄρθρα, καὶ ἅτε παντὶ βαίνοντες τῷ ποδί· οὐδὲν γὰρ ἧσσον τῇ πτέρνῃ οὗτοι βαίνουσιν ἢ τῷ ἔμπροσθεν· εἰ δέ γε ἠδύναντο μέγα προβαίνειν, κἂν πάνυ πτερνοβάται ἦσαν· καὶ γὰρ οἱ ὑγιαίνοντες, ὅσῳ ἂν μέζον προβαίνοντες ὁδοιπορέωσι, τοσούτῳ μᾶλλον πτερνοβάται εἰσί, τιθέντες τὸν πόδα, αἴροντες τὸν ἐναντίον. ὁπόσοισι δὲ δὴ
20 οὕτως ἐκπέπτωκε, καὶ ἔτι μᾶλλον τῇ πτέρνῃ προσεγχρίμπτουσιν ἢ τῷ ἔμπροσθεν· τὸ γὰρ ἔμπροσθεν τοῦ ποδός, ὁπόταν ἐκτεταμένον ᾖ τὸ ἄλλο σκέλος, οὐχ ὁμοίως δύναται ἐς τὸ πρόσω καμπύλλεσθαι, ὥσπερ ὅταν συγκεκαμμένον ᾖ τὸ σκέλος· οὐκ αὖ σιμοῦσθαι δύναται ὁ πούς, συγκεκαμμένου[4] τοῦ σκέλεος, ὡς ὅταν ἐκτεταμένον ᾖ τὸ σκέλος. ὑγιαίνουσά τε οὖν ἡ φύσις οὕτω πέφυκεν, ὥσπερ εἴρηται· ὅταν δὲ ἐκπεσὸν μὴ ἐμπέσῃ τὸ ἄρθρον, οὕτως ὁδοιπορέουσιν ὡς
30 εἴρηται, διὰ τὰς προφάσιας ταύτας τὰς εἰρημένας· ἀσαρκότερον μέντοι τὸ σκέλος τοῦ ἑτέρου γίνεται, κατά τε τὸ πυγαῖον, κατά τε τὴν

LX. In cases where this dislocation occurs in those already adult and is not reduced, these patients, when their pain subsides and the head of the bone has got accustomed to turning in the locality where it was displaced, are able to walk almost at once erect without a crutch, and even quite straight up, so far as the injured part is concerned, seeing that it cannot easily bend either at the groin or ham. Thus, owing to the stiffness at the groin, they keep the whole leg straighter in walking than when it was sound. And sometimes they drag the foot along the ground, seeing that they cannot easily flex the upper joints, and that they walk on the whole foot. In fact, they walk as much on the heel as on the front part; and if they could take long strides, they would be purely heel-walkers. For those with sound limbs, the longer the strides they take in walking, the more they go on their heels when putting down one leg and raising the other; but those who have this form of dislocation press upon the heel even more than on the front of the foot. For the front of the foot cannot be so well bent down when the leg is extended as when it is flexed; nor, on the other hand, can the foot be bent upwards when the leg is flexed so well as when it is extended. This is what happens in the natural sound condition, as was said; but when the joint is dislocated and not reduced, they walk in the way described, for the reasons given above. The leg, however, becomes less fleshy than the other, both

---

[1] Kw. omits. [2] ἐπί γε.
[3] χρέωνται. [4] συγκεκλιμένου.

γαστροκνημίην, καὶ κατὰ τὴν ὄπισθεν ἴξιν. οἷσι δ' ἂν νηπίοισιν ἔτι ἐοῦσι τὸ ἄρθρον [οὕτως] ὀλισθὸν μὴ ἐμπέσῃ, ἢ καὶ ἐκ γενεῆς οὕτω γένηται, καὶ τούτοισι τὸ τοῦ μηροῦ ὀστέον μᾶλλόν τι μινύθει ἢ τὰ τῆς κνήμης καὶ τὰ τοῦ ποδός. ἥκιστα μὴν ἐν τούτῳ τῷ τρόπῳ τοῦ ὀλισθήματος ὁ μηρὸς μειοῦται. μινύθουσι μέντοι αἱ σάρκες 40 πάντῃ, μάλιστα δὲ κατὰ τὴν ὄπισθεν ἴξιν, ὥσπερ ἤδη καὶ πρόσθεν εἴρηται. ὁπόσοι μὲν οὖν ἂν τιθηνηθέωσιν ὀρθῶς, οὗτοι μὲν δύνανται προσχρῆσθαι τῷ σκέλει αὐξανόμενοι, βραχυτέρῳ μέν τινι τοῦ ἑτέρου ἐόντι, ὅμως δὲ ἐρειδόμενοι ξύλῳ ἐπὶ ταῦτα, ᾗ τὸ σιναρὸν σκέλος· οὐ γὰρ κάρτα δύνανται ἄνευ τῆς πτέρνης τῷ στήθει τοῦ ποδὸς χρῆσθαι, ἐπικαθιέντες ὥσπερ ἐν ἑτέροισι χωλεύμασι ἔνιοι δύνανται· αἴτιον δὲ τοῦ μὴ δύνασθαι τὸ ὀλίγῳ πρόσθεν εἰρημένον· διὰ οὖν τοῦτο 50 προσδέονται ξύλου. ὁπόσοι δ' ἂν καταμεληθέωσι καὶ μηδὲν χρέωνται ἐπὶ τὴν γῆν τῷ σκέλει, ἀλλὰ μετέωρον ἔχωσι, τούτοισι μινύθει μὲν τὰ ὀστέα ἐς αὔξησιν μᾶλλον ἢ τοῖσι χρεομένοισιν· μινύθουσι δὲ [καὶ] αἱ σάρκες πολὺ μᾶλλον ἢ τοῖσι χρεομένοισι· κατὰ δὲ τὰ ἄρθρα ἐς τὸ εὐθὺ πηροῦται τούτοισι τὸ σκέλος μᾶλλόν τι ἢ οἷσι 57 ἂν ἄλλως ἐκπεπτώκῃ.

LXI. Ὡς μὲν οὖν ἐν κεφαλαίῳ εἰρῆσθαι, τὰ ἄρθρα τὰ ἐκπίπτοντα καὶ τὰ ὀλισθάνοντα ἀνίσως αὐτὰ ἑωυτοῖσιν ἐκπίπτει καὶ ὀλισθάνει, ἄλλοτε μὲν πολὺ πλέον, ἄλλοτε δὲ πολὺ ἔλασσον· καὶ οἷσι μὲν ἂν [πολὺ][1] πλέον ὀλίσθῃ ἢ ἐκπέσῃ, χαλεπώτερα ἐμβάλλειν τὸ ἐπίπαν ἐστί, καὶ ἢν μὴ ἐμβιβασθῇ, μέζους καὶ ἐπιδηλοτέρας τὰς

at the buttock and calf and all down the back of it. In those cases too where it is dislocated in childhood and not reduced, or where dislocation occurs congenitally, the thigh-bone is rather more atrophied than the bones of the leg and foot; but atrophy of the thigh-bone is least in this form of dislocation. The tissues are atrophied in the whole limb, but especially down the back of it, as was said before. Those, then, who are properly cared for are able to use the leg when they grow up, though it is a little shorter than the other; yet they do it by having a support on the side of the injured limb, for they have not much ability to use the ball of the foot without the heel, bringing it down, as some can do in other forms of lameness. The reason of their not being able is that mentioned a little above; and this is why they require a staff. In those who are neglected, and never use the leg to walk with, but keep it in the air, the bones are more atrophied than in those who do use it; and the tissues are much more atrophied than in those who use the leg. As regards the joints, the lesion keeps the leg straighter in these patients than in those who have other forms of dislocation.

LXI. To sum up—dislocations and slipping [separation][1] of joints vary among themselves in amount, and are sometimes much greater, sometimes much less. In cases where the slipping or dislocation is greater, it is, in general, harder to reduce; and, if unreduced, the resulting lesions and disabilities are

[1] It is usual to make ὀλισθαίνω, ὀλίσθημα refer to "partial dislocation"; but this hardly suits the context, or the reference to shoulder and hip-joints.

[1] Kw. omits.

πηρώσιας καὶ κακώσιας ἴσχει τὰ τοιαῦτα, καὶ ὀστέων καὶ σαρκῶν καὶ σχημάτων· ὅταν δὲ μεῖον
10 ἐκπέσῃ καὶ ὀλίσθῃ, ῥηΐδιον μὲν ἐμβάλλειν τὰ τοιαῦτα τῶν ἑτέρων γίνεται· ἢν δὲ καταπορηθῇ ἢ ἀμεληθῇ ἐμπεσεῖν, μείους καὶ ἀσινέστεραι αἱ πηρώσιες γίνονται τούτοισιν ἢ οἷσιν ὀλίγῳ πρόσθεν εἴρηται. τὰ μὲν οὖν ἄλλα ἄρθρα καὶ πάνυ πολὺ διαφέρει ἐς τὸ ὁτὲ μὲν μεῖον, ὁτὲ δὲ μέζον τὸ ὀλίσθημα ποιεῖσθαι· μηροῦ δὲ καὶ βραχίονος κεφαλαὶ παραπλησιώτατα ὀλισθάνουσιν αὐτὴ ἑωυτῇ ἑκατέρη· ἅτε γὰρ στρογγύλαι μὲν αἱ κεφαλαὶ ἐοῦσαι, ἁπλῆν τὴν στρογγύλωσιν
20 καὶ φαλακρὴν ἔχουσι, κυκλοτερεῖς δὲ αἱ κοιλίαι ἐοῦσαι αἱ δεχόμεναι τὰς κεφαλάς, ἁρμόζουσι δὲ τῇσι κεφαλῇσιν· διὰ τοῦτο οὐκ ἔστιν αὐτῇσι τὸ ἥμισυ ἐκστῆναι τοῦ ἄρθρου· ὀλισθάνοι γὰρ ἂν διὰ τὴν περιφερείην, ἢ ἐς τὸ ἔξω ἢ ἐς τὸ ἔσω. περὶ οὗ οὖν ὁ λόγος, ἐκπίπτουσι τελέως ἤδη, ἐπεὶ ἄλλως γε οὐκ ἐκπίπτουσι· ὅμως δὲ καὶ ταῦτα ὁτὲ μὲν πλεῖον ἀποπηδᾷ ἀπὸ τῆς φύσιος, ὁτὲ δὲ ἔλασσον· μᾶλλον δέ τι μηρὸς τοῦτο βραχίονος
29 πέπονθεν.

LXII. Ἐπεὶ ἔνια καὶ τῶν ἐκ γενεῆς ὀλισθημάτων, ἢν μικρὸν ὀλίσθῃ, οἷά τε ἐς τὴν φύσιν ἄγεσθαι, καὶ μάλιστα τὰ παρὰ τοῦ ποδὸς ἄρθρα. ὁπόσοι ἐκ γενεῆς κυλλοὶ γίνονται, τὰ πλεῖστα τούτων ἰήσιμά ἐστιν, ἢν μὴ πάνυ μεγάλη ἡ ἔκκλισις ᾖ, ἢ καὶ προαυξέων γεγονότων ἤδη τῶν παιδίων συμβῇ. ἄριστον μὲν οὖν ὡς τάχιστα ἰητρεύειν τὰ τοιαῦτα, πρὶν πάνυ μεγάλην τὴν ἔνδειαν τῶν ὀστέων τῶν ἐν τῷ ποδὶ γενέσθαι,
10 πρίν τε πάνυ μεγάλην τὴν ἔνδειαν τῶν σαρκῶν

greater and more manifest in the bones, the soft parts, and the attitudes. When there is less displacement, either with dislocation or separation, reduction is easier than in other cases; and if they are not reduced, owing to inability or neglect, the resulting deformities are smaller and less serious than in the cases just mentioned. Joints in general, then, differ very much in having their displacements sometimes less and sometimes greater; but the heads of the thigh and arm-bones each slip out in very similar ways; for the heads, being rounded, have a smooth and regular spherical surface, and the cavities which receive them, being also circular, fit the heads. Wherefore it is impossible for them to be put half out; for owing to the circular rim, it would slip either out or in. As regards our subject, then, they are put quite out, since otherwise they are not put out at all. Yet even these joints spring away, sometimes more, sometimes less, from the natural position. This is more pronounced in the thigh-bone than in the arm.

LXII. There are certain congenital displacements which, when they are slight, can be reduced to their natural position, especially those at the foot-joints. Cases of congenital club-foot are, for the most part, curable, if the deviation is not very great or the children advanced in growth. It is therefore best to treat such cases as soon as possible, before there is any very great deficiency in the bones of the foot, and

τῶν κατὰ τὴν κνήμην εἶναι. τρόπος μὲν οὖν κυλλώσιος οὐχ εἷς, ἀλλὰ πλείονες, τὰ πλεῖστα μὴν οὐκ ἐξηρθρηκότα παντάπασιν, ἀλλὰ δι' ἔθος σχήματος ἔν τινι ἀπολήψει τοῦ ποδὸς κεκυλλωμένα. προσέχειν δὲ καὶ ἐν τῇ ἰητρείῃ τοισίδε χρή· ἀπωθεῖν μὲν καὶ κατορθοῦν τῆς κνήμης τὸ κατὰ τὸ σφυρὸν ὀστέον τὸ ἔξωθεν ἐς τὸ ἔσω μέρος, ἀντωθεῖν δὲ ἐς τὸ ἔξω μέρος τὸ τῆς πτέρνης τὸ κατὰ τὴν ἴξιν, ὅπως ἀλλήλοις ἀπαντήσῃ τὰ 20 ὀστέα τὰ ἐξίσχοντα κατὰ μέσον τε καὶ πλάγιον τὸν πόδα· τοὺς δ' αὖ δακτύλους ἀθρόους σὺν τῷ μεγάλῳ δακτύλῳ ἐς τὸ ἔσω μέρος ἐγκλίνειν καὶ περιαναγκάζειν οὕτως· ἐπιδεῖν δὲ κηρωτῇ ἐρρητινωμένῃ εὖ, καὶ σπλήνεσι καὶ ὀθονίοισι μαλθακοῖσι μὴ ὀλίγοισι, μηδὲ ἄγαν πιέζοντα· οὕτω δὲ τὰς περιαγωγὰς ποιεῖσθαι τῆς ἐπιδέσιος, ὥσπερ καὶ τῇσι χερσὶν ἡ κατόρθωσις ἦν τοῦ ποδός, ὅπως ὁ πους ὀλίγῳ μᾶλλον ἐς τὸ βλαισὸν ῥέπων φαίνηται. ἴχνος δέ τι χρὴ ποιεῖσθαι ἢ δέρματος μὴ 30 ἄγαν σκληροῦ, ἢ μολύβδινον,[1] προσεπιδεῖν δέ, μὴ πρὸς τὸν χρῶτα τιθέντα, ἀλλ' ὅταν ἤδη τοῖσι ὑστάτοισιν ὀθονίοισι μέλλῃς ἐπιδεῖν· ὅταν δὲ ἤδη ἐπιδεδεμένος ᾖ, ἑνός τινος τῶν ὀθονίων χρή, οἷσιν ἐπιδεῖται, τὴν ἀρχὴν προσράψαι πρὸς τὰ κατὰ τοῦ ποδὸς ἐπιδέσματα κατὰ τὴν ἴξιν τοῦ μικροῦ δακτύλου· ἔπειτα ἐς τὸ ἄνω τείνοντα ὅπως ἂν δοκῇ μετρίως ἔχειν, περιβάλλειν ἄνωθεν τῆς γαστροκνημίης, ὡς μόνιμον ᾖ, κατατεταμένον οὕτως. ἁπλῷ δὲ λόγῳ, ὥσπερ κηροπλαστέοντα, 40 χρὴ ἐς τὴν φύσιν τὴν δικαίην ἄγειν καὶ τὰ ἐκκεκλιμένα καὶ τὰ συντεταμένα παρὰ τὴν φύσιν,

[1] μολυβδίου.

before the like occurs in the tissues of the leg. Now the mode of club-foot is not one, but manifold; and most cases are not the result of complete dislocation, but are deformities due to the constant retention of the foot in a contracted position.[1] The things to bear in mind in treatment are the following: push back and adjust the bone of the leg at the ankle from without inwards, making counter-pressure outwards on the bone of the heel where it comes in line with the leg, so as to bring together the bones which project at the middle and side of the foot; at the same time, bend inwards and rotate the toes all together, including the big toe. Dress with cerate well stiffened with resin, pads and soft bandages, sufficiently numerous, but without too much compression. Bring round the turns of the bandaging in a way corresponding with the manual adjustment of the foot, so that the latter has an inclination somewhat towards splay-footedness.[2] A sole should be made of not too stiff leather or of lead, and should be bound on as well, not immediately on to the skin, but just when you are going to apply the last dressings. When the dressing is completed, the end of one of the bandages used should be sewn on to the under side of the foot-dressings, in a line with the little toe; then, making such tension upwards as may seem suitable, pass it round the calf-muscle at the top, so as to keep it firm and on the stretch.[3] In a word, as in wax modelling, one should bring the parts into their true natural position, both those that are twisted and

[1] *I.e.* "an unnatural contraction of the muscles, ligaments and fasciae."

[2] *I.e. valgus* (outward distortion).

[3] *I.e.* so as to hold up the outer side of the foot.

καὶ τῇσι χερσὶν οὕτω διορθοῦντα, καὶ τῇ ἐπιδέσει ὡσαύτως, προσάγειν δὲ οὐ βιαίως, ἀλλὰ παρηγορικῶς· προσράπτειν δὲ τὰ ὀθόνια, ὅπως ἂν συμφέρῃ τὰς ἀναλήψιας ποιεῖσθαι· ἄλλα γὰρ ἄλλης τῶν χωλωμάτων δεῖται ἀναλήψιος. ὑποδημάτιον δὲ ποιεῖσθαι[1] μολύβδινον, ἔξωθεν τῆς ἐπιδέσιος ἐπιδεδεμένον, οἷον αἱ Χῖαι [κρηπῖδες][2] ῥυθμὸν εἶχον· ἀλλ' οὐδὲν αὐτοῦ δεῖ, ἤν τις ὀρθῶς μὲν
50 τῇσι χερσὶ διορθώσῃ, ὀρθῶς δὲ τοῖσιν ὀθονίοισιν ἐπιδέῃ, ὀρθῶς δὲ καὶ τὰς ἀναλήψιας ποιοῖτο.[3] ἡ μὲν οὖν ἴησις αὕτη, καὶ οὔτε τομῆς οὔτε καύσιος οὐδὲν δεῖ, οὔτ' ἄλλης ποικιλίης· θᾶσσον γὰρ ἐνακούει τὰ τοιαῦτα τῆς ἰητρείης ἢ ὡς ἄν τις οἴοιτο. προσνικᾶν μέντοι χρὴ τῷ χρόνῳ, ἕως ἂν αὐξηθῇ τὸ σῶμα ἐν τοῖσι δικαίοισι σχήμασιν. ὅταν δὲ ἐς ὑποδήματος λόγον ἴῃ, ἀρβύλαι ἐπιτηδειόταται αἱ πηλοπατίδες καλεόμεναι· τοῦτο γὰρ ὑποδημάτων ἥκιστα κρατεῖται ὑπὸ τοῦ
60 ποδός, ἀλλὰ κρατεῖ μᾶλλον· ἐπιτήδειος δὲ καὶ ὁ
61 Κρητικὸς τρόπος τῶν ὑποδημάτων.

LXIII. Ὁπόσοισι δ' ἂν κνήμης ὀστέα ἐξαρθρήσαντα καὶ ἕλκος ποιήσαντα τελέως ἐξίσχῃ κατὰ τὰ παρὰ τὸν πόδα ἄρθρα, εἴτε ἔσω ῥέψαντα, εἴτε μέντοι καὶ ἔξω, τὰ τοιαῦτα μὴ[4] ἐμβάλλειν, ἀλλ' ἐᾶν τὸν βουλόμενον τῶν ἰητρῶν ἐμβάλλειν. σαφέως γὰρ εἰδέναι χρὴ ὅτι ἀποθανεῖται ᾧ ἂν ἐμβληθέντα ἐμμείνῃ, καὶ ἡ ζωὴ δὲ ὀλιγήμερος τούτοισι γενήσεται·[5] ὀλίγοι γὰρ ἂν αὐτῶν τὰς ἑπτὰ ἡμέρας ὑπερβάλλοιεν· σπασμὸς γὰρ ὁ κτείνων

[1] ποιεῖν.
[2] κρηπῖδες Galen : omit Kw. and MSS. As Kw. shows, it is inserted from the Commentary.
[3] ποιῆται.
[4] οὐ χρή.
[5] γίνεται.

those that are abnormally contracted, adjusting them in this way both with the hands and by bandaging in like manner; but draw them into position by gentle means, and not violently. Sew on the bandages so as to give the appropriate support; for different forms of lameness require different kinds of support. A leaden shoe shaped as the Chian[1] boots used to be might be made, and fastened on outside the dressing; but this is quite unnecessary if the manual adjustment, the dressing with bandages, and the contrivance for drawing up are properly done. This then is the treatment, and there is no need for incision, cautery, or complicated methods; for such cases yield to treatment more rapidly than one would think. Still, time is required for complete success, till the part has acquired growth in its proper position. When the time has come for footwear, the most suitable are the so-called "mud-shoes," for this kind of boot yields least to the foot; indeed, the foot rather yields to it. The Cretan form[2] of footwear is also suitable.[3]

LXIII. In cases where the leg-bones are dislocated and, making a wound, project right through at the ankle-joint, whether it be towards the inner or outer side, do not reduce such a lesion; but let any practitioner who chooses do so.[4] For you may be certain that where there is permanent reduction the patients will die, and life in such cases lasts only a few days. Few go beyond seven days. Spasm

[1] Erotian says it was a "woman's boot." In Galen's time it was quite forgotten.

[2] "Reaching to the middle of the leg." Galen.

[3] "The most wonderful chapter in ancient surgery." Adams.

[4] *I.e.* leave it to anyone reckless enough.

10 ἐστίν· ἀτὰρ καὶ γαγγραινοῦσθαι ἱκνεῖται τὴν κνήμην καὶ τὸν πόδα. ταῦτα βεβαίως εἰδέναι χρὴ οὕτως ἐσόμενα· καὶ οὐκ ἄν μοι δοκεῖ οὐδὲ ἑλλέβορος ὠφελήσειν[1] αὐθημερόν τε δοθεὶς καὶ αὖθις πινόμενος, ἄγχιστα δὲ εἴπερ τι τοιοῦτο[ν]·[2] οὐ μέντοι γε οὐδὲ τοῦτο δοκέω. ἢν δὲ μὴ ἐμβληθῇ, μηδὲ ἀπ' ἀρχῆς μηδεὶς πειρηθῇ ἐμβάλλειν, περιγίνονται οἱ πλεῖστοι αὐτῶν. χρὴ δὲ ἡρμόσθαι μὲν τὴν κνήμην καὶ τὸν πόδα οὕτως, ὡς αὐτὸς ἐθέλει, μοῦνον δὲ μὴ ἀπαιωρεύμενα μηδὲ 20 κινεύμενα ἔστω. ἰητρεύειν δὲ πισσηρῇ καὶ σπλήνεσιν οἰνηροῖσιν ὀλίγοισι, μὴ ἄγαν ψυχροῖσι· ψῦχος γὰρ ἐν τοῖσι τοιούτοισι σπασμὸν ἐπικαλεῖται. ἐπιτήδεια δὲ καὶ φύλλα σεύτλων ἢ βηχίου ἢ ἄλλου τινὸς τῶν τοιούτων ἐν οἴνῳ μέλανι αὐστηρῷ ἡμίεφθα ἐπιτιθέντα ἰητρεύειν ἐπί τε τὸ ἕλκος ἐπί τε τὰ περιέχοντα, κηρωτῇ δὲ χλιερῇ ἐπιχρίειν[3] αὐτὸ τὸ ἕλκος· ἢν δὲ ἡ ὥρη χειμερινὴ ᾖ, καὶ ἔρια ῥυπαρὰ οἴνῳ καὶ ἐλαίῳ καταῤῥαίνοντα χλιεροῖσιν ἄνωθεν ἐπιτέγγειν· 30 καταδεῖν δὲ μηδὲν μηδενί,[4] μηδὲ περιπλάσσειν μηδενί· εὖ γὰρ εἰδέναι χρὴ ὅτι πίεξις καὶ ἀχθοφορίη πᾶν κακὸν τοῖσι τοιούτοισίν ἐστιν. ἐπιτήδεια δὲ πρὸς τὰ τοιαῦτα καὶ τῶν ἐναίμων μετεξέτερα, ὅσοισιν αὐτῶν συμφέρει· ἔρια δὲ ἐπιτιθέντα, οἴνῳ ἐπιτέγγοντα, πολὺν χρόνον ἐᾷν· τὰ δὲ ὀλιγημερώτατα τῶν ἐναίμων καὶ ὅσα ῥητίνῃ προσκαταλαμβάνεται οὐχ ὁμοίως ἐπιτήδεια ἐκείνοισίν ἐστιν. χρονίη ἡ κάθαρσις τῶν ἑλκέων γίνεται τούτων· πολὺν γὰρ χρόνον πλαδαρὴ γίνε-40 ται· τινὰς δὲ τούτων χρηστὸν ἐπιδεῖν. εἰδέναι

[1] ὠφελῆσαι.

(tetanus) is the cause of death; but gangrene of the leg and foot is also a sequel. It should be well known that this will happen; and I do not suppose that even hellebore, given on the day of the accident and repeated, would do good. If anything would help, something of this kind would come nearest; but I have no confidence even in that. But if there is no reduction or attempt at reduction to begin with, most of them survive. The leg and foot should be disposed as the patient himself wishes, only avoiding an unsupported position or movement. Treat with pitch cerate and a few compresses steeped in wine, not too cold; for cold in such cases evokes spasm. Other suitable applications are leaves of beet or colt's-foot or something similar, half-boiled in dark astringent wine, and applied both to the wound and the parts around it. Anoint the wound itself with warm cerate, and, if it is winter, apply an upper moist dressing of crude wool, sprinkling it with warm wine and oil; but avoid all bandaging and dressing with plasters, for one must bear well in mind that pressure and weight do nothing but harm in such cases. Some of the applications for fresh wounds are also suitable for these injuries, in cases where they are useful. Cover with wool, moistening it with wine, and leave on a long time. The wound remedies which last a very short time, and those incorporated with resin, are not so suitable for those patients; for the cleansing of these wounds then takes more time, since the flabby moist stage is prolonged. Bandaging is good for some of these cases. Finally, one should bear

---

[2] τοιοῦτον Galen. [3] ὑποχρίειν.
[4] Omit Kw. and many MSS.

μὲν δή που σάφα χρὴ ὅτι ἀνάγκη τὸν ἄνθρωπον χωλὸν αἰσχρῶς γενέσθαι· καὶ γὰρ ὁ ποὺς ἐς τὸ ἄνω ἀνέσπασται τῶν τοιούτων, καὶ τὰ ὀστέα τὰ διολισθήσαντα ἔξω ἐξέχοντα φαίνεται· οὔτε γὰρ ψιλοῦται τῶν τοιούτων ὀστέων οὐδὲν ὡς ἐπιτοπολύ, εἰ μὴ κατὰ βραχύ τι, οὐδὲ ἀφίσταται, ἀλλὰ περιωτειλοῦται λεπτῇσιν ὠτειλῇσι καὶ ἀσθενέσι, καὶ ταῦτα ἢν ἀτρεμίζωσι πολὺν χρόνον· ἢν[1] δὲ μή, ἑλκύδριον ἐγκαταλειφθῆναι
50 κίνδυνος ἀναλθές. ὅμως δέ, περὶ οὗ ὁ λόγος, οὕτω μὲν ἰητρευόμενοι σώζονται, ἐμβληθέντος δὲ τοῦ
52 ἄρθρου καὶ ἐμμείναντος, ἀποθνήσκουσιν.

LXIV. Ὡυτὸς δὲ λόγος οὗτος, ἢν καὶ τὰ τοῦ πήχεος ὀστέα τὰ παρὰ τὸν καρπὸν τῆς χειρὸς ἕλκος ποιήσαντα ἐξίσχῃ, ἤν τε ἐς τὸ ἔσω μέρος τῆς χειρός, ἤν τε ἐς τὸ ἔξω. σάφα γὰρ ἐπίστασθαι χρὴ ὅτι ἀποθανεῖται ἐν ὀλίγῃσιν ἡμέρῃσι τοιούτῳ θανάτῳ, οἵῳπερ καὶ πρόσθεν εἴρηται, ὅτῳ ἂν ἐμβληθέντα τὰ ὀστέα ἐμμένῃ.[2] οἷσι δ' ἂν μὴ ἐμβληθῇ μηδὲ πειρηθῇ ἐμβάλλεσθαι, οὗτοι πολὺ πλείονες περιγίνονται. ἰητρείη δὲ τοιαύτη
10 τοῖσι τοιούτοισιν ἐπιτηδείη, οἵηπερ εἴρηται· τὸ δὲ σχῆμα αἰσχρὸν τοῦ χωλώματος ἀνάγκη εἶναι, καὶ τοὺς δακτύλους τῆς χειρὸς ἀσθενέας καὶ ἀχρείους· ἢν μὲν γὰρ ἐς τὸ ἔσω μέρος ὀλίσθῃ τὰ ὀστέα, συγκάμπτειν οὐ δύνανται τοὺς δακτύλους·
15 ἢν δὲ ἐς τὸ ἔξω μέρος, ἐκτανύειν οὐ δύνανται.

LXV. Ὅσοισι δ' ἂν κνήμης ὀστέον, ἕλκος ποιησάμενον παρὰ τὸ γόνυ, ἔξω ἐξίσχῃ, ἤν τε ἐς τὸ ἔξω μέρος, ἤν τε ἐς τὸ ἔσω, τούτοισιν ἢν μέν τις ἐμβάλῃ, ἔτι ἑτοιμότερος ὁ θάνατός ἐστιν ἤπερ τοῖσιν ἑτέροισιν, καίπερ κἀκείνοισιν ἑτοίμος

clearly in mind that the patient will necessarily be deformed and lame; for the foot is drawn up, and the projection of the dislocated bones is obvious. There is no denudation of the bones as a rule, except to a slight extent, nor do they come away; but they get scarred over with thin and weak tissue—that is, if the patients keep at rest for a long time; otherwise there is risk of a small incurable ulcer being left. However, to return to our subject, those thus treated are saved; but if the joint is reduced and keeps its place, they die.

LXIV. The same remarks apply to cases where the bones of the forearm make a wound and stick out at the wrist, whether on the inner or outer side of the hand.[1] For one should understand clearly that the patient will die in a few days in the way which was mentioned above, if the bones are reduced and keep in place; but if there is no reduction or attempt at reduction, the great majority survive. The suitable treatment in such cases is such as was described, but the lesion is necessarily a deformity, and the fingers are weak and useless; for if the bones are displaced inwards, they cannot flex the fingers, if outwards, they cannot extend them.[2]

LXV. In cases where a bone of the leg makes a wound at the knee and projects either to the outer or inner side, death is more imminent, if one reduces the dislocation, than in the other cases, though it is

[1] Our "forwards or backwards."
[2] See note on wrist dislocation.

---

[1] εἰ. [2] ἐμμείνῃ.

ἐών. ἢν δὲ μὴ ἐμβαλὼν ἰητρεύῃς, ἐλπίδες μὲν σωτηρίης οὕτω μόνως εἰσίν· κινδυνωδέστερα δὲ ταῦτα τῶν ἑτέρων γίνεται καὶ ὅσῳ ἂν ἀνωτέρω καὶ ὅσῳ ἂν ἰσχυρότερα ᾖ καὶ ἀπὸ ἰσχυροτέρων
10 ὠλισθήκῃ. ἢν δὲ τὸ ὀστέον τὸ τοῦ μηροῦ τὸ πρὸς τοῦ γόνατος ἕλκος ποιησάμενον ἐξολίσθῃ, ἐμβληθὲν μὲν καὶ ἐμμεῖναν, ἔτι βιαιότερον καὶ θᾶσσον τὸν θάνατον ποιήσει τῶν πρόσθεν εἰρημένων·[1] μὴ ἐμβληθὲν δὲ πολὺ κινδυνωδέστερον ἢ
15 τὰ πρόσθεν· ὅμως δὲ μούνη ἐλπὶς αὕτη σωτηρίης.

LXVI. Ωὑτὸς δὲ λόγος καὶ περὶ τῶν κατὰ τὸν ἀγκῶνα ἄρθρων, καὶ περὶ τῶν τοῦ πήχεος καὶ βραχίονος· ὅσα γὰρ ἂν τούτων ἐξαρθρήσαντα ἐξίσχῃ ἕλκος ποιησάμενα, πάντα, ἢν ἐμβληθῇ, θάνατον φέρει, μὴ ἐμβληθέντα[2] δέ, ἐλπίδα σωτηρίης· χώλωσις δὲ ἑτοίμη τοῖσι περιγινομένοισιν. θανατωδέστερα δὲ τοῖσιν ἐμβαλλομένοισίν ἐστι τὰ ἀνωτέρω τῶν ἄρθρων, ἀτὰρ καὶ τοῖσι μὴ ἐμβαλλομένοισι κινδυνωδέστερα αὐτὰ ταῦτα. εἰ
10 δέ τινι τὰ ἀνώτατα ἄρθρα ἐξαρθρήσαντα ἕλκος ποιήσαντα ἐξίσχοι, ταῦτα δ' ἂν ἔτι καὶ ἐμβαλλόμενα ταχυθανατώτατα ἂν[3] εἴη καὶ μὴ ἐμβαλλόμενα κινδυνωδέστατα· ἰητρείη δὲ ἤδη εἴρηται οἵη τις ἐμοὶ δοκεῖ ἐπιτηδειοτάτη εἶναι τῶν
15 τοιούτων.

LXVII. Ὅσοισι δὲ ἄρθρα δακτύλων, ἢ ποδὸς ἢ χειρός, ἐξαρθρήσαντα ἕλκος ποιησάμενα

[1] ἢ τὰ πρόσθεν εἰρημένα.
[2] ἐμβαλλόμενα.
[3] Use of double ἂν characteristic. Even a triple ἂν is found (*J.* XLVI). Cf. *Vul. Cap.* IV., *Acut.* I, *Fract.* XXVIII, and (for triple ἂν) Thuc. II. 94.—Pq.

imminent in them too. If you treat it without reduction, this method, and this only, gives hope of recovery. These cases are the more dangerous, the higher the joint is, and the stronger the dislocated parts and those from which they are dislocated. If the thigh-bone at the knee makes a wound and is dislocated through it, when reduced and kept in place it will cause still more prompt and violent death than in the cases mentioned above; when not reduced, there is far more danger than in the former cases, yet this is the only hope of safety.

LXVI. The same remarks apply to the bones forming the elbow-joints, both those of the forearm and upper arm; for if any one of them is dislocated and projects, making a wound, they all bring a fatal issue if reduced; but if not reduced, there is hope of recovery, though those who survive are certain to be maimed. More fatal when reduced are compound dislocations of the more proximal joints; and they too involve greater danger even when unreduced. If anyone has the uppermost joints dislocated and projecting through the wound made, it is there that reduction brings swiftest death; and there too is most danger, even without reduction.[1] The kind of treatment which seems to me most suitable in such cases has already been described.

LXVII. When the joints of the fingers or toes are dislocated and project through a wound, the

[1] These two sentences seem to be of general application, not confined to the elbow—as in Littré's and Petrequin's versions.

ἐξέσχε, μὴ κατεηγότος τοῦ ὀστέου, ἀλλὰ κατ' αὐτὴν τὴν σύμφυσιν ἀποσπασθέντος, τούτοισιν ἢν ἐμβληθέντα ἐμμείνῃ, ἔνι μέν τις κίνδυνος σπασμοῦ, ἢν μὴ χρηστῶς ἰητρεύωνται· ὅμως δέ τι ἄξιον ἐμβάλλειν, προειπόντα ὅτι φυλακῆς πολλῆς καὶ μελέτης δεῖται. ἐμβάλλειν μέντοι ῥήϊστον καὶ δυνατώτατον καὶ τεχνικώτατόν ἐστι
10 τῷ μοχλίσκῳ, ὥσπερ καὶ πρόσθεν εἴρηται ἐν τοῖσι καταγνυμένοισι καὶ ἐξίσχουσι ὀστέοισιν· ἔπειτα ἀτρεμεῖν ὡς μάλιστα χρή, καὶ κατακεῖσθαι καὶ ὀλιγοσιτεῖν· ἄμεινον δὲ καὶ φαρμακεῦσαι ἄνω κούφῳ τινὶ φαρμάκῳ, τὸ δὲ ἕλκος ἰητρεύειν [1] μὲν ἢ ἐναίμοισι τοῖσιν ἐπιτέγκτοισι ἢ πολυοφθάλμοισιν ἢ οἷσι κεφαλῆς ὀστέα κατεηγότα ἰητρεύεται, κατάψυχρον δὲ κάρτα μηδὲν προσφέρειν. ἥκιστα μὲν οὖν τὰ πρῶτα ἄρθρα κινδυνώδεά ἐστι, τὰ δὲ ἔτι ἀνωτέρω [2] κινδυνωδέστερα. ἐμβάλλειν δὲ
20 χρὴ αὐθημερὸν ἢ τῇ ὑστεραίῃ, τριταίῳ δὲ καὶ τεταρταίῳ ἥκιστα· τεταρταῖα γὰρ ἐόντα ἐπισημαίνει τῇσι παλιγκοτίῃσι μάλιστα. οἷσιν ἂν οὖν μὴ αὐτίκα ἐγγένηται ἐμβάλλειν, ὑπερβαίνειν χρὴ ταύτας τὰς εἰρημένας ἡμέρας· ὅ τι γὰρ ἂν ἔσω δέκα ἡμερέων ἐμβάλλῃς, σπᾷν καταληπτέον.[3] ἢν δὲ ἄρα ἐμβεβλημένῳ σπασμὸς ἐπιγένηται, ἐκβάλλειν τὸ ἄρθρον δεῖ ταχύ, καὶ θερμῷ τέγγειν ὡς πλειστάκις, καὶ τὸ ὅλον σῶμα θερμῶς καὶ λιπαρῶς καὶ μαλθακῶς ἔχειν, μάλιστα
30 κατὰ τὰ ἄρθρα· κεκάμφθαι δὲ μᾶλλον ἢ ἐκτετάσθαι πᾶν τὸ σῶμα χρή. προσδέχεσθαι μέντοι χρὴ κατὰ τοὺς δακτύλους τὰ ἄρθρα τὰ ἐμβαλλόμενα ἀποστατικὰ ἔσεσθαι· τὰ γὰρ πλεῖστα οὕτω γίνεται, ἢν καὶ ὁτιοῦν φλεγμονῆς ὑπογένηται, ὡς,

bone being not fractured, but torn away at the connection, in these cases reduction and fixation involve some danger of spasm, if they are not skilfully treated; still, it is worth while to reduce the dislocation, giving warning beforehand as to the necessity for great caution and care. The easiest and most powerful reduction, and that most in accord with art, is that with the small lever, as described before in relation to fractured and protruding bones. Afterwards the patient should keep as quiet as possible, lie down, and take little food. It is rather advantageous to give a mild emetic. Treat the wound either with moist applications for fresh cuts, chamomile,[1] or remedies used for head fractures; but do not apply anything very cold. The distal joints, then, are least dangerous, the higher ones more so. One should make reduction on the first or following day, but not on the third or fourth, since the onset of exacerbations occurs mostly on the fourth day. In cases, then, where immediate reduction fails, one should pass over the aforesaid days. Any case you reduce within ten days is liable to spasm. If spasm supervenes after reduction, one ought to dislocate the joint quickly, make frequent warm affusions, and keep the whole body warmly, comfortably and softly at rest, especially at the joints. The whole body should be rather flexed than extended. In any case one must expect the articular ends of the phalanges to come away after reduction; for this happens in most cases, if there is any amount of inflammation. So, were it not that the surgeon

[1] "Ox-eye." Galen.

---

[1] θεραπεύειν. [2] τὰ δ' ἐπάνω.
[3] πᾶν καταληπτόν Kw.: κάρτα ἐλπτόν Reinhold.

εἰ μὴ δι' ἀμαθίην τῶν δημοτέων ἐν αἰτίῃ ἔμελλεν ὁ ἰητρὸς ἔσεσθαι, οὐδὲν ἂν πάντως οὐδ' ἐμβάλλειν ἔδει. τὰ μὲν οὖν κατὰ τὰ ἄρθρα ὀστέα ἐξίσχοντα
68 ἐμβαλλόμενα οὕτω κινδυνώδεά ἐστιν, ὡς εἴρηται.

LXVIII. Ὅσα δὲ κατὰ τὰ ἄρθρα τὰ κατὰ τοὺς δακτύλους ἀποκόπτεται τελέως, ταῦτα ἀσινέα τὰ πλεῖστά ἐστιν, εἰ μή τις ἐν αὐτῇ τῇ τρώσει λειποθυμήσας βλαβείη· καὶ ἰητρείη φαύλη ἀρκέσει τῶν τοιούτων ἑλκέων. ἀτὰρ καὶ ὅσα μὴ κατὰ τὰ ἄρθρα, ἀλλὰ κατ' ἄλλην τινὰ ἴξιν τῶν ὀστέων ἀποκόπτεται, καὶ ταῦτα ἀσινέα ἐστί, καὶ ἔτι εὐαλθέστερα τῶν ἑτέρων· καὶ ὅσα κατὰ τοὺς δακτύλους ὀστέα κατεηγότα[1] ἐξίσχει μὴ κατὰ
10 τὸ ἄρθρον, καὶ ταῦτα ἀσινέα ἐστὶν ἐμβαλλόμενα. ἀποκόψιες δὲ τέλειαι ὀστέων καὶ κατὰ τὰ ἄρθρα καὶ ἐν ποδὶ καὶ ἐν χειρὶ καὶ ἐν κνήμῃ, τοῖσι παρὰ τὰ σφυρὰ καὶ ἐν πήχει, τοῖσι παρὰ τοὺς καρπούς, τοῖσι πλείστοισιν ἀποκοπτομένοισιν ἀσινέα γίνεται, ὅσα ἂν μὴ αὐτίκα λειποθυμίη ἀνατρέψῃ ἢ τεταρταίοισιν ἐοῦσι πυρετὸς συνε-
17 χὴς ἐπιγένηται.

LXIX. Ἀποσφακελίσιες μέντοι σαρκῶν, καὶ ἐν τρώμασιν αἱμοῤῥόοισι γενομένοισιν ἢ ἀποσφίγξεσιν ἰσχυραῖς, καὶ ἐν ὀστέων κατήγμασι γενομένοισι[2] πιεχθεῖσι μᾶλλόν τι τοῦ καιροῦ, καὶ ἐν ἄλλοισι δεσμοῖσι βιαίοισιν, ἀποληφθέντα[3] ἀποπίπτει πολλοῖσι, καὶ οἱ πολλοὶ περιγίνονται τῶν τοιούτων, καὶ οἷσι μηροῦ μέρος τι ἀποπίπτει καὶ τῶν σαρκῶν καὶ τοῦ ὀστέου, καὶ οἷσι βραχίονος, ἧσσον[4] δέ· πήχεός τε καὶ

[1] καταγέντα.
[2] Kw. omits.
[3] ἀπομελανθέντα.
[4] ἡσσόνως.

is likely to incur blame owing to the ignorance of the vulgar, he should by no means make the reduction. The dangers, then, of reducing bones which project through the skin at the joints are such as have been described.[1]

LXVIII. Cases of complete amputation of fingers or toes at the joints are usually without danger—unless a patient suffers from collapse at the time of injury—and ordinary treatment will suffice for such wounds. Again, where the amputation is not at a joint, but somewhere in the line of the bones, these cases also are not dangerous, and heal even more readily than the former; and if the projection of fractured finger-bones is not at a joint, reduction is without danger in these cases also. Complete amputations even at the joints both of the foot and hand, or of the leg at the ankle, and of the forearm at the wrist, are in most cases without danger, unless syncope overcomes them at once, or continuous fever supervenes on the fourth day.[2]

LXIX. As for gangrene of the tissues occurring in wounds with supervening haemorrhage, or much strangulation, and in fractures which undergo greater compression than is opportune, and in other cases of tight bandaging, the intercepted[3] parts come away in many cases. The majority of such patients survive, even when a part of the thigh comes away with the soft parts and the bone, also part of the arm, but these less frequently. When the forearm or leg

[1] Surgeons such as Antyllus and Heliodorus probably performed amputation or resection in these cases. Even Paulus (VI. 121) is surprised at the timidity of Hippocrates.

[2] This chapter seems to refer to cases of injury, not surgical "resection" as Adams.

[3] Or "blackened" (ἀπομελανθέντα, Kw.).

10 κνήμης ἀποπεσούσης, καὶ ἔτι εὐφορωτέρως περιγίνονται. οἷσι μὲν οὖν κατεαγέντων τῶν ὀστέων ἀποσφίγξιες αὐτίκα ἐγένοντο καὶ μελασμοί, τούτοισι μὲν ταχεῖαι αἱ περιῤῥήξιες γίνονται τοῦ σώματος, καὶ τὰ ἀποπίπτοντα ταχέως ἀποπίπτει, ἤδη τῶν ὀστέων προενδεδωκότων· οἷσι δὲ ὑγιέων ἐόντων τῶν ὀστέων οἱ μελασμοὶ γίνονται, αἱ μὲν σάρκες ταχέως θνήσκουσι καὶ τούτοισι, τὰ δὲ ὀστέα βραδέως ἀφίσταται, ᾗ ἂν τὰ ὅρια τοῦ μελασμοῦ γένηται καὶ ἡ ψίλωσις τοῦ ὀστέου. 20 χρὴ δέ, ὅσα ἂν κατωτέρω τοῦ σώματος τῶν ορίων τοῦ μελασμοῦ ᾖ, ταῦτα, ὅταν ἤδη πάμπαν τεθνήκῃ καὶ ἀναλγέα ᾖ, ἀφαιρεῖν κατὰ τὸ ἄρθρον, προμηθεόμενον ὅπως μή τι τρώσῃς· ἢν γὰρ ὀδυνηθῇ ἀποταμνόμενος καὶ μήπω κυρήσῃ τὸ σῶμα τεθνεὸς ταύτῃ ᾗ ἀποτέμνεται, κάρτα κίνδυνος ὑπὸ τῆς ὀδύνης λειποθυμῆσαι· αἱ δὲ τοιαῦται λειποθυμίαι πολλοὺς παραχρῆμα ἤδη ἀπώλεσαν. μηροῦ μὲν οὖν ὀστέον, ψιλωθὲν ἐκ τοιούτου τρόπου, ὀγδοηκοσταῖον εἶδον ἐγὼ ἀπο-30 στάν· ἡ μέντοι κνήμη τούτῳ τῷ ἀνθρώπῳ κατὰ τὸ γόνυ ἀφῃρέθη εἰκοσταίη, ἐδόκει δέ μοι καὶ ἐγγυτέρω· οὐ γὰρ ἅμα, ἀλλ᾽ ἐπὶ τὸ προμηθέστερον ἔδοξέ μοί τι ποιεῖν.[1] κνήμης δὲ ὀστέα ἐκ τοιούτου μελασμοῦ, μάλα κατὰ μέσην τὴν κνήμην ἐόντα, ἑξηκοσταῖά μοι ἀπέπεσεν, ὅσα ἐψιλώθη αὐτῶν. διενέγκοι μὲν γὰρ ἄν τι καὶ ἰητρείη ἰητρείης ἐς τὸ θᾶσσόν τε καὶ βραδύτερον τὰ ὀστέα ψιλούμενα ἀποπίπτειν· διενέγκοι δ᾽

[1] Kw. ἐδόκει; omit ἅμα and μοι. Reinhold's emendation: οὐ γὰρ εἴα με . . . ἔταξέ μοι.

comes away, they survive still more easily. Now, in cases of fractured bones, when strangulation sets in at once with lividity, lines of demarcation are rapidly developed on the part, and that which is coming away does so quickly, the bones having already yielded; but in cases where the lividity comes on while the bones are sound, the flesh dies rapidly here also, but the bones separate slowly along the border of the lividity and denudation of the bone. As regards parts of the limb which are below the limit of mortification, when they are quite dead and painless, they should be taken off at the joint, taking care not to wound any live part. For if the patient suffers pain during the amputation, and the limb happens to be not yet dead at the place where it is cut away, there is great risk of collapse from pain; and collapses of this kind have brought sudden death to many. I have seen a thigh-bone, denuded in this way, separate on the eightieth day. The leg in this patient was removed at the knee on the twentieth day, and I thought it might have been done higher up—not all at once, of course—but I resolved to act rather on the safe side.[1] The bones of the leg in a similar case which I had of gangrene just in the middle of the leg came away on the sixtieth day, so far as they were denuded. One or another kind of treatment would make a great difference in the rapidity or slowness with which the denuded bones come away. So too pressure, if

[1] Seems to be the sense of a very obscure passage. "Sooner" gives best sense, but is a curious meaning for ἐγγυτέρω. "Too early, for it appeared to me that this should be done more guardedly" (Adams, Littré) does violence to the text. Galen apparently understood "higher up"; for he says H. means that it is safer to amputate at a joint.

ἄν τι καὶ πίεξις πιέξιος καὶ ἐπὶ τὸ ἰσχυρότερόν
40 τε καὶ ἀσθενέστερον, καὶ ἐς τὸ θᾶσσόν τε καὶ
βραδύτερον ἀπομελανθέντα ἀποθανεῖν τὰ νεῦρα
καὶ τὰς σάρκας καὶ τὰς ἀρτηρίας καὶ τὰς φλέβας·
ἐπεὶ ὅσα μὴ ἰσχυρῶς ἀποληφθέντων θνήσκει,
ἔνια τῶν τοιούτων οὐκ ἀφικνεῖται ἐς ὀστέων
ψιλώματα, ἀλλ' ἐπιπολαιότερα ἐκπίπτει· ἔνια
δὲ οὐδὲ ἐς νεύρων ψιλώματα ἀφικνεῖται, ἀλλ'
ἐπιπολαιότερα ἐκπίπτει. διὰ οὖν ταύτας τὰς
εἰρημένας προφάσιας οὐκ ἔστιν ἓν οὔνομα ἀριθ-
μοῦ τῷ χρόνῳ θέσθαι, ἐν ὁπόσῳ ἕκαστα τούτων
50 κρίνεται.

Προσδέχεσθαι δὲ μάλα χρὴ τοιαῦτα ἰήματα·
ἐσιδεῖν γὰρ φοβερώτερά ἐστίν τινι ἢ ἰητρεύειν·
καὶ ἰητρείη πραείη ἀρκεῖ πᾶσι τοιούτοισιν· αὐτὰ
γὰρ ἑωυτὰ κρίνει μοῦνον. τῆς δὲ διαίτης ἐπιμελεῖσθαι χρὴ ὡς κατὰ δύναμιν ἀπύρετος ᾖ, καὶ
ἐν σχήμασι δικαίοισι εὐθετίζειν τὸ σῶμα· δίκαια
δὲ ταῦτα μηδὲ μετέωρον ποιεῖν, μηδὲ ἐς τὸ κάτω
ῥέπον, ἀλλὰ μᾶλλον ἐς τὸ ἄνω, ποτὶ καὶ ἔστ' ἂν
τελέως περιῤῥαγῇ· αἱμοῤῥαγιέων γὰρ ἐν τούτῳ
60 τῷ χρόνῳ κίνδυνος· διὰ τοῦτο οὖν οὐ χρὴ κατάρ-
ῥοπα τὰ τρώματα ποιεῖν, ἀλλὰ τἀναντία. ἐπεὶ
ὅταν γε χρόνος ἐγγένηται πλείων καὶ καθαρὰ
τὰ ἕλκεα γένηται, οὐκ ἔτι τὰ αὐτὰ[1] σχήματα
ἐπιτήδειά ἐστιν, ἀλλ' ἡ εὐθεῖα θέσις, καὶ ἐνίοτε
ἐπὶ τὸ κατάῤῥοπον ῥέποντα· ἀνὰ χρόνον γὰρ
ἐνίοισι τούτων ἀποστάσιες πύου γίνονται, καὶ
ὑποδεσμίδων δέονται. προσδέχεσθαι δὲ χρὴ
τοὺς τοιούτους ἀνὰ χρόνον ὑπὸ δυσεντερίης
πιέζεσθαι· καὶ γὰρ ἐπὶ τοῖσι μελαινομένοισι,
70 τοῖσι πλείστοισιν ἐπιγίνεται δυσεντερίη, καὶ ἐπὶ

stronger or weaker, would make a difference in the rapidity or slowness of the blackening and mortification of the ligaments, flesh, arteries and veins. For where the parts perish without great strangulation, the denudation sometimes does not extend to the bones, but the more superficial tissues are thrown off; sometimes the denudation does not even extend to the ligaments, but the more superficial parts are thrown off. For the said reasons, then, one cannot fix on one definite time in which each of these cases is determined.

One should be quite ready to treat such cases, for they are more formidable to look at than to cure; and mild treatment is sufficient, for they determine their own process. One must be careful as to diet, so that the patient may be, so far as possible, without fever, and place the limb in a correct attitude. Correct attitudes are neither elevated nor sloping downwards, but rather upwards, especially before the line of demarcation is fully developed; for there is danger of haemorrhage in this period. Wherefore do not keep the injured part dependent, but the reverse. When a considerable time has elapsed, and the wounds are cleansed, the suitable attitude is no longer the same as before, but the horizontal position, and sometimes one sloping downwards; for in time purulent collections form in some of these cases, and they require under-bandages.[1] One must expect such patients to be troubled, after a time, with dysentery; for dysentery supervenes in most cases

[1] See Introduction.

---

[1] ταῦτα.

τῇσιν αἱμοῤῥαγίῃσιν[1] ἐξ ἑλκέων· ἐπιγίνεται δὲ ὡς ἐπὶ τὸ πολὺ κεκριμένων ἤδη τῶν μελασμῶν καὶ τῆς αἱμοῤῥαγίης, καὶ ὁρμᾶται μὲν λαύρως καὶ ἰσχυρῶς· ἀτὰρ οὔτε πολυήμερος γίνεται οὔτε θανατώδης· οὔτε γὰρ μάλα ἀπόσιτοι γίνονται οἱ τοιοῦτοι, οὔτε ἄλλως συμφέρει
76 κενεαγγεῖν.

LXX. Μηροῦ δὲ ὀλίσθημα κατ' ἰσχίον ὧδε χρὴ ἐμβάλλειν, ἢν ἐς τὸ ἔσω μέρος ὠλισθήκῃ· ἀγαθὴ μὲν ἥδε καὶ δικαίη καὶ κατὰ φύσιν ἡ ἐμβολή, καὶ δή τι καὶ ἀγωνιστικὸν ἔχουσα, ὅστις γε τοῖσι τοιούτοισιν ἥδεται κομψευόμενος. κρεμάσαι χρὴ τὸν ἄνθρωπον τῶν ποδῶν πρὸς μεσόδμην δεσμῷ δυνατῷ μέν, μαλθακῷ δὲ καὶ πλάτος ἔχοντι· τοὺς δὲ πόδας διέχειν χρὴ ὅσον τέσσαρας δακτύλους ἀπ' ἀλλήλων, ἢ καὶ ἔλασ-
10 σον· χρὴ δὲ καὶ ἐπάνωθεν τῶν ἐπιγουνίδων προσπεριβεβλῆσθαι πλατεῖ ἱμάντι καὶ μαλθακῷ, ἀνατείνοντι ἐς[2] τὴν μεσόδμην· τὸ δὲ σκέλος τὸ σιναρὸν ἐντετάσθαι χρὴ ὡς δύο δακτύλους μᾶλλον τοῦ ἑτέρου· ἀπὸ τῆς γῆς τὴν κεφαλὴν ἀπεχέτω ὡς δύο πήχεας, ἢ ὀλίγῳ πλέον ἢ ἔλασσον· τὰς δὲ χεῖρας παρατεταμένας παρὰ τὰς πλευρὰς προσδεδεμένος ἔστω μαλθακῷ τινί· πάντα δὲ ταῦτα ὑπτίῳ κατακειμένῳ κατασκευασθήτω, ὡς ὅτι ἐλάχιστον χρόνον κρέμηται. ὅταν δὲ κρε-
20 μασθῇ, ἄνδρα χρὴ εὐπαίδευτον καὶ μὴ ἀσθενέα, ἐνείραντα τὸν πῆχυν μεσηγὺ τῶν μηρῶν, εἶτα θέσθαι τὸν πῆχυν μεσηγὺ τοῦ τε περιναίου καὶ τῆς κεφαλῆς τοῦ μηροῦ τῆς ἐξεστηκυίης, ἔπειτα συνάψαντα τὴν ἑτέρην χεῖρα πρὸς τὴν διῃρμένην, παραστάντα ὀρθὸν παρὰ τὸ σῶμα τοῦ κρεμα-

of mortification, and in haemorrhage from wounds. It comes on as a rule when the mortification or haemorrhage has been determined, and is copious and violent at the start, but neither lasts long nor is dangerous to life. The patients in such cases do not lose their appetite much, nor is there any advantage in a restricted diet.

LXX. Dislocation of the thigh at the hip should be reduced as follows, if it is dislocated inwards. It is a good and correct method, and in accord with nature, and one too that has something striking about it, which pleases a dilettante in such matters. One should suspend the patient by his feet from a cross-beam with a band, strong, but soft, and of good breadth. The feet should be about four fingers apart, or even less. He should also be bound round above the knee-caps with a broad, soft band stretching up to the beam; and the injured leg should be extended about two fingers' breadth further than the other. Let the head be about two cubits, more or less, from the ground. The patient should have his arms extended along the sides and fastened with something soft. Let all these preparations be made while he is lying on his back, that the period of suspension may be as short as possible. When he is suspended, let an assistant who is skilful and no weakling insert his forearm between the patient's thighs, and bring it down between the perineum and the head of the dislocated bone. Then, clasping the inserted hand with the other, while standing erect beside the suspended patient, let him suddenly

---

[1] τοῖσι αἱμοῤῥαγήσασιν.
[2] πρὸς.

μένου, ἐξαπίνης ἐκκρεμασθέντα μετέωρον αἰωρηθῆναι ὡς ἰσοῤῥοπώτατον. αὕτη δὲ ἡ ἐμβολὴ παρέχεται πάντα ὅσα χρὴ κατὰ φύσιν· αὐτό τε γὰρ τὸ σῶμα κρεμάμενον τῷ ἑωυτοῦ βάρει κατά-
30 τασιν ποιεῖται, ὅ τε ἐκκρεμασθεὶς ἅμα μὲν τῇ κατατάσει ἀναγκάζει ὑπεραιωρεῖσθαι τὴν κεφαλὴν τοῦ μηροῦ ὑπὲρ τῆς κοτύλης, ἅμα δὲ τῷ ὀστέῳ τοῦ πήχεος ἀπομοχλεύει καὶ ἀναγκάζει ἐς τὴν ἀρχαίην φύσιν ὀλισθάνειν. χρὴ δὲ παγκάλως μὲν τοῖσι δεσμοῖσιν ἐσκευάσθαι, φρονέοντα δὲ καὶ ὡς ἰσχυρότατον[1] τὸν ἐξαιω-
37 ρούμενον εἶναι.

LXXI. Ὡς μὲν οὖν καὶ πρόσθεν εἴρηται, μέγα τὸ διαφέρον ἐστὶ τῶν φυσίων τοῖσι ἀνθρώποισιν ἐς τὸ εὐέμβλητα εἶναι καὶ δυσέμβλητα [τὰ ἄρθρα].[2] καὶ διότι μέγα διαφέρει, εἴρηται πρόσθεν ἐν τοῖσι περὶ ὤμου. ἐνίοισι γὰρ ὁ μηρὸς ἐμπίπτει ἀπ' οὐδεμιῆς παρασκευῆς, ἀλλ' ὀλίγης μὲν κατατάσιος, ὅσον τῇσι χερσὶ κατιθῦναι, βραχείης δὲ κιγκλίσιος· πολλοῖσι δὲ συγκάμψασι τὸ σκέλος κατὰ τὸ ἄρθρον ἐνέπεσεν, ἤδη ἀμφίσφαλ-
10 σιν ποιησάμενον. ἀλλὰ γὰρ τὰ πολὺ πλείω οὐκ ἐνακούει τῆς τυχούσης παρασκευῆς· διὰ τοῦτο ἐπίστασθαι μὲν χρὴ τὰ κράτιστα περὶ ἑκάστου ἐν πάσῃ τῇ τέχνῃ· χρῆσθαι δὲ οἷσιν ἂν δόξῃ ἑκάστοτε. εἴρηνται μὲν οὖν τρόποι κατατασίων καὶ ἐν τοῖσιν ἔμπροσθεν γεγραμμένοισιν, ὥστε χρῆσθαι τούτων ὅστις ἂν παρατύχῃ. δεῖ γὰρ

[1] According to Littré and Petrequin, the patient is meant; but Littré emends to ἐχυρώτατον. The καὶ favours reference to the assistant; as in the Latin interpreters and Ermerins.

[2] Omit Galen, Littré.

suspend himself from him, and keep himself in the air as evenly balanced as possible. This mode of reduction provides everything requisite according to nature, for the body itself when suspended makes extension by its own weight; the assistant who is suspended, while making extension, forces the head of the bone to a position above the socket, and at the same time levers it out with the bone of his forearm, and makes it slip into its old natural place. But the bandages must be perfectly arranged, and care taken that the suspended assistant is the strongest available.[1]

LXXI. Now, as was said before, there is a great difference in the constitution of individuals, as regards ease and difficulty in reducing their dislocated joints; and the reason of this great difference was given before in the part about the shoulder. Thus in some, the thigh is put in without any apparatus, by the aid of slight extension, such as can be managed with the hands, and a little jerking; while in many, flexion of the leg at the joint and making a movement of circumduction is found to reduce it. But the great majority do not yield to ordinary apparatus; wherefore one should know the most powerful methods which the whole art provides for each case, and use them severally where they seem appropriate. Now methods of extension have been described in previous chapters, so that one may use any one of them which happens to be available.[2]

[1] Pq. renders, "the patient very strongly suspended," so also Littré; but there are surely two injunctions. Adams, "the person suspended along with the patient [should] have a sufficiently strong hold." Littré's ἐχυρώτατον applied to the assistant.

[2] Cf. VII.

ἀντικατατετάσθαι ἰσχυρῶς, ἐπὶ θάτερα μὲν τοῦ σκέλεος, ἐπὶ θάτερα δὲ τοῦ σώματος· ἢν γὰρ εὖ καταταθῇ, ὑπεραιωρηθήσεται ἡ κεφαλὴ τοῦ
20 μηροῦ ὑπὲρ τῆς ἀρχαίης ἕδρης· καὶ ἢν μὲν ὑπεραιωρηθῇ οὕτως, οὐδὲ κωλῦσαι ἔτι ῥηΐδιον ἵζεσθαι αὐτὴν ἐς τὴν ἑωυτῆς ἕδρην, ὥστε ἤδη πᾶσα ἀρκεῖ μόχλευσίς τε καὶ κατόρθωσις· ἀλλὰ γὰρ ἐλλείπουσιν ἐν τῇ κατατάσει· διὰ τοῦτο ὄχλον πλείω παρέχει ἡ ἐμβολή. χρὴ οὖν[1] οὐ μοῦνον παρὰ τὸν πόδα τὰ δεσμὰ ἐξηρτῆσθαι, ἀλλὰ καὶ ἄνωθεν τοῦ γούνατος, ὅπως[2] μὴ κατὰ τὸ τοῦ γούνατος ἄρθρον ἐν τῇ τανύσει ἡ ἐπίδοσις[3] ᾖ μᾶλλον ἢ κατὰ τὸ τοῦ ἰσχίου ἄρθρον. οὕτω μὲν οὖν χρὴ τὴν κατάτα-
30 σιν τὴν πρὸς τὸ τοῦ ποδὸς μέρος ἐσκευάσθαι· ἀτὰρ καὶ τὴν ἐπὶ θάτερα κατάτασιν, μὴ μοῦνον ἐκ τῆς περὶ τὸ στῆθος καὶ τὰς μασχάλας περιβολῆς ἀντιτείνεσθαι, ἀλλὰ καὶ ἱμάντι μακρῷ, διπτύχῳ, ἰσχυρῷ, προσηνεῖ, παρὰ τὸν περίναιον βεβλημένῳ, παρατεταμένῳ, ἐπὶ μὲν τὰ ὄπισθεν παρὰ τὴν ῥάχιν, ἐπὶ δὲ τὰ ἔμπροσθεν παρὰ τὴν κληῖδα, προσηρτημένῳ πρὸς τὴν ἀρχὴν τὴν ἀντικατατείνουσαν, οὕτω διαναγκάζεσθαι, τοῖσι μὲν ἔνθα διατειναμένοισι, τοῖσι δὲ ἔνθα, ὅπως δὲ ὁ ἱμὰς ὁ
40 παρὰ τὸν περίναιον μὴ περὶ τὴν κεφαλὴν τοῦ μηροῦ παρατεταμένος ἔσται, ἀλλὰ μεσηγὺ τῆς κεφαλῆς καὶ τοῦ περιναίου, ἐν δὲ τῇ κατατάσει κατὰ μὲν τὴν κεφαλὴν τοῦ μηροῦ ἐρείσας τὴν πυγμὴν ἐς τὸ ἔξω ὠθείτω. ἢν δὲ μετεωρίζηται ἑλκόμενος, διέρσας τὴν χεῖρα καὶ ἐπισυνάψας τῇ ἑτέρῃ χειρὶ ἅμα συγκατατεινέτω, ἅμα δὲ ἐς τὸ ἔξω συναναγκαζέτω· ἄλλος δέ τις τὸ παρὰ τὸ γόνυ
48 τοῦ μηροῦ ἡσύχως ἐς τὸ ἔσω μέρος κατορθούτω.

There must be strong extension both ways, of the leg in one direction, and of the body in the other; for if good extension is made, the head of the thigh-bone will be lifted over its old seat, and when so brought up, it becomes difficult even to prevent it from settling into its position, so that any leverage and adjustment suffices; but it is in extension that operators fail, and that is why the reduction gives more trouble. One should attach the bands, not only at the foot, but also above the knee, so that, in stretching, the giving way may not occur at the knee-joint rather than at the hip. This then is how the extension towards the foot end should be arranged; but there should be also counter-extension in the other direction, not only from a band round the chest and under the armpits, but also from a long double strap, strong and soft, passed round the perineum and stretched behind along the spine, and in front by the collar-bone attached to the source of the counter-extension. With the cords so arranged, some are stretched in one direction, some in the other, taking care that the strap at the perineum is not stretched over the head of the thigh-bone but between it and the perineum. During extension, let the fist be pressed against the head of the thigh-bone and thrust it outwards. If the pulling lifts up the patient, insert one hand between the thighs and, clasping it with the other, combine extension with pressure outwards. Let another person make adjustment by pushing the knee end of the bone gently inwards.

---

[1] δὲ. [2] ἵνα.

[3] ἐπίδεσις Littré, Petrequin, and codd., except B. ἐπίδοσις B, Erm., Kw.

LXXII. Εἴρηται δὲ καὶ πρόσθεν ἤδη ὅτι ἐπάξιον, ὅστις ἐν πόλει πολυανθρώπῳ ἰητρεύει, ξύλον κεκτῆσθαι τετράγωνον ὡς ἑξάπηχυ, ἢ ὀλίγῳ μέζον, εὖρος δὲ ὡς δίπηχυ, πάχος δὲ ἀρκεῖ σπιθαμιαῖον· ἔπειτα κατὰ μῆκος μὲν ἔνθεν καὶ ἔνθεν ἐντομὴν ἔχειν χρή, ὡς μὴ ὑψηλοτέρη τοῦ καιροῦ ἡ μηχάνησις ᾖ· ἔπειτα φλιὰς βραχείας, ἰσχυρὰς καὶ ἰσχυρῶς ἐνηρμοσμένας, ὀνίσκον ἔχειν ἑκατέρωθεν· ἔπειτα ἀρκεῖ μὲν ἐν τῷ ἡμίσει τοῦ
10 ξύλου—οὐδὲν δὲ κωλύει καὶ διὰ παντός—ἐντετμῆσθαι ὡς καπέτους μακρὰς πέντε ἢ ἕξ, διαλειπούσας ἀπ' ἀλλήλων ὡς τέσσαρας δακτύλους, αὐτὰς δὲ ἀρκεῖ εὖρος τριδακτύλους εἶναι καὶ βάθος οὕτως. ἔχειν δὲ κατὰ μέσον τὸ ξύλον καὶ καταγλυφὴν χρὴ βαθυτέρην, ἐπὶ τετράγωνον, ὡς τριῶν δακτύλων· καὶ ἐς μὲν τὴν καταγλυφὴν ταύτην, ὅταν δοκῇ προσδεῖν, ξύλον ἐμπηγνύναι ἐνάρμοζον τῇ καταγλυφῇ, τὸ δὲ ἄνω στρογγύλον· ἐμπηγνύναι δέ, ἐπήν ποτε δοκῇ συμφέρειν, μεσηγὺ
20 τοῦ περιναίου καὶ τῆς κεφαλῆς τοῦ μηροῦ. τοῦτο τὸ ξύλον ἑστεὸς κωλύει τὴν ἐπίδοσιν ἐπιδιδόναι τὸ σῶμα τοῖσι πρὸς ποδῶν ἕλκουσιν· ἐνίοτε γὰρ ἀρκεῖ αὐτὸ τὸ ξύλον τοῦτο ἀντὶ τῆς ἄνωθεν ἀντικατατάσιος· ἐνίοτε δὲ καὶ κατατεινομένου τοῦ σκέλεος ἔνθεν καὶ ἔνθεν, αὐτὸ τὸ ξύλον τοῦτο, χαλαρὸν ἐγκείμενον ᾗ τῇ ᾗ τῇ, ἐκμοχλεύειν ἐπιτήδειον ἂν εἴη τὴν κεφαλὴν τοῦ μηροῦ ἐς τὸ ἔξω μέρος. διὰ τοῦτο γὰρ καὶ αἱ κάπετοι ἐντετμέαται, ὡς καθ' ὁποίην ἂν αὐτέων ἁρμόσῃ, ἐμβαλλόμενος
30 ξύλινος μοχλὸς μοχλεύοι, ἢ παρὰ τὰς κεφαλὰς τῶν ἄρθρων, ἢ κατὰ κεφαλὰς τελέως ἐρειδόμενος ἅμα τῇ κατατάσει, ἤν τε ἐς τὸ ἔξω μέρος συμφέρῃ

LXXII. It was said before[1] that it is worth while for one who practises in a populous city to get a quadrangular plank, six cubits long or rather more, and about two cubits broad; while for thickness a span is sufficient. Next, it should have an incision at either end of the long sides, that the mechanism may not be higher than is suitable.[2] Then let there be short strong supports, firmly fitted in, and having a windlass at each end. It suffices, next, to cut out five or six long grooves about four fingers' breadth apart; it will be enough if they are three fingers broad and the same in depth, occupying half the plank, though there is no objection to their extending the whole length. The plank should also have a deeper hole cut out in the middle, about three fingers' breadth square; and into this hole insert, when requisite, a post, fitted to it, but rounded in the upper part. Insert it, whenever it seems useful, between the perineum and the head of the thigh-bone. This post, when fixed, prevents the body from yielding when traction is made towards the feet; in fact, sometimes the post of itself is a substitute for counter-extension upwards. Sometimes also, when the leg is extended in both directions, this same post, so placed as to have free play to either side, would be suitable for levering the head of the thigh-bone outwards. It is for this purpose, too, that the grooves are cut, that a wooden lever may be inserted into whichever may suit, and brought to bear either at the side of the joint-heads or right upon them, making pressure simultaneously with the extension, whether the leverage is required

[1] *Fract.* XIII. The *Scamnum* or "Bench" of Hippocrates.
[2] *I.e.* the supports should be "let in," not fixed on the top.

ἐκμοχλεύεσθαι, ἤν τε ἐς τὸ ἔσω, καὶ ἤν τε στρογγύλον τὸν μοχλὸν συμφέρῃ εἶναι, ἤν τε πλάτος ἔχοντα· ἄλλος γὰρ ἄλλῳ τῶν ἄρθρων ἁρμόζει. εὔχρηστος δέ ἐστιν ἐπὶ πάντων τῶν ἄρθρων ἐμβολῆς τῶν κατὰ τὰ σκέλεα αὕτη ἡ μόχλευσις σὺν τῇ κατατάσει. περὶ οὗ οὖν ὁ λόγος ἐστί, στρογγύλος ἁρμόζει ὁ μοχλὸς εἶναι· τῷ μέντοι 40 ἔξω ἐκπεπτωκότι ἄρθρῳ πλατὺς ἁρμόσει εἶναι. ἀπὸ τούτων τῶν μηχανέων καὶ ἀναγκέων οὐδὲν ἄρθρον μοι δοκεῖ οἷόν τε εἶναι ἀπορηθῆναι ἐμ-43 πεσεῖν.

LXXIII. Εὕροι δ' ἄν τις καὶ ἄλλους τρόπους τούτου τοῦ ἄρθρου ἐμβολῆς· εἰ γὰρ τὸ ξύλον τὸ μέγα τοῦτο ἔχοι κατὰ μέσον καὶ ἐκ πλαγίων φλιὰς δύο ὡς ποδιαίας,[1] ὕψος δὲ ὅπως ἂν δοκέοι συμφέρειν, τὴν μὲν ἔνθεν, τὴν δὲ ἔνθεν· ἔπειτα ξύλον πλάγιον ἐνείη ἐν τῇσι φλιῇσιν ὡς κλιμακτήρ, ἔπειτα διέρσαι [2] τὸ ὑγιὲς σκέλος μεσηγὺ τῶν φλιέων, τὸ δὲ σιναρὸν ἄνωθεν τοῦ κλιμακτῆρος ἔχειν [3] ἐνάρμοζον ἀπαρτὶ πρὸς τὸ ὕψος καὶ πρὸς 10 τὸ ἄρθρον, ᾗ ἐκπέπτωκεν· ῥηΐδιον δὲ [χρὴ] [4] ἁρμόζειν· τὸν γὰρ κλιμακτῆρα ὑψηλότερόν τινι χρὴ ποιεῖν τοῦ μετρίου, καὶ ἱμάτιον πολύπτυχον, ὡς ἂν ἁρμόσῃ, ὑποτείνειν ὑπὸ τὸ σῶμα. ἔπειτα χρὴ ξύλον ἔχον τὸ πλάτος μέτριον, καὶ μῆκος ἄχρι τοῦ σφυροῦ ὑποτεταμένον, ὑπὸ τὸ σκέλος εἶναι, ἱκνεύμενον ἐπέκεινα τῆς κεφαλῆς τοῦ μηροῦ

[1] ποδὸς μῆκος Paulus VI. 118.
[2] εἰ διέρσειεν Kw., ἐρείσειε Apoll.
[3] ἔχοι.
[4] Omit.

outwards or inwards, and whether the lever should be rounded or broad, for one form suits one joint, another another. This leverage, combined with extension, is very efficacious in all reductions of the leg-joints. As regards our present subject, it is proper that the lever be rounded; but for an external dislocation of the joint, a flat one will be suitable. It seems to me that no joint is incapable of reduction with these mechanical forces.

LXXIII. One might find other ways of reducing this joint. This big plank might have two props at the middle and to the sides,[1] about a foot long —height as may seem suitable—one on one side, the other on the other; then a crossbar of wood should be inserted in the props like a ladder-step. One might then insert[2] the sound leg between the props, and have the injured one on the top of the bar, fitting exactly to its height and to the joint where it is dislocated. This is easily arranged; for the crossbar should be put somewhat higher than is sufficient, and a folded garment spread under the patient, so that it fits. Then a piece of wood of suitable breadth and of a length sufficient to reach to the ankle should be extended under the leg, going up as far as possible beyond the head of the thigh-

[1] These props seem to have been removable and at the sides of the hole for the perineal post, which was *κατὰ μέσον*; not fixtures at the sides of the "bench," as usually figured. See the description in Paulus (VI. 118). The wooden crosspiece must have been either very thick or much shorter than three feet, to stand the pressure required. It could be put either at the top, when the whole resembled the letter pi, or lower down, when it resembled êta (H). This also shows that the arrangement was not very wide.

[2] *διέρσειεν* surely implies that the props were not far apart.

ὡς οἷόν τε· προσκαταδεδέσθαι δὲ χρὴ πρὸς τὸ σκέλος, ὅπως ἂν μετρίως ἔχῃ. κἄπειτα κατατεινομένου τοῦ σκέλεος, εἴτε ξύλῳ ὑπεροειδεῖ, εἴτε
20 τούτων τινὶ τῶν κατατασίων, ὁμοῦ χρὴ καταναγκάζεσθαι τὸ σκέλος περὶ τὸν κλιμακτῆρα ἐς τὸ κάτω μέρος σὺν τῷ ξύλῳ τῷ προσδεδεμένῳ· τὸν δέ τινα κατέχειν τὸν ἄνθρωπον ἀνωτέρω τοῦ ἄρθρου κατὰ τὸ ἰσχίον. καὶ γὰρ οὕτως ἅμα μὲν ἡ κατάτασις ὑπεραίροιτο[1] τὴν κεφαλὴν τοῦ μηροῦ ὑπὲρ τῆς κοτύλης, ἅμα δὲ ἡ μόχλευσις ἀπωθέοι τὴν κεφαλὴν τοῦ μηροῦ ἐς τὴν ἀρχαίην φύσιν. αὗται πᾶσαι αἱ εἰρημέναι ἀνάγκαι ἰσχυραὶ καὶ πᾶσαι κρέσσους τῆς συμφορῆς, ἤν τις
30 ὀρθῶς καὶ καλῶς σκευάζῃ.[2] ὥσπερ δὲ καὶ πρόσθεν ἤδη εἴρηται, πολύ τι ἀπὸ ἀσθενεστέρων κατατασίων καὶ φαυλοτέρης κατασκευῆς τοῖσι
33 πλείοσιν[3] ἐμπίπτει.

LXXIV. Ἢν δὲ ἐς τὸ ἔξω κεφαλὴ μηροῦ ὀλίσθῃ, τὰς μὲν κατατάσιας ἔνθα καὶ ἔνθα οὕτω χρὴ ποιεῖσθαι ὥσπερ εἴρηται, ἢ τοιουτοτρόπως· τὴν δὲ μόχλευσιν πλάτος ἔχοντι μοχλῷ μοχλεύειν χρὴ ἅμα τῇ κατατάσει, ἐκ τοῦ ἔξω μέρους ἐς τὸ ἔσω ἀναγκάζοντα, κατά γε αὐτὸν τὸν γλουτὸν τιθέμενον τὸν μοχλὸν καὶ ὀλίγῳ ἀνωτέρω· ἐπὶ τὸ ὑγιὲς ἰσχίον κατὰ τὸν γλουτὸν ἀντιστηριζέτω τις τῇσι χερσὶν ὡς μὴ ὑπείκῃ τὸ σῶμα, ἢ ἑτέρῳ τινὶ
10 τοιούτῳ μοχλῷ ὑποβάλλων καὶ ἐρείσας, ἐκ[4] τῶν καπέτων τὴν ἁρμόζουσαν ἀντικατεχέτω· τοῦ δὲ μηροῦ τοῦ ἐξηρθρηκότος τὸ παρὰ τὸ γόνυ ἔσωθεν ἔξω παραγέτω ἡσύχως. ἡ δὲ κρέμασις οὐχ

[1] ὑπεραιωρέοι ἄν.

bone; it should be attached to the leg in a suitable manner. Then, while the leg is being extended either by a pestle-shaped rod or any of the above modes of extension, one should simultaneously force the leg with the wood attached to it downwards over the crossbar; while an assistant holds down the patient at the hip above the joint. For thus the extension will raise the head of the thigh-bone over its socket, while the leverage will thrust it back into its natural place.[1] All these forcible methods of reduction are strong, and all are able to overcome the lesion, if one makes a proper and good application of them; but, as was said before, in the majority of cases the joint is put in with much weaker extensions and more ordinary apparatus.

LXXIV. When a thigh-bone head slips outwards, extension should be made in both directions as described, or in similar fashion. The leverage should be done with a broad lever simultaneously with the extension, forcing it from without inwards, the lever being applied to the buttock itself and a little above it. Let someone give counter-support to the hip on the sound side at the buttock with his hands, that the body may not yield, or make counter-pressure by slipping a similar lever under the joint, using a suitable groove as fulcrum. Let the bone of the dislocated thigh be gently brought from within outwards at the knee. The suspension method will

[1] An imitation of the method of reducing the shoulder-joint (VII).

---

[2] σκευάζηται, as Apollonius. [3] πλείστοισιν.
[4] ἐς for ἐκ Kw., following Erm.'s conjecture.

ἁρμόσει τούτῳ τῷ τρόπῳ τῆς ὀλισθήσιος τοῦ ἄρθρου· ὁ γὰρ πῆχυς τοῦ ἐκκρεμαμένου ἀπωθέοι[1] ἂν τὴν κεφαλὴν τοῦ μηροῦ ἀπὸ τῆς κοτύλης. τὴν μέντοι σὺν τῷ ξύλῳ τῷ ὑποτεινομένῳ μόχλευσιν μηχανήσαιτ' ἄν τις ὥστε ἁρμόζειν καὶ τούτῳ τῷ τρόπῳ τοῦ ὀλισθήματος, ἔξωθεν προσαρτέων.
20 ἀλλὰ τί καὶ δεῖ [πλείω λέγειν]; [2] ἢν γὰρ ὀρθῶς μὲν καὶ εὖ κατατείνηται, ὀρθῶς δὲ μοχλεύηται, τί
22 οὐκ ἂν ἐμπέσοι ἄρθρον οὕτως ἐκπεπτωκός;

LXXV. Ἢν δὲ ἐς τοὔπισθεν μέρος ἐκπεπτώκῃ ὁ μηρός, τὰς μὲν κατατάσιας καὶ ἀντιτάσιας οὕτω δεῖ ποιεῖσθαι, καθάπερ[3] εἴρηται· ἐπιστορέσαντα δὲ ἐπὶ τὸ ξύλον ἱμάτιον πολύπτυχον, ὡς μαλακώτατον ᾖ, πρηνέα κατακλίναντα τὸν ἄνθρωπον, οὕτω κατατείνειν· ἅμα δὲ τῇ κατατάσει χρὴ τῇ σανίδι καταναγκάζειν τὸν αὐτὸν τρόπον ὡς τὰ ὑβώματα, κατ' ἴξιν τοῦ πυγαίου ποιησάμενον τὴν σανίδα, καὶ μᾶλλον ἐς τὸ κάτω μέρος ἢ ἐς τὸ
10 ἄνω τῶν ἰσχίων· καὶ ἡ ἐντομὴ ἡ ἐν τῷ τοίχῳ τῇ σανίδι μὴ εὐθεῖα ἔστω, ἀλλ' ὀλίγον καταφερὴς πρὸς τὸ τῶν ποδῶν μέρος. αὕτη ἡ ἐμβολὴ κατὰ φύσιν τε μάλιστα τῷ τρόπῳ τούτῳ τοῦ ὀλισθήματός ἐστι καὶ ἅμα ἰσχυροτάτη. ἀρκέσειε δ' ἂν ἴσως ἀντὶ τῆς σανίδος καὶ ἐφεζόμενόν τινα, ἢ τῇσι χερσὶν ἐρεισάμενον ἢ ἐπίβαντα ἐξαπίνης ὁμοίως ἐπαιωρηθῆναι ἅμα τῇ κατατάσει. ἄλλη δὲ οὐδεμίη ἐμβολὴ τῶν πρόσθεν εἰρημένων κατὰ
19 φύσιν ἐστὶ τῷ τρόπῳ τούτῳ τοῦ ὀλισθήματος.

LXXVI. Ἢν δὲ ἐς τὸ ἔμπροσθεν ὀλίσθῃ, τῶν μὲν κατατασίων ὁ αὐτὸς τρόπος ποιητέος· ἄνδρα δὲ χρὴ ὡς ἰσχυρότατον ἀπὸ τῶν χειρῶν καὶ ὡς εὐπαιδευτότατον, ἐνερείσαντα τὸ θέναρ τῆς χειρὸς

not suit this form of dislocation, for the forearm of the person who hangs himself on would push the head of the thigh-bone away from its socket; but one might arrange the leverage with the board attached so as to suit this form of dislocation also, fitting it to the outside. But what need is there [to say more]? For if the extension is correct and good, and the leverage correct, what dislocation of this kind would not be reduced?

LXXV. If the thigh is dislocated backwards, extension and counter-extension should be made in the way described. Spreading a folded cloak on the plank, so that it may be as soft as possible, with the patient lying prone, one should make extension thus, and simultaneously make downward pressure with the plank, as in cases of hump-back, putting the board in a line with the buttock, and rather below than above the hip. Let the groove in the wall for the board be not level, but sloping a little down towards the feet. This mode of reduction is most naturally in accord with this form of dislocation, and at the same time very powerful. Instead of the board it would, perhaps, suffice for someone to sit on the part, or make pressure with his hands or with the foot, in each case bringing his weight suddenly to bear at the moment of extension. None of the other modes of reduction mentioned above is in natural conformity with this dislocation.

LXXVI. In dislocation forwards, the same extensions are to be used; and the strongest-handed and best-trained assistant available should make pressure

---

[1] ἀπωθοίη. [2] Omit Kw. and a few MSS.
[3] ὡς.

τῆς ἑτέρης παρὰ τὸν βουβῶνα, καὶ τῇ ἑτέρῃ χειρὶ τὴν ἑωυτοῦ χεῖρα προσκαταλαβόντα, ἅμα μὲν ἐς τὸ κάτω ὠθεῖν τὸ ὀλίσθημα, ἅμα δὲ ἐς τὸ ἔμπροσθεν τοῦ γόνατος μέρος. οὗτος γὰρ ὁ τρόπος τῆς ἐμβολῆς μάλιστα κατὰ φύσιν τούτῳ τῷ ὀλισ-
10 θήματί ἐστιν. ἀτὰρ καὶ ὁ κρεμασμὸς ἐγγύς τι τοῦ κατὰ φύσιν· δεῖ μέντοι τὸν ἐκκρεμάμενον ἔμπειρον εἶναι, ὡς μὴ ἐκμοχλεύῃ τῷ πήχει τὸ ἄρθρον, ἀλλὰ περὶ μέσον τὸν περίναιον καὶ
14 κατὰ τὸ ἱερὸν ὀστέον τὴν ἐκκρέμασιν ποίηται.

LXXVII. Εὐδοκιμεῖ δὲ δὴ καὶ [ὁ πειραθεὶς][1] ἀσκῷ τοῦτο τὸ ἄρθρον ἐμβάλλεσθαι· καὶ ἤδη μέν τινας εἶδον οἵτινες ὑπὸ φαυλότητος καὶ τὰ ἔξω ἐκκεκλιμένα καὶ τὰ ὄπισθεν ἀσκῷ ἐπειρῶντο ἐμβάλλειν, οὐ γιγνώσκοντες ὅτι ἐξέβαλλον αὐτὸ μᾶλλον ἢ ἐνέβαλλον· ὁ μέντοι πρῶτος ἐπινοήσας δῆλον ὅτι πρὸς τὰ ἔσω ὠλισθηκότα ἀσκῷ ἐμβάλλειν ἐπειρήσατο. ἐπίστασθαι μὲν οὖν χρὴ ὡς χρηστέον ἀσκῷ, εἰ δέοι χρῆσθαι·
10 διαγινώσκειν δὲ χρὴ[2] ὅτι ἕτερα πολλὰ ἀσκοῦ κρέσσω ἐστίν. χρὴ δὲ τὸν μὲν ἀσκὸν καταθεῖναι[3] ἐς τοὺς μηροὺς ἀφύσητον ἐόντα, ὡς ἂν δύναιτο ἀνωτάτω πρὸς τὸν περίναιον ἀνάγοντα· ἀπὸ δὲ τῶν ἐπιγουνίδων ἀρξάμενον, ταινίῃ πρὸς ἀλλήλους τοὺς μηροὺς καταδῆσαι ἄχρι τοῦ ἡμίσεος τῶν μηρῶν· ἔπειτα ἐς ἕνα τῶν ποδῶν,[4] τὸν λελυμένον, ἐνθέντα αὐλὸν ἐκ χαλκείου, φῦσαν ἐσαναγκάζειν ἐς τὸν ἀσκόν· τὸν δὲ ἄνθρωπον πλάγιον κατακεῖσθαι, τὸ σιναρὸν σκέλος ἐπι-
20 πολῆς ἔχοντα. ἡ μὲν οὖν παρασκευὴ αὕτη

[1] Omit Kw. and most MSS. [2] δεῖ.

at the groin with the palm of one hand, grasping it with the other, and pushing the dislocated part downwards, while at the same time the part at the knee is brought forwards.[1] This mode of reduction is in most natural accord with this dislocation. For the rest, suspension rather approaches the natural method; but the man who hangs himself on must be experienced, so as not to lever out the joint with his arm, but make the suspension weight act at the middle of the perineum, and over the sacrum.

LXXVII. Finally, there is an approved method of reducing this joint also with a bag;[2] and I have seen some who, through incompetence, kept trying to reduce even external and posterior dislocations with a bag, not knowing that they were putting it out rather than putting it in. The first inventor of the method, however, obviously used the bag in trying to reduce inward dislocations. One ought, therefore, to know how to use it, if required, while bearing in mind that many other methods are more effective. The bag should be applied to the thighs uninflated, and brought up as close as possible to the perineum. Bind the thighs to one another with a band extending from above the knee-caps half-way up the thighs; then, inserting a brass tube into one of the feet[3] which has been untied, force air into the bag. The patient should lie on his side with the injured leg on top. This, then, is the arrangement;

[1] In the "Apollonius" illustration he makes pressure with one hand on top of the other.

[2] *I.e.* wine-skin. Cf. use for spine (XLVII).

[3] Of the wine-skin.

---

[3] ἐνθεῖναι. [4] ποδεώνων Weber, Kw.

ἐστίν· σκευάζονται δὲ κάκιον οἱ πλεῖστοι ἢ ὡς
ἐγὼ εἴρηκα· οὐ γὰρ καταδέουσι τοὺς μηροὺς ἐπὶ
συχνόν, ἀλλὰ μοῦνον τὰ γόνατα, οὐδὲ κατα-
τείνουσι· χρὴ δὲ καὶ προσκατατείνειν· ὅμως δὲ
ἤδη τινὲς ἐνέβαλον ῥηϊδίου πρήγματος ἐπιτυ-
χόντες. εὐφόρως δὲ οὐ πάνυ ἔχει διαναγκάζεσθαι οὕτως· ὅ τε γὰρ ἀσκὸς ἐμφυσώμενος οὐ τὰ
ὀγκηρότατα αὐτοῦ ἔχει πρὸς τῷ ἄρθρῳ τῆς
κεφαλῆς, ἣν δεῖ μάλιστα ἐκμοχλεύσασθαι, ἀλλὰ
30 καθ' ἑωυτὸν αὐτὸς μέσος καὶ τῶν μηρῶν ἴσως
ἢ κατὰ τὸ μέσον ἢ ἔτι κατωτέρω· οἵ τε αὖ μηροὶ
φύσει γαυσοὶ πεφύκασιν, ἄνωθεν γὰρ σαρκώδεές
τε καὶ σύμμηροι, ἐς δὲ τὸ κάτω ὑπόξηροι, ὥστε καὶ
ἡ τῶν μηρῶν φύσις ἐπαναγκάζει τὸν ἀσκὸν ἀπὸ
τοῦ ἐπικαιροτάτου χωρίου. εἴ τε οὖν τις σμικρὸν
ἐνθήσει τὸν ἀσκόν, σμικρὴ ἡ ἰσχὺς ἐοῦσα ἀδύνατος ἔσται ἀναγκάζειν τὸ ἄρθρον. εἰ δὲ δεῖ
ἀσκῷ χρῆσθαι, ἐπὶ πολὺ οἱ μηροὶ συνδετέοι
πρὸς ἀλλήλους, καὶ ἅμα τῇ κατατάσει τοῦ
40 σώματος ὁ ἀσκὸς φυσητέος· τὰ δὲ σκέλεα ἀμφότερα ὁμοῦ καὶ καταδεῖν ἐν τούτῳ τῷ τρόπῳ
42 τῆς ἐμβολῆς ἐπὶ τὴν τελευτήν.

LXXVIII. Χρὴ δὲ περὶ πλείστου μὲν ποιεῖσθαι ἐν πάσῃ τῇ τέχνῃ ὅπως ὑγιέα ποιήσῃς τὸν
νοσέοντα· εἰ δὲ πολλοῖσι τρόποισι οἷόν τε εἴη
ὑγιέα ποιεῖν, τὸν ἀοχλότατον χρὴ αἱρεῖσθαι·
καὶ γὰρ ἀνδραγαθικώτερον τοῦτο καὶ τεχνικώ-
τερον, ὅστις μὴ ἐπιθυμεῖ δημοειδέος κιβδηλίης.
περὶ οὗ οὖν ὁ λόγος ἐστί, τοιαίδε ἄν τινες
κατοικίδιοι κατατάσιες εἶεν τοῦ σώματος, ὥστε
ἐκ τῶν παρεόντων τὸ εὔπορον εὑρίσκειν· τοῦτο
10 μὲν εἰ τὰ δεσμὰ τὰ ἱμάντινα μὴ παρείη τὰ

but most operators make less suitable preparation than that which I have described. They do not fasten the thighs together over a good space, but only at the knees; nor do they make extension, though there should be extension as well. Still, some are found to have made reduction, chancing upon an easy case. But the forcible separation is by no means lightly accomplished thus; for the inflated bag does not present its largest part at the articular head of the bone, which it is especially requisite to get levered out, but at its own middle, and perhaps at the middle of the thighs, or still lower down. The thighs, too, have a natural curve; for at the top they are fleshy and close together, but taper off downwards, so that the natural disposition of the thighs also forces the bag away from the most opportune place. If one inserts a small bag, its power being small, it will be unable to reduce the joint. So, if one must use a bag, the thighs are to be bound together over a large space, and the bag inflated simultaneously with the extension of the body; also tie both legs together at their extremity, in this form of reduction.

LXXVIII. What you should put first in all the practice of our art is how to make the patient well; and if he can be made well in many ways, one should choose the least troublesome. This is more honourable and more in accord with the art for anyone who is not covetous of the false coin of popular advertisement. To return to our subject—there are certain homely means of making extension, such as might readily be found among things at hand. First, supposing no soft supple leather holdfasts are

μαλθακὰ καὶ προσηνέα, ἀλλ' ἢ σιδήρεα[1] ἢ ὅπλα ἢ σχοινία, ταινίῃσι χρὴ ἢ ἐκρήγμασι τρυχίων ἐρινέων περιελίσσειν ταύτῃ μάλιστα ᾗ μέλλει τὰ δεσμὰ καθέξειν, καὶ ἔτι ἐπὶ πλέον· ἔπειτα οὕτω δεῖν τοῖσι δεσμοῖσιν· τοῦτο δέ, ἐπὶ κλίνης χρὴ ἥτις ἰσχυροτάτη καὶ μεγίστη τῶν παρεουσέων κατατετάσθαι καλῶς τὸν ἄνθρωπον· τῆς δὲ κλίνης τοὺς πόδας, ἢ τοὺς πρὸς κεφαλῆς ἢ τοὺς πρὸς ποδῶν, ἐρηρεῖσθαι πρὸς τὸν οὐδόν, εἴ
20 τε ἔξωθεν συμφέρει, εἴ τε ἔσωθεν· παρὰ δὲ τοὺς ἑτέρους πόδας παρεμβεβλῆσθαι ξύλον τετράγωνον πλάγιον, διῆκον ἀπὸ τοῦ ποδὸς πρὸς τὸν πόδα, καὶ ἢν μὲν λεπτὸν ᾖ τὸ ξύλον, προσδεδέσθω πρὸς τοὺς πόδας τῆς κλίνης, ἢν δὲ παχὺ ᾖ, μηδέν·[2] ἔπειτα τὰς ἀρχὰς χρὴ τῶν δεσμῶν καὶ τῶν πρὸς τῆς κεφαλῆς καὶ τῶν πρὸς τῶν ποδῶν προσδῆσαι ἑκατέρας πρὸς ὕπερον ἢ πρὸς ἄλλο τι τοιοῦτον· ὁ δὲ δεσμὸς ἐχέτω ἰθυωρίην κατὰ τὸ σῶμα ἢ καὶ ὀλίγῳ ἀνωτέρω, συμμέτρως δὲ
30 ἐκτετάσθω πρὸς τὰ ὕπερα, ὡς, ὀρθὰ ἑστεῶτα, τὸ μὲν παρὰ τὸν οὐδὸν ἐρείδηται, τὸ δὲ παρὰ τὸ ξύλον τὸ παραβεβλημένον· κἄπειτα οὕτω τὰ ὕπερα ἀνακλῶντα χρὴ τὴν κατάτασιν ποιεῖν. ἀρκεῖ δὲ καὶ κλῖμαξ ἰσχυροὺς ἔχουσα τοὺς κλιμακτῆρας, ὑποτεταμένη ὑπὸ τὴν κλίνην, ἀντὶ τοῦ οὐδοῦ τε καὶ ξύλου τοῦ παρατεταμένου, ὡς τὰ ὕπερα, πρὸς τῶν κλιμακτήρων τοὺς ἁρμόζοντας ἔνθεν καὶ ἔνθεν προσερηρεισμένα, ἀνακλώμενα, οὕτω τὴν κατάτασιν ποιῆται τῶν
40 δεσμῶν.

Ἐμβάλλεται δὲ μηροῦ ἄρθρον καὶ τόνδε τὸν

[1] σειραί.

available, one might still wrap up iron chains, ship's tackle, or cords, in scarves, or torn woollen rags, especially at the part where they are fastened on, and somewhat further, and then proceed to bind them on as holdfasts. Again, one should use a bed, the strongest and largest available, for making good extension;[1] the legs of the bed either at the head or foot should press against the threshold, outside or inside, as is opportune, and a quadrangular plank should be laid crosswise against the other legs, reaching from one to the other. If the plank is thin, let it be fastened to the legs of the bed; but if thick, this is unnecessary. Next, one should tie the ends of the bands, both those at the head and those at the feet respectively, to a pestle, or some other such piece of wood. Let the bands be in line with the body, or slanting a little upwards, and evenly stretched to the pestles, so that, when they are vertical, one is pressed against the threshold, the other against the plank laid across; and then one should make the extension by drawing back the pestles thus arranged. A ladder with strong crossbars stretched under the bed is a good substitute for the threshold and cross-beam, so arranged that the pestles may get their fulcra at either end against suitable crossbars, and, when drawn back, may thus make extension on the bands.

The thigh-joint is also reduced in the following

[1] Littré and Petrequin render κατατετάσθαι simply "coucher"; but the word is used throughout for surgical "extension." Adams: "the patient should be comfortably laid."

---

[a] οὐ δεῖ (Kw.'s conjecture from οὐδὲν of BMV).

τρόπον, ἢν ἐς τὸ ἔσω ὠλισθήκῃ καὶ ἐς τὸ ἔμπροσθεν· κλίμακα γὰρ χρὴ κατορύξαντα ἐπικαθίσαι τὸν ἄνθρωπον, ἔπειτα τὸ μὲν ὑγιὲς σκέλος ἡσύχως κατατείναντα προσδῆσαι, ὅπου ἂν ἁρμόσῃ· ἐκ δὲ τοῦ σιναροῦ ἐς κεράμιον ὕδωρ ἐγχέας ἐκκρεμάσαι, ἢ ἐς σφυρίδα λίθους ἐμβαλών. ἕτερος τρόπος ἐμβολῆς, ἢν ἐς τὸ ἔσω ὠλισθήκῃ· στρωτῆρα χρὴ καταδῆσαι μεταξὺ δύω στύλων ὕψος
50 ἔχοντα σύμμετρον· προεχέτω δὲ τοῦ στρωτῆρος κατὰ τὸ ἓν μέρος ὁπόσον τὸ πυγαῖον·[1] περιδήσας δὲ περὶ τὸ στῆθος τοῦ ἀνθρώπου ἱμάτιον, ἐπικαθίσαι τὸν ἄνθρωπον ἐπὶ τὸ προέχον τοῦ στρωτῆρος· εἶτα προσλαβεῖν τὸ στῆθος πρὸς τὸν στῦλον πλατεῖ τινί· ἔπειτα τὸ μὲν ὑγιὲς σκέλος κατεχέτω τις, ὡς μὴ περισφάλληται· ἐκ δὲ τοῦ σιναροῦ ἐκκρεμάσαι βάρος, ὅσον ἂν ἁρμόζῃ, ὡς
58 καὶ πρόσθεν ἤδη εἴρηται.

LXXIX. Πρῶτον μὲν οὖν δεῖ εἰδέναι ὅτι πάντων τῶν ὀστέων αἱ συμβολαί εἰσιν ὡς ἐπὶ πολὺ ἡ κεφαλὴ καὶ ἡ κοτύλη· ἐφ' ὧν δὲ καὶ ἡ χώρα κοτυλοειδὴς καὶ ἐπίμακρος· ἔνιαι δὲ τῶν χωρέων γληνοειδέες εἰσίν. ἀεὶ δὲ ἐμβάλλειν δεῖ πάντα τὰ ἐκπίπτοντα ἄρθρα, μάλιστα μὲν εὐθὺς παραχρῆμα ἔτι θερμῶν ἐόντων· εἰ δὲ μή, ὡς τάχιστα· καὶ γὰρ τῷ ἐμβάλλοντι ῥηΐτερον καὶ θᾶσσόν ἐστιν ἐμβάλλειν, καὶ τῷ ἀσθενέοντι πολὺ ἀπο-
10 νωτέρη ἡ ἐμβολὴ ἡ πρὶν διοιδεῖν ἐστίν. δεῖ δὲ

[1] πηχυαῖον Littré; πυγμαῖον Pq.; πυγαῖον vulg., Kw.

manner, if it is dislocated inwards or forwards. One should fix a ladder in the ground, and seat the patient upon it; then, gently extending the sound leg, fasten it at a suitable point, and from the injured limb suspend a jar and pour in water, or a basket and put in stones. Another way of reducing it, if dislocated inwards :– Fasten a crossbar between two props at a moderate height, and let one end of it project a buttock's length.[1] After passing a cloak round the patient's chest, seat him on the projecting crossbar, and then fasten his chest to the upright with a broad band. Let an assistant hold the sound leg, to prevent him from slipping round, and hang a suitable weight from the injured one, as has already been described.[2]

LXXIX. One must know, to begin with, that the connections between all bones are as a rule the head and the socket. In some, the cavity is large and cup-shaped; but in others, the cavities are shallowly concave. One must always reduce any dislocated joint, preferably at once, and while the parts are still warm; failing that, as soon as possible, for reduction before swelling sets in is accomplished much more easily and quickly by the operator, and is much less painful for the patient. When you are

[1] "What a measure!" says Petrequin, and suggests πυγμαῖον. Littré reads πηχυαῖον, "a cubit." The reading of the MSS. is supported by Apollonius (both text and illustration), though it is hard to see why the patient should not sit between the posts.

[2] According to Galen, the treatise ended here. The rest is a sort of appendix of fragments, some of them (*e.g.* LXXX) perhaps genuine parts which were lost and subsequently rediscovered. Most is from *Mochlicon*, as explained in the Introduction.

ἀεὶ παντα τὰ ἄρθρα, ὁπόταν μέλλῃς ἐμβάλλειν, προαναμαλάξαι καὶ διακιγκλίσαι· ῥᾷον γὰρ ἐθέλει ἐμβάλλεσθαι. παρὰ πάσας δὲ τὰς τῶν ἄρθρων ἐμβολὰς ἰσχναίνειν δεῖ τὸν ἄνθρωπον, μάλιστα μὲν περὶ τὰ μέγιστα ἄρθρα καὶ χαλεπώτατα ἐμβάλλεσθαι, ἥκιστα δὲ περὶ τὰ ἐλάχιστα
17 καὶ ῥηΐδια.

LXXX. Δακτύλων δὲ ἢν ἐκπεσῃ ἄρθρον τι τῶν τῆς χειρός, ἤν τε τὸ πρῶτον, ἤν τε τὸ δεύτερον, ἤν τε τὸ τρίτον, ωὑτὸς [καὶ ἴσος][1] τρόπος τῆς ἐμβολῆς· χαλεπώτερα μέντοι ἀεὶ τὰ μέγιστα τῶν ἄρθρων ἐμβάλλειν. ἐκπίπτει δὲ κατὰ τέσσαρας τρόπους, ἢ ἄνω ἢ κάτω ἢ ἐς τὸ πλάγιον ἑκατέρωθεν, μάλιστα μὲν ἐς τὸ ἄνω, ἥκιστα δὲ ἐς τὰ πλάγια, ἐν τῷ σφόδρα κινεῖσθαι. ἑκατέρωθεν δὲ τῆς χώρης, οὗ ἐκβέβηκεν, ὥσπερ ἄμβη ἐστίν. ἢν μὲν οὖν ἐς τὸ
10 ἄνω ἐκπέσῃ ἢ ἐς τὸ κάτω διὰ τὸ λειοτέρην εἶναι ταύτην τὴν χώρην, ἢ ἐκ τῶν πλαγίων, καὶ ἅμα μικρῆς ἐούσης τῆς ὑπερβάσιος, ἢν μεταστῇ τὸ ἄρθρον, ῥηΐδιόν ἐστιν ἐμβάλλειν. τρόπος δὲ τῆς ἐμβολῆς ὅδε· περιελίξαι τὸν δάκτυλον ἄκρον ἢ ἐπιδέσματί τινι ἢ ἄλλῳ τρόπῳ τοιούτῳ τινί, ὅπως, ὁπόταν κατατείνῃς ἄκρου λαβόμενος, μὴ ἀπολισθάνῃ· ὅταν δὲ περιελίξῃς, τὸν μέν τινα διαλαβέσθαι ἄνωθεν τοῦ καρποῦ τῆς χειρός, τὸν δὲ τοῦ κατειλημμένου·[2] ἔπειτα κατατείνειν πρὸς
20 ἑωυτὸν ἀμφοτέρους εὖ μάλα, καὶ ἅμα ἀπῶσαι τὸ ἐξεστηκὸς ἄρθρον ἐς τὴν χώρην. ἢν δὲ ἐς τὰ πλάγια ἐκπέσῃ, τῆς μὲν κατατάσιος ωὑτὸς τρόπος· ὅταν δὲ δὴ δοκῇ σοι ὑπερβεβηκέναι τὴν γραμμήν,[3] ἅμα χρὴ κατατείναντας ἀπῶσαι ἐς τὴν χώρην εὐθύς, ἕτερον δέ τινα ἐκ τοῦ ἑτέρου

going to put in any joint, you must always first make it supple and move it about, for it will thus be more easily reduced. In all cases of reduction, the patient must be put on restricted diet, especially when the joints are very large and very difficult to put in, and least so when they are very small and easy.

LXXX. If any of the finger-joints, whether first, second, or third, is dislocated, the mode of reduction is identically the same, though the largest joints are always the hardest to put in. Dislocation takes place in four ways, up or down[1] or to either side; chiefly upwards, most rarely to the sides, in some violent movement. On each side of the part whence it is displaced there is a sort of rim. Thus, if the displacement is upwards or downwards, it is easier to reduce, because this part is smoother than that at the sides, and the obstacle to get over is small, if the joint is dislocated. The mode of reduction is as follows:—Wrap a bandage or something of the kind round the end of the finger, in such a way that it will not slip off when you grasp the end and make extension. When it is applied, let one person take hold of the wrist from above, the other of the part wrapped up. Next, let each make vigorous extension in his own direction, and at the same time push back the projecting joint into place. In case of lateral dislocation, the mode of extension is the same. When you think it has passed over the line of the joint, push it at once into place, while keeping up the extension; an assistant should keep guard over

[1] Or "backwards" or "forwards."

---

[1] Omit B, Kw. [2] κατειλυμένου Weber.
[3] ἄμβην (Kw.'s conjecture).

μέρεος τοῦ δακτύλου φυλάσσειν καὶ ἀνωθεῖν, ὅπως μὴ πάλιν ἐκεῖθεν ἀπολίσθῃ. ἐμβάλλουσι δὲ ἐπιεικέως καὶ αἱ σαῦραι αἱ ἐκ τῶν φοινίκων πλεκόμεναι, ἢν κατατείνῃς ἔνθεν καὶ ἔνθεν τὸν
30 δάκτυλον, λαβόμενος τῇ μὲν ἑτέρῃ τῆς σαύρης, τῇ δὲ ἑτέρῃ τοῦ καρποῦ τῆς χειρός. ὅταν δὲ ἐμβάλλῃς, ἐπιδεῖν δεῖ ὀθονίοισιν ὡς τάχιστα, λεπτοτάτοισι κεκηρωμένοισι κηρωτῇ μήτε λίην μαλακῇ μήτε λίην σκληρῇ, ἀλλὰ μετρίως ἐχούσῃ. ἡ μὲν γὰρ σκληρὴ ἀφέστηκεν ἀπὸ τοῦ δακτύλου, ἡ δὲ ἁπαλὴ καὶ ὑγρὴ διατήκεται καὶ ἀπόλλυται, θερμαινομένου τοῦ δακτύλου. λύειν δὲ ἄρθρον δακτύλου τριταῖον ἢ τεταρταῖον· τὸ δὲ ὅλον, ἢν μὲν φλεγμήνῃ, πυκνότερον λύειν, ἢν δὲ μή, ἀραιό-
40 τερον· κατὰ πάντων δὲ τῶν ἄρθρων ταῦτα λέγω. καθίσταται δὲ τοῦ δακτύλου τὸ ἄρθρον τεσσαρεσκαιδεκαταῖον. ὁ αὐτὸς δέ ἐστι θεραπείης
43 τρόπος δακτύλων χειρός τε καὶ ποδός.

LXXXI. Παρὰ πάσας δὲ τὰς τῶν ἄρθρων ἐμβολὰς δεῖ ἰσχναίνειν καὶ λιμαγχονεῖν καὶ ἄχρι ἑβδόμης· καὶ εἰ φλεγμαίνοι, πυκνότερον λύειν, εἰ δὲ μή, ἀραιότερον· ἡσυχίην δὲ δεῖ ἔχειν ἀεὶ τὸ πόνεον ἄρθρον, καὶ ὡς κάλλιστα
6 ἐσχηματισμένον κεῖσθαι.

LXXXII. Γόνυ δὲ εὐηθέστερον ἀγκῶνος διὰ τὴν εὐσταλίην καὶ τὴν εὐφυΐην, διὸ καὶ ἐκπίπτει καὶ ἐμπίπτει ῥᾷον· ἐκπίπτει δὲ πλειστάκις ἔσω, ἀτὰρ καὶ ἔξω καὶ ὄπισθεν. ἐμβολαὶ δέ, ἐκ τοῦ

the other side of the finger and make counter-pressure, to prevent another dislocation to that side. The "lizards"[1] woven out of palm tissue are satisfactory means of reduction, if you make extension of the finger both ways, grasping the "lizard" at one end and the wrist at the other. After reduction you must apply at once very light bandages soaked in cerate, neither too soft nor too hard, but of medium consistency; for the hard gets detached from the finger, while the soft and moist is melted and disappears as the finger gets warm. Change the dressing of a finger-joint on the third or fourth day; in general, if there is inflammation, change it oftener; if not, more rarely. I apply this rule to all joints. A finger-joint is healed in fourteen days. The mode of treatment is the same for fingers and toes.

LXXXI.[2] In all reductions of joints, the patient should have attenuating and starvation diet up to the seventh day; if there is inflammation, change the dressing oftener; if not, more rarely. The injured joint should be kept always at rest, and be placed in the best possible attitude.

LXXXII.[3] The knee is more favourable for treatment than the elbow, because of its compact and regular form, whence it is both dislocated and reduced more easily. It is most often dislocated inwards, but also externally and backwards. Modes

[1] Hollow cylinders of plaited material which contract on being pulled out. Once a well-known toy. Also mentioned by Diocles, who calls them "the lizards which the children plait." Aristotle (*P.A.* IV. 9) calls them πλεγμάτια, and compares them with the suckers of cuttle-fish.

[2] An insertion repeated from §§ LXXIX (end) and LXXX.

[3] From *Fract.* XXXVIII and *Mochl.* XXVI

συγκεκάμφθαι ἢ ἐκλακτίσαι ὀξέως, ἢ συνελίξας ταινίης ὄγκον, ἐν τῇ ἰγνύῃ θείς, ἀμφὶ τοῦτον ἐξαίφνης ἐς ὄκλασιν ἀφιέναι τὸ σῶμα. δύναται δὲ καὶ κατατεινόμενον μετρίως, ὥσπερ ἀγκών, ἐμπίπτειν τὰ ὄπισθεν· τὰ δὲ ἔνθα καὶ ἔνθα, ἐκ
10 τοῦ συγκεκάμφθαι ἢ ἐκλακτίσαι, ἀτὰρ καὶ ἐκ κατατάσιος μετρίης. ἡ διόρθωσις ἅπασι κοινή. ἢν δὲ μὴ ἐμπέσῃ τοῖσι μὲν ὄπισθεν, συγκάμπτειν οὐ δύνανται, ἀτὰρ οὐδὲ τοῖσι ἄλλοισι πάνυ. μινύθει δὲ μηροῦ καὶ κνήμης τοὔμπροσθεν· ἢν δὲ ἐς τὸ ἔσω, βλαισότεροι, μινύθει δὲ τὰ ἔξω. ἢν δὲ ἐς τὸ ἔξω, γαυσότεροι, χωλοὶ δὲ ἧσσον· κατὰ γὰρ τὸ παχύτερον ὀστέον ὀχεῖ, μινύθει δὲ τὰ ἔσω. ἐκ γενεῆς δὲ καὶ ἐν αὐξήσει κατὰ λόγον
19 τὸν πρόσθεν.

LXXXIII. Τὰ δὲ κατὰ τὰ σφυρὰ κατατάσιος ἰσχυρῆς δεῖται, ἢ τῇσι χερσὶν ἢ ἄλλοισι τοιούτοισι,[1] κατορθώσιος δὲ ἅμα ἀμφότερα ποιεούσης·
4 κοινὸν δὲ τοῦτο ἅπασιν.

LXXXIV. Τὰ δὲ ἐν ποδὶ ὡς καὶ τὰ ἐν χειρὶ
2 ὑγιέες.[2]

LXXXV. Τὰ δὲ τῆς κνήμης συγκοινωνέοντα καὶ ἐκπεσόντα[3] ἐκ γενεῆς, ἢ καὶ ἐν αὐξήσει
3 ἐξαρθρήσαντα, ταὐτὰ ἃ καὶ ἐν χειρί.

LXXXVI. Ὁκόσοι δὲ πηδήσαντες ἄνωθεν

[1] τοῖσι. [2] ὑγιῆ *Mochl.*
[3] μὴ ἐμπεσόντα *Mochl.*

of reduction: by flexion or a sharp kick upwards[1] (? jerking the leg upwards), or placing a rolled bandage in the ham, on which the patient brings the weight of his body by crouching suddenly. Suitable extension can reduce backward dislocations, as with the elbow. Those to one or the other side are put in by flexion or leg-jerking, and also by suitable extension. Adjustment[2] is the same for all. If there is no reduction, in posterior cases patients cannot flex the limb, but they can hardly do so in the others; there is atrophy of the thigh and leg in front. If inwards, they are more knock-kneed, and there is atrophy of the outer side; if outwards, they are more bandy, but not so lame, for the weight comes on the larger bone; the inner side atrophies. Cases which occur congenitally or during adolescence follow the rule given above.

LXXXIII.[3] Dislocations at the ankle require strong extension, either with the hands or other such means, and a rectification involving the two[4] combined. This is common to all.

LXXXIV. Dislocations in the foot heal in the same way as those in the hand.

LXXXV. The bones connecting the foot with the leg, whether dislocated from birth or put out during adolescence, follow the same course as those in the hand.

LXXXVI. Those who in leaping from a height

[1] In Hippocrates *Coacae Prenotiones* 108 it is applied to involuntary "jerking of the legs."

[2] The slight variation in *Mochl.* XXVI seems to favour Pq.'s rendering. "This (*i.e.* extension) is common to all cases."

[3] Partly repeated in § LXXXVII.

[4] Extension and counter-extension? Extension and adjustment? It seems an obscure summary of *Fract.* XIII.

ἐστηρίξαντο τῇ πτέρνῃ, ὥστε διαστῆναι τὰ ὀστέα καὶ φλέβας ἐκχυμωθῆναι καὶ νεῦρα ἀμφιφλασθῆναι, ὁπόταν γένηται οἷα τὰ δεινά, κίνδυνος μὲν σφακελίσαντα τὸν αἰῶνα πρήγματα παρασχεῖν· ῥοιώδη μὲν τὰ ὀστέα, τὰ δὲ νεῦρα ἀλλήλοισι κοινωνέοντα. ἐπεὶ καὶ οἷσιν ἂν μάλιστα καταγεῖσιν ἢ ὑπὸ τρώματος ἢ ἐν κνήμῃ ἢ ἐν μηρῷ, ἢ νεύρων ἀπολυθέντων ἃ κοινωνεῖ
10 τούτων, ἢ ἐκ κατακλίσιος ἀμελέος, ἐμελάνθη ἡ πτέρνη, καὶ τούτοισι τὰ παλιγκοτέοντα ἐκ τῶν τοιούτων. ἔστιν ὅτε καὶ πρὸς τῷ σφακελισμῷ γίνονται πυρετοὶ ὀξέες λυγμώδεες, γνώμης ἁπτόμενοι, ταχυθάνατοι, καὶ ἔτι φλεβῶν αἱμορροιέων πελιώσιες. σημεῖα δὲ τῶν παλιγκοτησάντων, ἢν τὰ ἐκχυμώματα καὶ τὰ μελάσματα καὶ τὰ περὶ ταῦτα ὑπόσκληρα καὶ ὑπέρυθρα·[1] ἢν δὲ σὺν σκληρύσματι πελιδνωθῇ, κίνδυνος μελανθῆναι· ἢν δὲ ὑποπέλια ᾖ, ἢ καὶ πέλια
20 μάλα καὶ ἐκχυμώμενα,[2] ἢ ὑπόχλωρα καὶ μαλακά, ταῦτα ἐπὶ πᾶσι τοῖσι τοιούτοισιν ἀγαθά. ἴησις, ἢν μὲν ἀπύρετος ᾖ, ἐλλέβορον· ἢν δὲ μή, μή· ἀλλὰ ποτὸν ὀξύγλυκυ, εἰ δέοι. ἐπίδεσις δὲ ἄρθρων· ἐπὶ δὲ πάντα, μᾶλλον τοῖσι φλάσμασιν, ὀθονίοισι πλείοσι καὶ μαλθακωτέροισιν· πίεξις ἧσσον· προσπεριβάλλειν δὲ τὰ πλεῖστα τῇ πτέρνῃ. τὸ σχῆμα, ὅπερ ἡ ἐπίδεσις, ὡς μὴ ἐς τὴν
28 πτέρνην ἀποπιέζηται· νάρθηξι δὲ μὴ χρῆσθαι.

LXXXVII. Οἷσι δ᾿ ἂν ἐκβῇ ὁ πούς ἢ αὐτὸς ἢ σὺν τῇ ἐπιφύσει, ἐκπίπτει μὲν μᾶλλον ἐς τὸ ἔσω· ἢν δὲ μὴ ἐμπέσῃ, λεπτύνεται ἀνὰ χρόνον

[1] ὑπέρυθρα ᾖ *Mochl.*

come down on the heel, so that the bones are separated, and there is extravasation of blood and contusion of ligaments—when grave injuries such as these occur, there is danger of necrosis and life-long trouble; for the bones slip easily, and the ligaments are in connection with one another. Further, when in cases of fracture especially, or a wound either of leg or thigh, or when the ligaments joining up with these parts are torn away, or from carelessness as to position in bed, mortification of the heel has set in, in these patients also such causes give rise to exacerbations. Sometimes acute fevers follow the necrosis, with hiccoughs, affecting the mind and rapidly fatal; there are also lividities from haemorrhage. Signs of exacerbation are ecchymoses, blackenings of the skin with some induration and redness of the surrounding parts. If the lividity is accompanied with hardness, there is danger of mortification; but if the part is sublivid or even very livid after ecchymosis, or greenish yellow and soft, these are good signs in all such cases. Treatment: if there is no fever, hellebore, otherwise not, but let him drink oxymel, if required. Bandaging: that used for joints; over all, especially in contusions, use plenty of soft bandages; pressure, rather slight; additional bandaging, especially round the heel. Attitude: the same object as in bandaging, so as to avoid pressure on the heel. Do not use splints.

LXXXVII. In cases where the foot is dislocated, either by itself or with the epiphysis, it is usually displaced inwards; and if not reduced, the hip,

---

[2] ἐκκεχυμωμένα.

τό τε ἰσχίον καὶ ὁ μηρός, καὶ κνήμης τὸ ἀντίον τοῦ ὀλισθήματος. ἐμβολὴ δὲ ἄλλη,[1] ὥσπερ καρποῦ, κατάτασις δὲ ἰσχυρή· ἴησις δέ, νόμος ἄρθρων. παλιγκοτεῖ, ἧσσον δὲ καρποῦ, ἢν ἡσυχάσωσιν. δίαιτα μείων· ἐλινύουσι. τὸ δὲ
9 ἐκ γενεῆς ἢ ἐν αὐξήσει, κατὰ λόγον τὸν πρότερον.

[1] δὲ ἄλλη omit *Mochl.* and translators, except Pq.

thigh and leg become in time attenuated on the side opposed to the dislocation. Reduction in other respects as for the wrist; but strong extension is required. Treatment: that customary for joints. Exacerbation occurs, but less than in wrist cases, if the patients keep at rest. Diet more reduced; they do no work. Congenital and adolescent cases follow the rule given before.[1]

[1] See notes on these chapters in *Mochlicon*, pp. 425–429.

ΜΟΧΛΙΚΟΝ[1]

I. Ὀστέων φύσις· δακτύλων μὲν ἁπλᾶ καὶ ὀστέα καὶ ἄρθρα, χειρὸς δὲ καὶ ποδὸς πολλά, ἄλλα ἀλλοίως συνηρθρωμένα· μέγιστα δὲ τά ἀνωτάτω. πτέρνης δὲ ἕν, οἷον ἔξω φαίνεται, πρὸς δὲ αὐτὴν οἱ ὀπίσθιοι τένοντες τείνουσιν. κνήμης δὲ δύο, ἄνωθεν καὶ κάτωθεν συνεχόμενα, κατὰ μέσον δὲ διέχοντα σμικρόν· τὸ ἔξωθεν, κατὰ τὸν σμικρὸν δάκτυλον λεπτότερον βραχεῖ, πλεῖστον δὲ ταύτῃ διεχούσῃ καὶ σμικροτέρῃ ῥοπῇ κατὰ 10 γόνυ, καὶ ὁ τένων ἐξ αὐτοῦ πέφυκεν, ὁ παρὰ τὴν ἰγνύην ἔξω. ἔχουσι δὲ κάτωθεν κοινὴν ἐπίφυσιν πρὸς ἣν ὁ πούς κινεῖται· ἄλλην δὲ ἄνωθεν ἔχουσιν ἐπίφυσιν, ἐν ᾗ τὸ τοῦ μηροῦ ἄρθρον κινεῖται, ἁπλόον καὶ εὐσταλὲς ὡς ἐπὶ μήκει· εἶδος κονδυλῶδες, ἔχον ἐπιμυλίδα· αὐτὸς δὲ ἔγκυρτος ἔξω καὶ ἔμπροσθεν· ἡ δὲ κεφαλὴ ἐπίφυσίς ἐστι στρογγύλη, ἐξ ἧς τὸ νεῦρον τὸ ἐν τῇ κοτύλῃ τοῦ ἰσχίου πέφυκεν· ὑποπλάγιον δὲ καὶ τοῦτο προσήρτηται, ἧσσον δὲ βραχίονος. 20 τὸ δὲ ἰσχίον προσίσχεται πρὸς τῷ μεγάλῳ σπονδύλῳ τῷ παρὰ τὸ ἱερὸν ὀστέον χονδρονευρώδει δεσμῷ.

[1] ΜΟΧΛΙΚΟΣ Littré ; and the word is used as a synonym for *μοχλίσκος* in XLII. : but ΜΟΧΛΙΚΟΝ is supported by the MSS., and by the analogy of ΠΡΟΓΝΩΣΤΙΚΟΝ and ΠΡΟΡΡΗΤΙΚΟΝ. Cf. also Galen XVIII.(2) 327.

INSTRUMENTS OF REDUCTION

I. Nature of bones. In the fingers and toes, both bones and joints are simple; but in hand and foot they are diverse and diversely articulated, the uppermost being largest. The heel has a single bone which appears as a projection, and the hind tendons pull upon it. There are two leg-bones joined together above and below, but slightly separated in the middle. The outer one, towards the little toe, is rather more slender, most so in the separated part, and in the smaller inclination at the knee;[1] and the tendon on the outer side of the ham has its origin from it. They have below a common epiphysis on which the foot moves; and above they have another epiphysis, in which the articular end of the thigh-bone moves. This is simple and compact, considering the length of the bone; it is knuckle-shaped, and has a knee-cap. The bone itself is curved outwards and forwards; its head is a spherical epiphysis, from which the ligament arises which has its attachment in the cavity[2] of the hip; this (tendon)[3] is inserted rather obliquely, but less so than that of the arm.[4] The hip-bone is attached to the great vertebra[5] next the sacrum by a fibro-cartilaginous ligament.

[1] Or, "with the greatest deviation (from the vertical) at this point, and less at the knee"; but the passage is obscure.
[2] Acetabulum.
[3] Ligamentum teres.
[4] Long head of the biceps.
[5] Fifth lumbar.

Ῥάχις δὲ ἀπὸ μὲν τοῦ ἱεροῦ ὀστέου μέχρι τοῦ μεγάλου σπονδύλου κυφή. κύστις τε καὶ γονὴ καὶ ἀρχοῦ τὸ ἐγκεκλιμένον ἐν τούτῳ. ἀπὸ δὲ τούτου ἄχρι φρενῶν ἦλθεν ἡ ἰθύλορδος, καὶ αἱ ψόαι κατὰ τοῦτο· ἐντεῦθεν δὲ ἄχρι τοῦ μεγάλου σπονδύλου τοῦ ὑπὲρ τῶν ἐπωμίδων ἰθυκυφής· ἔτι δὲ μᾶλλον δοκεῖ ἢ ἐστιν· αἱ γὰρ ὄπισθεν τῶν
30 σπονδύλων ἀποφύσιες ταύτῃ ὑψηλόταται· τὸ δὲ τοῦ αὐχένος ἄρθρον λορδόν. σπόνδυλοι δὲ ἔσωθεν ἄρτιοι πρὸς ἀλλήλους, ἀπὸ δὲ τῶν ἔξωθεν χόνδρων νεύρῳ συνεχόμενοι· ἡ δὲ συνάρθρωσις αὐτῶν ἐν τῷ ὄπισθεν τοῦ νωτιαίου· ὄπισθεν δὲ ἔχουσιν ἔκφυσιν ὀξεῖαν ἔχουσαν ἐπίφυσιν χονδρώδεα· ἔνθεν νεύρων ἀπόφυσις καταφερής, ὥσπερ καὶ οἱ μῦες παραπεφύκασιν ἀπὸ αὐχένος ἐς ὀσφύν, πληροῦντες δὲ πλευρέων καὶ ἀκάνθης τὸ μέσον. πλευραὶ δὲ κατὰ τὰς διαφύσιας τῶν
40 σπονδύλων νευρίῳ προσπεφύκασιν ἀπ' αὐχένος ἐς ὀσφὺν ἔσωθεν, ἐπίπροσθεν δὲ κατὰ τὸ στῆθος χαῦνον καὶ μαλθακὸν τὸ ἄκρον ἔχουσαι· εἶδος ῥαιβοειδέστατον τῶν ζώων· στενότατος γὰρ ταύτῃ ὁ ἄνθρωπος ἐπ' ὄγκον· ᾗ δὲ μὴ πλευραί εἰσιν, ἔκφυσις πλαγίη, βραχεῖα καὶ πλατεῖα· ἐφ' ἑκάστῳ σπονδύλῳ νευρίῳ προσπεφύκασιν.

Στῆθος δὲ συνεχὲς αὐτὸ ἑωυτῷ, διαφύσιας ἔχον πλαγίας, ᾗ πλευραὶ προσήρτηνται, χαῦνον δὲ καὶ χονδρῶδες. κληῖδες δὲ περιφερέες ἐς
50 τοὔμπροσθεν, ἔχουσαι πρὸς μὲν τὸ στῆθος βραχείας κινήσιας, πρὸς δὲ τὸ ἀκρώμιον συχνοτέρας. ἀκρώμιον δὲ ἐξ ὠμοπλατέων πέφυκεν, ἀνομοίως δὲ τοῖσι πλείστοισι. ὠμοπλάτη δὲ

[1] "The ensemble of the articulations." Pq.

The spine from the end of the sacrum to the great vertebra is convex backwards. The bladder, generative organs, and inclined portion of the rectum are in this part. From here to the diaphragm it ascends in a forward curve, and there are the psoa-muscles; but thence up to the great vertebra above the shoulders it rises in a curve backwards, and seems more convex than it is, for the backward processes of the vertebrae are here at their highest. The neck-joint[1] is concave behind. The vertebrae on the inside are fitted to one another, being held together by a ligament from the outer side of the cartilages; but their jointing (synarthrosis) is behind the spinal cord, and they have posteriorly a sharp process with a cartilaginous epiphysis. Hence arise the ligaments which pass downwards, just as muscles also are disposed at the side from neck to loins, filling up the part between the ribs and the spinal ridge. The ribs are attached by a ligament at the intervals between the vertebrae from neck to loins behind, but in front to the breast-bone, having the termination spongy and soft. In shape they are the most curved of any animal; for man is flattest here in proportion to his size. Where there are no ribs, there is a short and broad lateral process; they are connected with each vertebra by a small ligament.

The sternum is a continuous bone, having lateral interstices where the ribs are inserted; it is spongy and cartilaginous. The collar-bones are rounded in front, having slight movements at the sternal end, but more extensive ones at the acromion. The acromion has its origin from the shoulder-blades in a different way from that in most animals.[2] The

[2] See notes on *Joints* XIII.

χονδρώδης τὸ πρὸς ῥάχιν, τὸ δ' ἄλλο χαύνη, τὸ ἀνώμαλον ἔξω ἔχουσα, αὐχένα δὲ καὶ κοτύλην ἔχουσα χονδρώδεα, ἐξ ἧς αἱ πλευραὶ κίνησιν ἔχουσι, εὐαπόλυτος ἐοῦσα ὀστέων, πλὴν βραχίονος. τούτου δὲ ἐκ τῆς κοτύλης νευρίῳ ἡ κεφαλὴ ἐξήρτηται, χόνδρου χαύνου περιφερῆ
60 ἐπίφυσιν ἔχουσα· αὐτὸς δ' ἔγκυρτος ἔξω καὶ ἔμπροσθεν πλάγιος, οὐκ ὀρθὸς πρὸς κοτύλην· τὸ δὲ πρὸς ἀγκῶνα αὐτοῦ πλατὺ καὶ κονδυλῶδες καὶ βαλβιδῶδες καὶ στερεόν, ἔγκοιλον ὄπισθεν, ἐν ᾧ ἡ κορώνη ἡ ἐκ τοῦ πήχεος, ὅταν ἐκταθῇ ἡ χείρ, ἔνεστιν· ἐς τοῦτο καὶ τὸ ναρκῶδες νεῦρον, ὃ [1] ἐκ τῆς διαφύσιος τῶν τοῦ πήχεος ὀστέων, ἐκ
67 μέσων ἐκπέφυκε καὶ περαίνεται.

II. Ῥὶς δὲ κατεαγεῖσα ἀναπλάσσεσθαι οἵη τε αὐθωρόν. κἢν μὲν οὖν ὁ χόνδρος, ἐντίθεσθαι [2] ἄχνην ὀθονίου, ἐναποδέοντα λοπῷ Καρχηδονίῳ, ἢ ἐν ἄλλῳ ὃ μὴ ἐρεθιεῖ· τῷ λοπῷ δὲ τὰς παραλλάξιας παρακολλᾷν καὶ ἀναλαμβάνειν· ταῦτα δὲ ἐπίδεσις κακὰ ποιεῖ.[3] ἴησις ἄλλη· ἅμα δὲ τῷ συμβαλεῖν σὺν μάννῃ [4] ἢ θείῳ σὺν κηρωτῇ· αὐτίκα ἀναπλάσσειν, ἔπειτα ἀνακωχήσειν, τοῖσι δακτύλοισι ἐσματευόμενον καὶ παραστρέφοντα·
10 καὶ τὸ Καρχηδόνιον· πωροῖτο ἂν καὶ ἢν ἕλκος ἐνῇ· καὶ ἢν ὀστέα ἀπιέναι μέλλῃ—οὐ γὰρ
12 παλιγκοτώτατα—οὕτω ποιητέα.

[1] τὸ. [2] ἐντιθέναι Littré, Kw.
[3] καταποιεῖ codd.; κακοποιεῖ M marg.; κακὰ ποιέει Lit. conj.
[4] ἀλήτῳ σὺν μάννῃ.

---

[1] Long tendon of the biceps.
[2] Galen *U.P.* II. 14. Our "olecranon." Both processes of the ulna were called κορωνόν, because of their semicircular shape.

shoulder-blade is cartilaginous in the part towards the spine, and spongy elsewhere; it has an irregular shape on the outer side, and the neck and articular cavity are cartilaginous. Its disposition allows free movement to the ribs, since it is not closely connected with the bones, except that of the upper arm. The head of this bone is attached to its socket by a small ligament,[1] and has a rounded epiphysis of spongy cartilage. The bone itself is convex outwards and oblique in front, and does not meet the cavity at right angles. Its elbow end is broad, knuckle-shaped, and grooved; it is also solid, and has a hollow at the back, in which the coronoid process[2] of the ulna is lodged when the arm is extended. Here too the cord which stupefies,[3] arising from the interstice between the bones of the forearm, has its issue and termination.

II. A fractured nose is a thing to be adjusted at once. If the cartilage is the part affected, introduce lint, rolling it up in thin Carthaginian leather, or in some other non-irritant substance. Glue strips of the leather to the distorted parts, and raise them up. Bandaging does harm[4] in these cases. Another treatment: while bringing the parts together, apply frankincense or sulphur with cerate; adjust at once. Afterwards keep it up by inserting the fingers, feeling for and reducing the deviation; also the Carthaginian leather. It will consolidate, even though there be a wound; and if bones are going to come away—for there are no very grave exacerbations—this is the treatment to use.

[3] Surely our ulnar nerve (funny-bone), though Foës and others call it "a ligament void of sensation."

[4] Pq. renders "depresses," reading καταποιεῖ, as opposed to ἀναπλάσσειν.

III. Οὖς κατεαγὲν μὴ ἐπιδεῖν, μηδὲ καταπλάσσειν· ἢν δέ τι δέῃ, ὡς κουφότατον, ἡ κηρωτὴ καὶ θείῳ κατακολλᾶν. ὧν δὲ ἔμπυα τὰ ὦτα διὰ παχέος εὑρίσκεται, πάντα δὲ τὰ ὑπόμυξα καὶ τῇ ὑγρῇ σαρκὶ πλήρεα ἐξαπατᾷ· οὐ μὴ βλάβη [γένηται][1] στομωθὲν τὸ τοιοῦτον· ἔστι γὰρ ἄσαρκα καὶ ὑδατώδεα, μύξης πλέα· ὅπου δὲ καὶ οἷα ἐόντα θανατώδεά ἐστι, παρεθέντα.[2] ὤτων καῦσις πέρην, τάχιστα ὑγιάζει· κυλλὸν δὲ καὶ
10 μεῖον γίνεται τὸ οὖς, ἢν πέρην καυθῇ. ἢν δὲ
11 στομωθῇ, κούφῳ ἐναίμῳ δεήσει χρῆσθαι.

IV. Γνάθοι δὲ κατασπῶνται μὲν πολλάκις καὶ καθίστανται· ἐκπίπτουσι δὲ ὀλιγάκις, μάλιστα μὲν χασμωμένοισιν· οὐ γὰρ ἐκπίπτει, ἢν μή τις χανὼν μέγα παραγάγοι· ἐκπίπτει δὲ μᾶλλον, ὅτι τὰ νεῦρα ἐν πλαγίῳ καὶ λελυγισμένα συνδιδοῖ. σημεῖα· προΐσχει ἡ κάτω γνάθος καὶ παρέστραπται τἀναντία τοῦ ἐκπτώματος· συμβάλλειν οὐ δύνανται· ἢν δὲ ἀμφότεραι, προΐσχουσι μᾶλλον, συμβάλλουσιν ἧσσον, ἀστραβέες· δηλοῖ δὲ τὰ
10 ὅρια τῶν ὀδόντων τὰ ἄνω τοῖσι κάτω κατ᾽ ἶξιν. ἢν οὖν ἀμφότεραι ἐκπεσοῦσαι μὴ αὐτίκα ἐμπέσωσι, θνήσκουσι δεκαταῖοι οὗτοι μάλιστα πυρετῷ συνεχεῖ νωθρῇ τε καρώσει· οἱ γὰρ μῦες οὗτοι τοιοῦτοι. γαστὴρ ἐπιταράσσεται ὀλίγα ἄκρητα· καὶ ἢν ἐμέωσι, τοιαῦτα ἐμέουσιν· ἡ δ᾽ ἑτέρη ἀσινεστέρη. ἐμβολὴ δὲ ἡ αὐτὴ ἀμφοτέρων· κατακειμένου ἢ καθημένου τοῦ ἀνθρώπου, τῆς

[1] Kw. omits.
[2] Cf. *Art.* XL. παρεῖται.

III. Do not bandage a broken ear, and do not apply a plaster. If one is required, let it be cerate plaster as light as possible, and agglutinate with sulphur. When there is suppuration of the ears, it is found at a depth; for all pulpy tissues and those full of moisture are deceptive. There is certainly no harm in opening such an abscess, for the parts are fleshless and watery, full of mucus; but the position and nature of abscesses which cause death are not mentioned. Perforating cautery of the ears cures a case very quickly; but the ear becomes mutilated and smaller if it is burnt through. If an abscess is opened, a light wound application must be used.

IV. The jaw is often partially displaced, and reduces itself. It is rarely put out, and that chiefly when yawning; for it is not put out unless it is drawn to one side during a wide yawn; and dislocation occurs the more because the ligaments, being oblique and twisted, give way. Symptoms: the lower jaw projects and deviates to the side opposite the dislocation; patients cannot close the mouth. If both sides are dislocated, the projection is greater, ability to close the mouth less, no deviation; this is shown by the upper row of teeth corresponding in line with the lower. If, then, bilateral dislocation is not reduced immediately, these patients usually die in ten days with continuous fever, stupor and coma; for such is the influence of the muscles in this region. The bowels are affected, and there are scanty, undigested motions; if there is vomiting, it is of a similar nature. One-sided dislocation is less harmful. Reduction is the same in both cases; the patient being either

κεφαλῆς ἐχόμενον, περιλαβόντα τὰς γνάθους ἀμφοτέρας ἀμφοτέρῃσι χερσὶν ἔσωθεν καὶ ἔξωθεν,
20 τρία ἅμα ποιῆσαι· ὦσαι ἐς ὀρθὸν καὶ ἐς τοὐπίσω, καὶ συσχεῖν τὸ στόμα. ἴησις· μαλάγμασι καὶ σχήμασι καὶ ἀναλήψει γενείου· ποιοῦσι ταῦτα[1]
23 τῇ ἐμβολῇ.

V. Ὦμος δὲ ἐκπίπτει κάτω· ἄλλῃ δὲ οὔπω ἤκουσα. δοκεῖ μὲν γὰρ ἐς τοὔμπροσθεν ἐκπίπτειν, ὧν αἱ σάρκες αἱ περὶ τὸ ἄρθρον μεμινυθήκασι διὰ τὴν φθίσιν,[2] οἷον καὶ τοῖσι βουσὶ χειμῶνος φαίνεται διὰ λεπτότητα. καὶ ἐκπίπτει μᾶλλον τοῖσι δὲ λεπτοῖσιν ἢ ἰσχνοῖσιν ἢ ξηροῖσι καὶ τοῖσιν ὑγράσματα περὶ τὰ ἄρθρα ἔχουσιν ἄνευ φλεγμονῆς· αὕτη γὰρ συνδεῖ· οἱ δὲ καὶ βουσὶν ἐμβάλλοντες καὶ ἀποπερονῶντες ἐξαμαρ-
10 τάνουσι, καὶ ὅτι διὰ τὴν χρῆσιν, ὡς χρῆται βοῦς σκέλει, λήθει, καὶ ὅτι κοινὸν καὶ ἀνθρώπῳ οὕτως ἔχοντι τὸ σχῆμα τοῦτο· τό τε Ὁμήρειον· καὶ διότι λεπτότατοι βόες τηνικαῦτα. ὅσα τε τὸν πῆχυν πλάγιον ἀπὸ πλευρέων ἄραντες δρῶσιν, οὐ πάνυ δύνανται δρᾶν, οἷσιν ἂν μὴ ἐμπέσῃ. οἷσι μὲν οὖν ἐκπίπτει μάλιστα, καὶ ὡς ἔχουσιν, εἴρηται. οἷσι δὲ ἐκ γενεῆς, τὰ ἐγγύτατα μᾶλλον βραχύνεται ὀστέα, οἷον ἐν τούτῳ οἱ γαλιάγκωνες· πῆχυς δὲ ἧσσον, χεὶρ δὲ ἔτι ἧσσον, τὰ δ᾽ ἄνωθεν
20 οὐδέν· καὶ ἀσαρκότατα ἐγγύς· μινύθει δὲ μάλιστα

[1] ταὐτὰ.
[2] Littré's correction. φύσιν MSS. would give sense, but the writer is evidently copying *Joints* I.

---

[1] The safety-pin was a very ancient instrument. Cf. *Iliad* XIV. 180. It is strange that there is no other mention

lying down or seated, his head fixed, take hold of both sides of the jaw with both hands, inside and out, and perform three actions at once—get it straight, thrust it back, and shut the mouth. Treatment: with emollients, position, and support of the chin; these things co-operate in the reduction.

V. The shoulder is dislocated downwards. I have no knowledge of any other direction. It appears indeed to be dislocated forwards in cases where the tissues about the joint have diminished through wasting disease, as one observes also with cattle in winter, because of their leanness. Dislocation occurs preferably in thin and slight subjects, or those of dry habit; also those who have the region of the joints charged with moisture without inflammation, for this braces them up. Those who use reductions and fixations with fibulae [1] in oxen are in error, and forget that the appearance is due to the way the ox uses its leg, and that this attitude is common also to man in the same condition—also the Homeric quotation, and the reason why oxen are very thin at that time. Actions requiring lateral elevation of the arm from the ribs are quite impossible for patients in whom the joint is not reduced. The subjects, then, most liable to dislocation, and their condition, have been described. In congenital cases, the proximal bones are shortened most, as is the case with the weasel-armed; the forearm less than the arm, the hand still less, and parts above the lesion not at all; the most fleshless parts are near the lesion. Atrophy occurs especially on the side

of it in the Hippocratic surgical works. That it was then in surgical use for closing wounds seems indicated by Eur. *Bacchae* 97.

τὰ ἐναντία τῶν ὀλισθημάτων, καὶ τὰ ἐν αὐξήσει, ἧσσον δέ τινι τῶν ἐκ γενεῆς. καὶ τὰ παραπυήματα, τὰ κατ' ἄρθρον βαθέα, νεογενέσι μάλιστα παρ' ὦμον γίνεται, καὶ τούτοισιν ὥσπερ τὰ ἐξαρθρήσαντα ποιεῖ. ἢν δὲ ηὐξημένοισι, τὰ μὲν ὀστέα οὐ μειοῦται, οὐδὲ γὰρ ἔχει ᾗ ἄλλα οὐ συναύξεται ὁμοίως, αἱ δὲ μινυθήσιες τῶν σαρκῶν. τοῦτο γὰρ καθ' ἡμέρην καὶ αὔξεται καὶ μειοῦται, καὶ καθ' ἡλικίας. καὶ ἃ δύναται σχήματα, καὶ 30 αὖ σημεῖον τὸ παρὰ τὸ ἀκρώμιον κατεσπασμένον καὶ κοῖλον, διότι ὅταν τὸ ἀκρώμιον ἀποσπασθῇ καὶ κοῖλον ᾖ, οἴονται τὸν βραχίονα ἐκπεπτωκέναι· κεφαλὴ δὲ τοῦ βραχίονος ἐν τῇ μασχάλῃ φαίνεται· αἴρειν [γὰρ][1] οὐ δύνανται, οὐδὲ παράγειν ἔνθα καὶ ἔνθα ὁμοίως· ὁ ἕτερος ὦμος μηνύει. ἐμβολαὶ δέ· αὐτὸς μὲν τὴν πυγμὴν ὑπὸ μασχάλην ὑποθεὶς τὴν κεφαλὴν ἀνωθεῖν, τὴν δὲ χεῖρα ἐπιπαράγειν ἐπὶ τὸ στῆθος. ἄλλη· ἐς τοὐπίσω περιαναγκάσαι, ὡς ἀμφισφαλῇ. ἄλλη· κεφαλὴ 40 μὲν πρὸς τὸ ἀκρώμιον, χερσὶ δὲ ὑπὸ μασχάλην, κεφαλὴν ὑπάγειν βραχίονος, γούνασι δὲ ἀγκῶνα ἀπωθεῖν, ἢ ἀντὶ τῶν γουνάτων τὸν ἀγκῶνα τὸν ἕτερον παράγειν ὡς τὸ πρότερον· ἢ κατ' ὦμον ἵζεσθαι, ὑποθεὶς τῇ μασχάλῃ τὸν ὦμον· ἢ τῇ πτέρνῃ ἐνθέντα ἐκπληρώματα τῇ μασχάλῃ, δεξιῇ δεξιόν· ἢ περὶ ὕπερον· ἢ περὶ κλιμακτῆρα· ἢ περίοδος σὺν τῷ ξύλῳ τῷ ὑπὸ χεῖρα τεινομένῳ. ἴησις· τὸ σχῆμα, πρὸς πλευρῇσι βραχίων, χεὶρ

opposite to the dislocations, and when they occur during adolescence, but is somewhat less than in congenital cases. Deep suppurations at a joint occur in infants, especially at the shoulder, and have the same effect as dislocations. In adults there is no shortening, for there is no opportunity for one bone to have less growth than another; but there is atrophy of the tissues; for in the young there is increase and decrease, both daily and according to age. [Consider] too the effect of attitudes, and also what is indicated by the hollow at the point of the shoulder, due to avulsion; for when the acromion is torn away and there is a hollow, people think the humerus has been dislocated. If so, the head of the humerus is found in the armpit, the patients cannot lift the arm, nor move it to either side equally;[1] the other shoulder is an index. Modes of reduction: let the patient put his fist in the armpit, push up the head of the bone, and bring the arm to the chest. Another method: force the arm backwards, so as to make a movement of circumduction. Another: with the head against the point of the shoulder, and the hands under the armpit, lift the head of the humerus, and push back the elbow with the knees, or, instead of using the knees, let the assistant bring the elbow to the side, as above; or suspend the patient on the shoulder, putting it under the armpit, or with the heel, putting plugs into the armpit, using the right heel for the right shoulder, or on a pestle or ladder; or make a circular movement with the wood (lever) fixed under the arm. Treatment; position; arm to

[1] Or, "as before."

[1] Omit.

ἄκρη ἄνω, ὦμος ἄνω· οὕτως ἐπίδεσις, ἀνάληψις.
50 ἢν δὲ μὴ ἐμπέσῃ, ἀκρώμιον προσλεπτύνεται.

VI. Ἀκρώμιον ἀποσπασθέν, τὸ μὲν εἶδος φαίνεται οἷόν περ ὤμου ἐκπεσόντος, στερίσκεται δὲ οὐδενός, ἐς δὲ τὸ αὐτὸ οὐ καθίσταται. σχῆμα τὸ αὐτὸ ᾧ[1] καὶ ἐκπεσόντι, ἐν ἐπιδέσει καὶ ἀνα-
5 λήψει· ἐπίδεσις καὶ ὡς νόμος.

VII. Ἀγκῶνος ἄρθρον παράλλαξαν μὲν[2] ἢ πρὸς πλευρὴν ἢ ἔξω, μένοντος τοῦ ὀξέος τοῦ ἐν τῷ κοίλῳ τοῦ βραχίονος, ἐς ἰθὺ[3] κατατείνοντα,
4 τὰ ἐξέχοντα ἀνωθεῖν[4] ὀπίσω καὶ ἐς τὸ πλάγιον.

VIII. Τὰ δὲ τελέως ἐκβάντα ἢ ἔνθα ἢ ἔνθα· κατάτασις μὲν ἐν ᾗ ὁ βραχίων[5] ἐπιδεῖται· οὕτω γὰρ τὸ καμπύλον τοῦ ἀγκῶνος οὐ κωλύσει. ἐκπίπτει δὲ μάλιστα ἐς τὸ πρὸς πλευρέα[6] μέρος. τὰς δὲ κατορθώσιας, ἀπάγοντα ὅτι πλεῖστον, ὡς μὴ ψαύσῃ τῆς κορώνης ἡ κεφαλή, μετέωρον δὲ περιάγειν καὶ περικάμψαι, καὶ μὴ ἐς ἰθὺ[7] βιάζεσθαι, ἅμα δὲ ὠθεῖν τἀναντία ἐφ' ἑκάτερα, καὶ παρωθεῖν ἐς χώρην. συνωφελοίη δ' ἂν καὶ
10 ἐπίστρεψις ἀγκῶνος ἐν τούτοισιν, ἐν τῷ μὲν ἐς τὸ ὕπτιον, ἐν τῷ δὲ ἐς τὸ πρηνές. ἐμβολὴ δέ·[8] σχήματος μὲν ὀλίγον[9] ἀνωτέρω ἄκρην χεῖρα ἀγκῶνος[10] ἔχειν, βραχίονα δὲ κατὰ τὰς[11] πλευράς· οὕτω δὲ ἡ ἀνάληψις,[12] καὶ εὔφορον, καὶ χρῆσις ἐν τῷ κοινῷ, ἢν ἄρα μὴ κακῶς πωρωθῇ· πωροῦται δὲ ταχέως. ἴησις·[13] ὀθονίοισι κατὰ τὸν νόμον τὸν
17 ἀρθριτικόν, καὶ τὸ ὀξὺ προσεπιδεῖν.

IX. Παλιγκοτώτατον δὲ ἀγκὼν[14] πυρετοῖσι, ὀδύνῃ,[15] ἀσώδει, ἀκρητοχόλῳ· ἀγκῶνος δὲ μάλιστα ὀπίσω διὰ τὸ ναρκῶδες, δεύτερον τὸ ἔμπροσθεν. ἴησις ἡ αὐτή·[16] ἐμβολαὶ δὲ τοῦ μὲν ὀπίσω ἐκ-

ribs, hand elevated, shoulder elevated; bandaging and support in this attitude. If not reduced, the point of the shoulder atrophies as well.

VI. Avulsion of the acromion (process of the shoulder-blade), appears in form like a dislocation of the shoulder, but there is no loss of function; yet it does not stay in place when reduced. Position as regards bandaging and support the same as in a case of dislocation; the bandaging follows the customary rule.

VII–XIX. *Mochlicon* VII–XIX corresponds verbally (except a few "various readings" such as occur in different MSS.)[1] with *Joints* XVII–XXIX. Instead of repeating the translation, we may, therefore, attempt a few explanatory notes; for dislocation of the elbow has always been an obscure subject, owing to the complicated form of the joint, and the presence of three bones.

All the chief surgical commentators, Apollonius, Adams, Petrequin, agree that VII represents dislocation of the radius only, in directions which we call "forwards" and "backwards"; though Galen says that *Fractures* XXXVIII, of which it is an epitome, refers to partial lateral dislocations of the ulna. "Diastasis" (X) can hardly mean anything else than dislocation of the radius in the other possible direction—outwards, or away from the ulna.

[1] These are given in the notes.

---

[1] ἃ. [2] Add ἢ παραρθρῆσαν. [3] εὐθὺ.
[4] ἀπωθεῖν. [5] Add καταγεὶς. [6] πλευρὰς.
[7] εὐθὺ. [8] ἴησις δέ (so Kw. here). [9] ὀλίγῳ.
[10] τοῦ ἀγκῶνος. [11] Omit τὰς. [12] Add καὶ θέσις.
[13] ἴησις δέ. [14] ὁ ἀγκὼν. [15] ὀδύνῃσι.
[16] ἴησις δὲ αὐτή.

τείνοντα[1] κατατεῖναι. σημεῖον δέ· οὐ γὰρ δύνανται ἐκτείνειν· τοῦ δὲ ἔμπροσθεν οὐ δύνανται συγκάμπτειν. τούτῳ δὲ ἐνθέντα τι σκληρὸν συνειλιγμένον, περὶ τοῦτο συγκάμψαι ἐξ ἐκτάσιος
9 ἐξαίφνης.

X. Διαστάσιος δὲ ὀστέων σημεῖον κατὰ τὴν φλέβα τὴν κατὰ τὸν βραχίονα σχιζομένην
3 διαψαύοντι.

XI. Ταῦτα δὲ ταχέως διαπωροῦται· ἐκ γενεῆς δέ, βραχύτερα τὰ κάτω ὀστέα τοῦ σίνεος,[2] πλεῖστον τὰ ἐγγύτατα πήχεος, δεύτερον χειρός, τρίτον δακτύλων. βραχίων δὲ καὶ ὦμος ἐγκρατέστερα διὰ τὴν τροφήν· ἡ δ' ἑτέρη χεὶρ διὰ τὰ ἔργα πλείω ἔτι ἐγκρατεστέρη. μινύθησις δὲ σαρκῶν, εἰ μὲν ἔξω ἐξέπεσεν, ἔσω·[3] εἰ δὲ μή, ἐς
8 τοὐναντίον ἢ ᾗ ἐξέπεσεν.

XII.[4] Ἀγκὼν δὲ ἢν μὲν[5] ἔσω ἢ ἔξω ἐκβῇ, κατάτασις μὲν ἐν σχήματι ἐγγωνίῳ, κοινῷ τῷ πήχει πρὸς βραχίονα· καὶ μασχάλην ἀναλαβὼν[6] ταινίῃ ἀνακρεμάσαι, ἀγκῶνι δὲ ἄκρῳ ὑποθείς[7] τι παρὰ τὸ ἄρθρον βάρος ἐκκρεμάσαι, ἢ χερσὶ καταναγκάσαι. ὑπεραιωρηθέντος δὲ τοῦ ἄρθρου, αἱ παραγωγαὶ τοῖσι θέναρσιν, ὡς τὰ ἐν χερσίν. ἐπίδεσις ἐν τούτῳ τῷ σχήματι, καὶ ἀνάληψις καὶ
9 θέσις.

XIII.[8] Τὰ δὲ ὄπισθεν, ἐξαίφνης ἐκτείνοντα διορθοῦν τοῖσι θέναρσιν· ἅμα δὲ δεῖ ἐν τῇ διορθώσει, καὶ τοῖσιν ἑτέροισιν. ἢν δὲ πρόσθεν, ἀμφὶ ὀθόνιον συνειλιγμένον, εὔογκον, συγκάμπ-
5 τοντα ἅμα διορθοῦσθαι.[9]

[1] ἐκτείναντα. [2] τοῦ σίνεος ὀστέα.
[3] ἔσωθεν.

As regards complete dislocations, Littré and Adams refer those in VIII to lateral cases, and those in IX to dislocation forwards and backwards; while Petrequin, turning the bend of the elbow inwards, takes the opposite view. The most frequent and mildest form of complete dislocation is that of the forearm backwards (or the humerus forwards), and the Hippocratic writers can only be got to agree with this by assuming the Petrequin attitude; for they evidently describe this form as a dislocation of the humerus inwards (cf. *Fract.* XL, XLI). The dislocation "backwards" which specially affects the ulnar nerve would thus be our external lateral dislocation of the forearm.

Still, the accounts remain obscure and often difficult to accommodate with facts; nor do we get much help from the existence of a sort of double epitome, XII and XIII repeating VIII and IX from a more practical standpoint, while XIV refers to the radius dislocations noticed above in VII and X.

The account of wrist dislocation (XVI, XVII) combines theoretic clearness with even greater practical obscurity. As Adams says, "in the wrist, nothing is more common than fracture, and nothing more rare than dislocation." Yet the epitomist gives us a neat schematic arrangement of dislocation in all four directions, and says nothing of fracture, unless we take "with the epiphysis" to imply this. The original account is lost; but its essence is doubtless contained in *Joints* LXIV, on compound dislocations of the wrist.

---

[4] Variant of VIII.
[5] Omit μὲν.
[6] ἀναλαβόντα.
[7] ὑποθέντα.
[8] Cf. IX.
[9] διορθοῦν.

XIV.[1] Ἢν δὲ ἑτεροκλινὲς ᾖ, ἐν τῇ διορθώσει ἀμφότερα χρὴ ποιεῖν· τῆς δὲ μελέτης [2] κοινὸν καὶ τὸ σχῆμα καὶ ἡ ἐπίδεσις· δύναται γὰρ [3] ἐκ τῆς
4 διατάσιος κοινῇ συμπίπτειν πάντα.[4]

XV. Τῶν δὲ ἐμβολέων αἱ μὲν ἐξ ὑπεραιωρήσιος ἐμβάλλονται, αἱ δὲ ἐκ κατατάσιος, αἱ δὲ ἐκ περισφάλσιος· αὗται δὲ ἐκ τῶν ὑπερβολέων τῶν
4 σχημάτων ᾗ τῇ ᾗ τῇ σὺν τῷ τάχει.

XVI. Χειρὸς δὲ ἄρθρον ὀλισθάνει ἢ ἔσω ἢ ἔξω, ἔσω δὲ τὰ πλεῖστα. σημεῖα δ' εὔσημα· ἢν μὲν ἔσω, συγκάμπτειν ὅλως σφῶν [5] τοὺς δακτύλους οὐ δύνανται· ἢν δὲ ἔξω, ἐκτείνειν. ἐμβολὴ δέ· ὑπὲρ τραπέζης τοὺς δακτύλους ἔχων, τοὺς μὲν τείνειν, τοὺς δὲ ἀντιτείνειν· τὸ δὲ ἐξέχον ἢ θέναρι ἢ πτέρνῃ ἅμα ἀπωθεῖν [6] πρόσω καὶ κάτωθεν,[7] κατὰ τὸ ἕτερον ὀστέον ὄγκον τε [8] μαλθακὸν ὑποθείς, κἢν [9] μὲν ἄνω, καταστρέψας τὴν χεῖρα, ἢν
10 δὲ κάτω, ὑπτίην. ἴησις,[10] ὀθονίοισιν.

XVII. Ὅλη δὲ χεὶρ ὀλισθάνει ἢ ἔσω ἢ ἔξω, μάλιστα δὲ ἔξω, ἢ ἔνθα ἢ ἔνθα.[11] ἔστι δ' ὅτε ἡ ἐπίφυσις [12] ἐκινήθη· ἔστι δ' ὅτε τὸ ἕτερον τῶν ὀστέων διέστη. τούτοισι κατάτασις ἰσχυρὴ ποιητέη, καὶ τὸ μὲν ἐξέχον ἀπωθεῖν, τὸ δὲ ἕτερον ἀντωθεῖν, δύο εἴδεα ἅμα καὶ ἐς τοὐπίσω καὶ ἐς τὸ πλάγιον, ἢ χερσὶν ἐπὶ τραπέζης ἢ πτέρνῃ. παλίγκοτα δὲ καὶ ἀσχήμονα, τῷ χρόνῳ δὲ κρατύνεται ἐς χρῆσιν. ἴησις, ὀθονίοισι σὺν τῇ χειρὶ καὶ τῷ
10 πήχει· καὶ νάρθηκας μέχρι δακτύλων τιθέναι· ἐν νάρθηξι δὲ τεθέντα [13] ταῦτα πυκνότερον λύειν ἢ τὰ
12 κατήγματα, καὶ καταχύσει πλέονι χρῆσθαι.

[1] Cf. VII. [2] Add τῆς θεραπείης.
[3] Add καί. [4] ἅπαντα.

Here the writer evidently describes dislocation of the bones of the forearm from the wrist; while the epitomist (unless, with Littré and Petrequin, we put some strain on the Greek) speaks of dislocation of the hand, but follows Hippocrates in saying that "when the dislocation is inwards (our 'forwards'), they cannot flex the fingers, when outwards, they cannot extend them."

This is the view of Celsus (VIII. 17), and is most in accordance with modern experience—when the hand is dislocated backwards, the flexor tendons are on the stretch and the fingers cannot be extended, and vice versa, though exceptions have been observed, and the accidents are too rare and complicated for the establishment of neat rules. The typical "dislocation" of the wrist is the fracture of the end of the radius, known as Colles's fracture.

The brief account of congenital dislocation (XVIII) may have been added to complete the picture. The results described are those of all congenital dislocations, as frequently given in *Joints*. Perhaps, however, "nothing can show more remarkably the attention which our author must have paid to the subject than his being acquainted with a case of such rarity" (Adams).[1]

[1] Littré treats these subjects at length in his Introductions, and Petrequin at still greater length in his Notes and Excursus. They confirm the observation of Adams that a full discussion would lead to no conclusion, and would be tedious even to professional readers.

---

[5] Omit ὅλως σφῶν. [6] Add καὶ ὠθεῖν.
[7] πρόσω κάτω, κάτωθεν. [8] δὲ. [9] ἣν.
[10] ἴησις δέ. [11] ἢ ἔνθα ἢ ἔνθα, μάλιστα δὲ ἔσω.
[12] καὶ ἡ ἐπίφυσις. [13] δεθέντα.

XVIII. Ἐκ γενεῆς δέ, βραχυτέρη ἡ χεὶρ γίνεται, καὶ ἡ[1] μινύθησις σαρκῶν μάλιστα τἀναντία ἢ ὡς[2] τὸ ἔκπτωμα· ηὐξημένῳ δὲ τὰ ὀστέα
4 μένει.

XIX. Δακτύλου δὲ ἄρθρον ὀλισθὸν μὲν εὔσημον [οὐ δεῖ γράφειν].[3] ἐμβολὴ δὲ αὐτοῦ ἥδε·[4] κατατείναντα ἐς ἰθὺ τὸ μὲν ἐξέχον ἀπωθεῖν, τὸ δὲ ἐναντίον ἀντωθεῖν. ἴησις δὲ ἡ προσήκουσα,[5] τοῖσι ὀθονίοισι[6] ἐπίδεσις.[7] μὴ ἐμπεσὸν γὰρ ἐπιπωροῦται ἔξωθεν. ἐκ γενεῆς δὲ ἢ ἐν αὐξήσει ἐξαρθρήσαντα τὰ ὀστέα βραχύνεται κάτω[8] τοῦ ὀλισθήματος· καὶ σάρκες μινύθουσι τἀναντία μάλιστα ἢ ὡς[9] τὸ ἔκπτωμα· ηὐξημένῳ δὲ τὰ
10 ὀστέα μένει.

XX. Μηροῦ ἄρθρον ἐκπίπτει κατὰ τρόπους τέσσαρας· ἔσω πλεῖστα, ἔξω δεύτερον, τὰ δὲ ἄλλα ὁμοίως. σημεῖα· κοινὸν μὲν τὸ ἕτερον σκέλος· ἴδιον δὲ τοῦ μὲν ἔσω. παρὰ τὸν περίναιον[10] ψαύεται ἡ κεφαλή· συγκάμπτουσι οὐχ ὁμοίως, δοκεῖ δὲ μακρότερον[11] τὸ σκέλος, καὶ πολύ, ἢν μὴ ἐς μέσον ἀμφότερα ἄγων παρατείνῃς· καὶ γὰρ οὖν ἔξω ὁ πούς καὶ τὸ γόνυ ῥέπει. ἢν μὲν οὖν ἐκ γενεῆς ἢ ἐν αὐξήσει ἐκπέσῃ, βραχύτερος ὁ μηρός,
10 ἧσσον δὲ κνήμη, κατὰ λόγον δὲ τἄλλα· μινύθουσι δὲ σάρκες, μάλιστα δὲ ἔξω. οὗτοι κατοκνέουσιν ὀρθοῦσθαι, καὶ εἰλέονται ἐπὶ τὸ ὑγιές· ἢν δὲ ἀναγκάζωνται, σκίμπονι ἑνὶ ἢ δυσὶν ὁδοιπορέουσι, τὸ δὲ σκέλος αἴρουσιν· ὅσῳ γὰρ μεῖον, τόσῳ ῥᾷον. ἢν δὲ ηὐξημένοισι, τὰ μὲν ὀστέα μένει, αἱ

[1] Omit ἡ. [2] ἢ ἧ. [3] Omit ("probably a gloss." Kw.).
[4] Omit αὐτοῦ ἥδε. [5] Omit ἡ προσήκουσα.

The problem of the knee (XXVI) seems insoluble. All writers, from the author of *Mochlicon* to Ambroise Paré, copy the statement of Hippocrates (*Fract.* XXXVII) that dislocation is frequent and of slight severity. We know that it is rare and requires great violence which usually has serious results. Suggestions such as confusion with "internal derangement," or displacement of the knee-cap, seem unsatisfactory. The existence of some peculiar grip in wrestling which dislocated the knee without further injury seems the most probable explanation. One of the modern causes—being dragged in the stirrup by a runaway horse—was absent in antiquity.

XX. The thigh-joint is dislocated in four ways, most frequently inwards, secondly outwards, in the other directions equally. Symptoms: in general, comparison with the other leg. Peculiar to internal dislocation: the head of the thigh-bone is felt towards the perineum; they do not flex the thigh as on the other side; the leg appears longer, especially if you do not bring both legs to the middle line for comparison, for the foot and knee incline outwards. If then the dislocation is congenital, or occurs during adolescence, the thigh is shortened, the lower leg less so, and the rest in proportion. There is atrophy of the tissues, especially on the outer side. These patients shrink from standing erect, and wriggle along on the sound leg. If they have to stand up, they walk with a crutch or two, and keep the leg up, which they do more easily the smaller it is. In adults the bones are unaltered, but

[6] ταινίοισι ὀθονίοισι.

[7] Omit ἐπίδεσις.

[8] τὰ κάτω.

[9] μάλιστα, ᾗ ῇ.

[10] περίνεον.

[11] πολὺ μακρότερον.

δὲ σάρκες μινύθουσι, ὡς προείρηται. ὁδοιπορέουσι δὲ περιστροφάδην, ὡς βόες, ἐν δὲ κενεῶνι[1] καμπύλοι, ἐπὶ τὸ ὑγιὲς ἐξίσχιοι ἐόντες· τῷ μὲν γὰρ ἀνάγκη ὑποβαίνειν ὡς ὀχῇ, τὸ[2] δὲ
20 ἀποβαίνειν (οὐ γὰρ δύναται ὀχεῖν), ὥσπερ οἱ ἐν ποδὶ ἕλκος ἔχοντες. κατὰ δὲ τὸ ὑγιές, πλάγιον[3] ξύλῳ τῷ σώματι ἀντικοντοῦσι, τὸ δὲ σιναρὸν τῇ χειρὶ ὑπὲρ τοῦ γόνατος καταναγκάζουσι ὡς ὀχεῖν ἐν τῇ μεταβάσει τὸ σῶμα· ἰσχίῳ κάτωθεν[4] εἰ χρῆται, κάτωθεν[5] ἧσσον μινύθει καὶ τὰ ὀστέα,
26 μᾶλλον δὲ σάρκες.

XXI. Τοῦ δὲ ἔξω τἀναντία καὶ τὰ σημεῖα καὶ αἱ στάσιες· καὶ τὸ γόνυ καὶ ὁ ποὺς ἔξω ῥέπει βραχύ. τοῖσι δὲ ἐν αὐξήσει ἢ ἐκ γενεῆς παθοῦσιν οὐχ ὁμοίως συναύξεται[6] κατὰ τὸν αὐτὸν λόγον· ἰσχίον ἀνωτέρω τινί, οὐχ ὁμοίως. οἷσι δὲ πυκινὰ ἐκπίπτει ἐς τὸ ἔξω ἄνευ φλεγμονῆς, ὑγροτέρῳ τῷ σκέλει χρῶνται, ὥσπερ ὁ μέγας τῆς χειρὸς δάκτυλος· μάλιστα δὲ οὗτος ἐκπίπτει φύσει· οἷς μὲν ἐκπίπτει μᾶλλον ἢ ἧσσον, καὶ οἷς μὲν ἐκπίπ-
10 τει χαλεπώτερον ἢ ῥήϊον, καὶ οἷσιν ἐλπὶς θᾶσσον ἐμπεσεῖν, καὶ οἷσιν οὐκ ἀκὴ τούτου, καὶ οἷσι πολλάκις ἐκπίπτει, ἴησις τούτου. ἐκ γενεῆς δὲ ἢ ἐπ' αὐξήσει ἢ ἐν νούσῳ (μάλιστα γὰρ ἐκ νούσου) ἔστι μὲν [οὖν][7] οἷσιν ἐπισφακελίζει τὸ ὀστέον, ἀτὰρ καὶ οἷσι μή, πάσχει μὲν πάντα, ἧσσον δὲ ἢ τὸ ἔσω, ἢν χρηστῶς ἐπιμεληθῶσιν, ὥστε καὶ ὅλῳ βαίνοντας τῷ ποδὶ διαῤῥίπτειν· διὰ μελέτης

[1] τῷ κενεῶνι. [2] τῷ. [3] πλάγιοι.
[4] ἰσχίων κατωτέρω. [5] κάτω τε.
[6] Kw. puts colon after συναύξεται. [7] Omit.

---

[1] Cf. J. LIV.

there is atrophy of the tissues in the way described. They walk with shambling gait, like oxen, bent in at the loin and projecting at the hip on the sound side; for they have to bring the leg under to serve as support, and keep the other leg out (for it cannot give support), like people with a wound on the foot. On the sound side they use a staff as a lateral prop, and press down the injured limb with the hand above the knee, so as to support the body in the change of step. If the part below the hip is used, there is less atrophy of the bones (below). It occurs more in the tissues.

XXI. In outward dislocation, both symptoms and attitudes are the reverse. Knee and foot incline slightly inwards. In adolescent or congenital patients there is inequality of growth, in the same proportion (as with inward dislocation). Hip somewhat elevated, not corresponding.[1] Those in whom outward dislocation is frequent without inflammation have the limb more charged with humours, as is the case with the thumb; for this is by its nature most liable to dislocation. In some the dislocation is more or less complete; in some it takes place with more or less difficulty; in some there is hope of speedy reduction: in some there is no cure for the condition; in cases of frequent dislocation there is a treatment. In congenital and adolescent cases, and those due to disease (for disease is the principal cause), in some cases there is necrosis of bone, but in others not. They have all the affections above mentioned, but to a less degree than those with internal dislocation, if they are well cared for, so as to balance themselves and walk on the whole foot. The youngest require the greatest care. Left to

πλείστης τοῖσι νηπιωτάτοισιν· ἐαθέντα κακοῦται, ἐπιμεληθέντα δὲ ὠφελεῖται· τοῖσιν ὅλοισιν, ἧσσον 20 δέ τι, μινύθουσι.

XXII. Οἷσι δ' ἂν ἀμφότερα οὕτως ἐκπέσῃ, τῶν ὀστέων ταὐτὰ παθήματα· εὔσαρκοι μέν, πλὴν ἔσωθεν, ἐξεχέγλουτοι, ῥοικοὶ μηροί, ἢν μὴ ἐπισφακελίσῃ. εἰ κυφοὶ τὰ ἄνωθεν ἰσχίων γένοιντο, ὑγιη- 5 ροὶ μέν, ἀναυξέες δὲ τὸ σῶμα, πλὴν κεφαλῆς.

XXIII. Οἷσι δὲ ὄπισθεν, σημεῖα· ἔμπροσθεν λαπαρώτερον, ὄπισθεν ἐξέχον, ποὺς ὀρθός· συγκάμπτειν οὐ δύνανται, εἰ μὴ μετ' ὀδύνης, ἐκτείνειν ἥκιστα· τούτοισι σκέλος βραχύτερον. ἀτὰρ οὐδ' ἐκτανύειν δύνανται κατ' ἰγνύην ἢ [1] κατὰ βουβῶνα, ἢν μὴ πάνυ αἴρωσιν, οὐδὲ συγκάμπτειν. ἡγεῖται ἐν τοῖσι πλείστοισι τὸ ἄνω ἄρθρον τὸ πρῶτον· κοινὸν τοῦτο ἄρθροισι, νεύροισι, μυσίν, ἐντέροισιν, ὑστέρῃσιν, ἄλλοισιν· τούτοις τοῦ ἰσχίου τὸ 10 ὀστέον καταφέρεται εἰς τὸν γλουτόν· διὰ τοῦτο βραχύ, καὶ ὅτι ἐκτείνειν οὐ δύνανται. σάρκες παντὸς τοῦ σκέλεος ἐν πᾶσι μινύθουσιν· ἐφ' οἷσι δὲ μάλιστα, καὶ οἷ,[2] εἴρηται· τὰ ἔργα τὰ ἑωυτοῦ ἕκαστον τοῦ σώματος ἐργαζόμενον μὲν ἰσχύει, ἄργεον δὲ κακοῦται, πλὴν κόπου, πυρετοῦ, φλεγμονῆς. καὶ τὸ ἔξω, ὅτι ἐς σάρκα ὑπείκουσαν, βραχύτερον· τὸ δὲ ἔσω, ὅτι ἐπ' ὀστέον προέχον, μακρότερον. ἢν μὲν οὖν ηὐξημένοισι μὴ ἐμπέσῃ, ἐπὶ βουβῶσι καμπύλοι ὁδοιπορέουσι, καὶ ἡ ἑτέρη

[1] ἢ = "and not" (cf. *Surg.* XIV); but Kw. reads ⟨ἢν⟩ μή, from *J.* LVII.

[2] *I.e.* "to what extent" (?); but Kw. (M) has ᾗ.

---

[1] Hardly intelligible without reference to *J.* LVII.

itself, the lesion gets worse; if cared for, it improves. There is atrophy of all the parts, but somewhat less (than in dislocation inwards).

XXII. When both hips are thus dislocated, the bones are similarly affected. The patients have well-nourished tissues, except on the outer side; they have prominent buttocks, and arched thighs, unless there is also necrosis of the bone. If they become hump-backed above the hips, they retain health; but the body ceases to grow, except the head.

XXIII. Symptoms of posterior dislocation: anterior region rather hollow, posterior projecting, foot straight; they cannot flex the thigh without pain, nor extend it at all; the limb is shorter in these cases. Note also that people cannot do extension at the knee and not at the groin unless they lift it quite high, nor can they flex.[1] In most cases the proximal joint takes precedence (in function); this applies to the joints, ligaments, muscles, intestines, uterus, and other organs.[2] In these dislocations, the hip-bone is carried to the buttock, which causes the shortening and inability to extend the joint. In all cases there is atrophy of the tissues throughout the leg; in which cases this occurs most, and where, has been explained. Each part of the body which performs its proper function gets strong; but when idle, it deteriorates, unless the inaction is due to fatigue, fever, or inflammation. External dislocation, because it is into yielding tissue, produces shortening: internal, because it is on to projecting bone, lengthening. If then it is unreduced in adults, they walk in a bent attitude at the groins,

[2] *I.e.* movements, including contractions, start from above.

20 ἰγνύῃ κάμπτεται· στήθεσι μόλις[1] καθικνεῖται·[2] χειρὶ τὸ σκέλος καταλαμβάνει, ἄνευ ξύλου, ἢν ἐθέλωσιν· ἢν μὲν γὰρ μακρότερον ᾖ, οὐ βήσεται· ἢν δὲ βαίνῃ, βραχύ. μινύθησις δὲ σαρκῶν, οἷσι πόνοι, καὶ ἡ ἴξις ἔμπροσθεν, καὶ τῷ ὑγιεῖ κατὰ λόγον· οἷσι δὲ ἐκ γενεῆς ἢ αὐξομένοισι ἢ ὑπὸ νούσου ἐνόσησε καὶ ἔξαρθρα ἐγένετο (ἐν αἷς, εἰρήσεται), οὗτοι μάλιστα κακοῦνται διὰ τὴν τῶν νεύρων καὶ ἄρθρων ἀργίην· καὶ τὸ γόνυ διὰ τὰ εἰρημένα συγκακοῦνται. συγκεκαμμένον οὗτοι 30 ἔχοντες ὁδοιπορέουσιν ἐπὶ ξύλου, ἑνὸς ἢ δύο· τὸ 31 δὲ ὑγιές, εὔσαρκον διὰ χρῆσιν.

XXIV. Οἷσι ἐς τοὔμπροσθεν, σημεῖα τἀναντία· ὄπισθεν λαπαρόν, ἔμπροσθεν ἐξέχον· ἥκιστα συγκάμπτουσιν οὗτοι τὸ σκέλος, μάλιστα δὲ ἐκτείνουσι· ὀρθὸς πούς, σκέλος ἴσον, πτέρνα· βραχεῖ ἄκρως ἀνέσταλται. [ἢ][3] πονέουσι μάλιστα οὗτοι αὐτίκα, καὶ οὖρον ἴσχεται μάλιστα ἐν τούτοισι τοῖσιν ἐξαρθρήμασιν· ἐν γὰρ τόνοισιν ἔγκειται τοῖσιν ἐπικαίροισιν. τὰ ἔμπροσθεν κατατέταται [ἀναυξέα, νοσώδεα, ταχύγηρα]·[4] τὰ 10 ὄπισθεν στολιδώδεες· οἷσιν ηὐξημένοισιν, ὁδοιπορέουσι ὀρθοί, πτέρνῃ μᾶλλον βαίνοντες· εἰ δὲ ἠδύναντο μέγα προβαίνειν, κἂν πάνυ· σύρουσι δέ. μινύθει δὲ ἥκιστα, τούτοισι δὲ ἡ χρῆσις αἰτία· μάλιστα δὲ ὄπισθεν· διὰ παντὸς τοῦ σκέλεος, ὀρθότεροι τοῦ μετρίου, ξύλου δέονται κατὰ τὸ

[1] μόγις.
[2] κινεῖται codd.; ἱκνεῖται Littré.
[3] Kw. deletes. Perhaps ἦ emphatic.
[4] Words from *J.* LVIII referring to effects of disuse, evidently out of place here.

and the sound knee is flexed. The ball of the foot barely reaches the ground; they hold the leg with the hand if they choose to walk without a crutch. A crutch for walking should be short; if too long, he will not use the foot. There is wasting of the flesh in painful cases[1] down the front, and on the sound side in proportion. In congenital and adolescent patients, or where the dislocation follows disease (what the diseases are will be explained), these cases especially go to the bad through disuse of the sinews and joints; and the knee shares in the deterioration, for the reasons given. They walk with the leg flexed, on one or two crutches; but the sound limb is well nourished, because it is used.

XXIV. In cases of dislocation forwards the symptoms are reversed; hind region depressed, front projecting. These patients are least able to flex the leg, but have most power to extend it. The foot is straight, and the leg equal to the other, if measured to the heel; the foot is a little drawn up at the tip. Now these patients suffer especially at first, and there is a special liability to retention of urine in these dislocations; for the bone lies upon cords of vital importance. The parts in front are stretched [cease to grow, and are liable to disease and premature age]; the hinder parts are wrinkled. In the case of adults, they walk erect, chiefly on the heel, and, if they could take long strides, would do so entirely; but they drag the leg. There is very little atrophy in these cases on account of the exercise, and it is chiefly in the hinder parts. Because the whole leg is straighter than it should be, they require a crutch

[1] Pq. renders "in those who exercise the limb" (!); surely the sense is, "where it is too painful to use."

σιναρόν. οἷσι δὲ ἐκ γενεῆς ἢ αὐξομένοισι, χρηστῶς μὲν ἐπιμεληθεῖσιν ἡ χρῆσις, ὥσπερ τοῖσιν ηὐξημένοισιν· ἀμεληθεῖσι δὲ βραχύ, ἐκτεταμένον· πωροῦται[1] γὰρ τούτοισι, μάλιστα δὲ ἐς ἰθὺ τὰ
20 ἄρθρα· αἱ δὲ τῶν ὀστέων μειώσιες καὶ αἱ τῶν
21 σαρκῶν μινυθήσιες κατὰ λόγον.

XXV. Μηροῦ δὲ κατάτασις μὲν ἰσχυρή· καὶ ἡ διόρθωσις κοινή, ἢ χερσὶν ἢ σανίδι ἢ μοχλῷ, τὰ μὲν ἔσω στρογγύλῳ, τὰ δὲ ἔξω πλατεῖ, μάλιστα δὲ τὰ ἔξω. καὶ τὰ μὲν ἔσω ἀσκοῖσιν ἀκεσάμενον ἐς τὸ ὑπόξηρον τοῦ μηροῦ, κατατάσιος δὲ καὶ συνδέσιος σκελέων· κρεμάσαι διαλείποντα σμικρὸν τοὺς πόδας, ἔπειτα πλέξαντα ἐκκρεμασθῆναί τινα, ἐν τῇ διορθώσει ἀμφότερα ἅμα ποιεῦντα. καὶ τῷ ἔμπροσθεν τοῦτο ἱκανὸν
10 καὶ τοῖσιν ἑτέροισιν, ἥκιστα δὲ τῷ ἔξω. ἡ τοῦ ξύλου ὑπόστασις,[2] ὥσπερ ὤμῳ, ὑπὸ τὴν χεῖρα, οἷς ἔσω· τοῖσι γὰρ ἄλλοισιν ἧσσον· καταναγκάσεις δὲ μετὰ διατάσιος, μάλιστα τῶν ἔμπροσθεν ἢ ὄπισθεν, ἢ ποδὶ ἢ χειρὶ ἐφίζεσθαι
15 ἢ σανίδι.

XXVI. Γόνυ δὲ εὐηθέστερον ἀγκῶνος διὰ τὴν εὐσταλίην καὶ εὐφυΐην, διὸ καὶ ἐκπίπτει καὶ ἐμπίπτει ῥᾷον. ἐκπίπτει δὲ πλειστάκις ἔσω, ἀτὰρ καὶ ἔξω καὶ ὄπισθεν. ἐμβολαὶ δέ· ἢ ἐκ τοῦ συγκεκάμφθαι, ἢ ἐκλακτίσαι ὀξέως, ἢ συνελίξας ταινίης ὄγκον, ἐν ἰγνύῃ θείς, ἀμφὶ τοῦτον ἐξαίφνης ἐς ὄκλασιν ἀφεῖναι τὸ σῶμα, [μάλιστα

[1] πηροῦται, perhaps the correct reading, as in *J.* LX. Foës, Littré, Kw.
[2] ὑπότασις.

on the injured side. In congenital and adolescent cases, if exercise is well managed, they get on like adults; but in neglected patients, the leg is short and extended. Ankylosis occurs in these cases, with the joints usually in an extended position. The shortening of the bones and atrophy of the tissues are according to rule.

XXV. For the thigh strong extension is required, and the adjustment in all cases is with the hands or a board or lever, rounded for internal, flat for external dislocations. The external cases want it most. As to internal cases, there is a treatment with bags to the tapering part of the thigh, with extension and binding together of the legs. Suspend the patient with his legs slightly parted; then let someone be suspended from him, twisting [his arms between the patient's legs],[1] performing both acts of adjustment at once (extension and leverage outwards). This suffices in anterior dislocation and the rest, but is no good in the external form. The plan with wood beneath the limb, as under the arm in shoulder dislocation, suits internal cases, but is not so good in the others; you will succeed in reducing anterior and posterior cases especially by double extension, using foot or hand or a plank to make pressure from above.

XXVI–XXXI. In these chapters we have an epitome of an obscure subject already given verbally (with a few various readings) in *Joints* LXXXII–LXXXVII. Instead of repeating the English version, we may therefore attempt some explanation of the difficulties.[2] The chief of these are:—Why is there no mention of the astragalus in ankle dis-

[1] Cf. *J.* LXX. [2] For note on § XXVI, see p. 417.

ἐν τῇ τῶν ὄπισθεν·][1] δύναται δὲ καὶ κατατεινόμενα μετρίως, ὥσπερ ἀγκών, ἐμπίπτειν τὰ
10 ὄπισθεν· τὰ δὲ ἔνθα ἢ ἔνθα, ἐκ τοῦ συγκεκάμφθαι ἢ ἐκλακτίσαι ἢ [ἐν] κατατάσει, [μάλιστα δὲ αὐτὴ[2] τὸ ὄπισθεν]. ἀτὰρ καὶ ἐκ κατατάσιος μετρίης, ἡ διόρθωσις ἅπασι κοινή. ἢν δὲ μὴ ἐμπέσῃ, τοῖσι μὲν ὄπισθεν συγκάμπτειν οὐ δύνανται, ἀτὰρ οὐδὲ τοῖσιν ἄλλοισιν πάνυ τι. μινύθει δὲ μηροῦ καὶ κνήμης τὸ ἔμπροσθεν. ἢν δὲ ἐς τὸ ἔσω, βλαισότεροι, μινύθει δὲ τὰ ἔξω· ἢν δὲ ἐς τὸ ἔξω, γαυσότεροι, χωλοὶ δὲ ἧσσον· κατὰ γὰρ τὸ παχύτερον ὀστέον ὀχεῖ· μινύθει δὲ τὰ ἔσω. ἐκ
20 γενεῆς δὲ ἢ ἐν αὐξήσει, κατὰ λόγον τὸν ἔμπροσθεν.

XXVII. Τὰ δὲ κατὰ σφυρὰ κατατάσιος ἰσχυρῆς δεῖται, ἢ τῇσι χερσὶν ἢ ἄλλοισι τοιούτοισι, κατορθώσιος δὲ ἅμα ἀμφότερα ποιεύσης· κοινὸν
4 δὲ πᾶσιν.

1 XXVIII. Τὰ δὲ ἐν ποδί, ὡς τὰ ἐν χειρί, ὑγιῆ.

XXIX. Τὰ δὲ ἐν τῇ κνήμῃ συγκοινωνέοντα καὶ μὴ ἐμπεσόντα, ἐκ γενεῆς καὶ ἐν αὐξήσει
3 ἐξαρθρήσαντα, ταὐτὰ ἃ καὶ ἐν χειρί.

XXX. Ὅσοι δὲ πηδήσαντες ἄνωθεν ἐστηρίξαντο τῇ πτέρνῃ, ὥστε διαστῆναι τὰ ὀστέα καὶ φλέβας ἐκχυμωθῆναι καὶ νεῦρα ἀμφιφλασθῆναι, ὅταν γένηται οἷα τὰ δεινότατα, κίνδυνος μὲν σφακελίσαντα τὸν αἰῶνα πρήγματα παρασχεῖν· καὶ ῥοικώδη[3] μὲν τὰ ὀστέα, τὰ δὲ νεῦρα ἀλλήλοισι κοινωνέοντα. ἐπεὶ καὶ οἷσιν ἂν καταγεῖσιν ἢ ὑπὸ τρώματος, οἷα ἐν κνήμῃ, ἢ μηρῷ, νεύρων ἀπολυθέντων ἃ κοινωνεῖ τούτοισιν, ἢ ἐξ
10 ἄλλης κατακλίσιος ἀμελέος ἐμελάνθη[4] ἡ πτέρνη, καὶ τούτοισι παλίγκοτα ἐκ τοιούτων. ἔστιν ὅτε

locations? and, What is meant by the epiphysis of the foot and leg?

We are told (*Fract.* XII, *Mochl.* I) that the leg-bones towards the foot have "a common epiphysis" against which (πρὸς ἥν) the foot moves. The bones may be dislocated with the epiphysis, or the epiphysis only may be displaced (*Fract.* XIII). In the epitome, however, the epiphysis is considered part of the foot, which may be dislocated either with or without it. Littré discusses the subject at great length,[1] and concludes, somewhat doubtfully, that the epiphysis is "la réunion des deux malléoles considérées comme une seule pièce." Its dislocation is the separation of the two bones. But Hippocrates has a special word for each of these, συμφυάς for the union and διάστασις for the separation; and he uses neither here. Adams,[2] following a suggestion by Gardeil, confines the term to the lower end of the fibula; dislocation of the epiphysis is fracture or displacement of the fibula. He admits, however, that a full discussion would be futile and tedious even to the professional reader. The chief argument in favour of this view is that fracture of the lower end of the fibula frequently accompanies ankle dislocation. On the other hand *Fract.* XIII seems to distinguish clearly between the epiphysis and either of the leg-bones.

A third view, hardly bolder than that of Adams,

[1] III. 393 ff.; IV. 45 ff. Petrequin agrees with Littré.
[2] II. 522, also 504.

---

[1] *J.* LXXXII omits here and below. [2] αὐτῇ.
[3] ῥοιώδεα. [4] μελανθῇ.

πρὸς σφακελισμῷ γίνονται πυρετοὶ ὑπερόξεες, λυγγώδεες, τρομώδεες, γνώμης ἁπτόμενοι, ταχυθάνατοι, καὶ ἔτι φλεβῶν αἱμοῤῥόων πελιώσιες καὶ γαγγραινώσιες. σημεῖα τῶν παλιγκοτησάντων· ἢν τὰ ἐκχυμώματα καὶ τὰ μελάσματα καὶ τὰ περὶ ταῦτα ὑπόσκληρα καὶ ὑπέρυθρα ᾖ· ἢν γὰρ σὺν σκληρύσματι πελιωθῇ, κίνδυνος μελανθῆναι· ἢν δὲ ὑποπέλια ᾖ, καὶ πελιὰ μάλα καὶ κεχυμένα,[1] 20 ἢ ὑπόχλωρα καὶ μαλθακά, ταῦτα ἐν[2] πᾶσι τοῖσι τοιούτοισιν ἀγαθά. ἴησις δέ· ἢν μὲν ἀπύρετοι ἔωσιν, ἐλλεβορίζειν·[3] ἢν δὲ μή, μή· ἀλλὰ ποτὸν διδόναι ὀξύγλυκυ, εἰ δέοι. ἐπίδεσις δὲ ἡ ἄρθρων σύνθεσις· ἔτι δὲ[4] πάντα μᾶλλον τοῖσι φλάσμασι· καὶ ὀθονίοισι πλέοσι καὶ μαλθακωτέροισι χρῆσθαι· πίεξις ἧσσον· ὕδωρ πλέον·[5] προσπεριβάλλειν τὰ πλεῖστα τῇ πτέρνῃ· τὸ σχῆμα ὅπερ ἡ ἐπίδεσις, ὡς μὴ ἐς τὴν πτέρνην ἀποπιέξηται· ἀνωτέρω γούνατος ἔστω εὔθετος· νάρθηξι μὴ 30 χρήσασθαι.[6]

XXXI. Ὅταν δὲ ἐκστῇ ὁ πούς, ἢ μοῦνος[7] ἢ σὺν τῇ ἐπιφύσει, ἐκπίπτει μᾶλλον ἐς τὸ ἔσω· εἰ[8] δὲ μὴ ἐμπέσῃ, λεπτύνεται ἀνὰ χρόνον ἰσχίου καὶ μηροῦ καὶ κνήμης τὸ ἀντίον τοῦ ὀλισθήματος. ἐμβολή, ὡς ἡ καρποῦ, κατάτασις δὲ ἰσχυροτέρη. ἴησις, νόμος ἄρθρων· παλιγκοτεῖ ἧσσον καρποῦ, ἢν ἡσυχάσῃ. δίαιτα μείων, ἐλινύουσι γάρ. τὰ δὲ ἐκ γενεῆς μὲν ἢ ἐν αὐξήσει, κατὰ λόγον τὸν 9 πρότερον.

XXXII. Ἐπεὶ τὰ σμικρὸν ὠλισθηκότα ἐκ γενεῆς, ἔνια οἷά τε διορθοῦσθαι. μάλιστα δὲ

[1] ἐκκεχυμωμένα. [2] ἐπὶ.
[3] ἀπύρετος ᾖ, ἐλλέβορον.

is that the epiphysis is our astragalus, looked upon either as an annex to the leg-bones or an epiphysis of the foot. This would explain much, *e.g.*, the fact that Hippocrates speaks of dislocation of the leg from the foot (*Fract.* XIII, *Joints* LIII, LXIII); for, with the astragalus, the leg-bones would have a convex end; so too the foot is said to move *on* (πρός) not *in* this joint. We may also note that the epitomist, taking the epiphysis as part of the foot, adopts the modern view, dislocating the foot from the leg, yet retains the language of his original (*Fract.* XIV) in saying that the commonest dislocation is inwards. The commonest dislocation is that of the leg inwards and the foot outwards, so we can only make him correct by a bold translation such as that of Gardeil, who renders ὁ ποὺς ἐκπίπτει μᾶλλον ἐς τὸ ἔσω, "la partie supérieure de l'astragale se place communément en dedans."

The other Hippocratic account of the ankle-joint (*Loc. Hom.* VI) says, "towards the foot the leg has a joint at the ankles and another below the ankles." The part between is the astragalus; and it is left doubtful whether this belongs to the foot or the leg.[1]

XXXII. Among slight congenital dislocations, some can be put straight, and especially club-foot.[2]

[1] So, too, in *Joints* LIII, we hear of a "bone of the leg at the ankle" which seems distinct from the leg-bones proper, and more closely connected with those of the foot.

[2] An almost ludicrous epitome of *J.* LXII.

---

[4] ἐπίδεσις δέ, ἄρθρων σύνδεσις· ἐπιδεῖν Kw.
[5] Omit. [6] χρῆσθαι.
[7] αὐτὸς. [8] ἦν.

ποδὸς κύλλωσις· κυλλώσιος γὰρ οὐχ εἷς ἐστὶ τρόπος. ἡ δὲ ἴησις τούτου, κηροπλαστεῖν· κηρωτὴ ῥητινώδης,[1] ὀθόνια συχνά, ἢ πέλμα ἢ μολύβδιον προσεπιδεῖν, μὴ χρωτί· ἀνάληψις, τά
7 τε σχήματα ὁμολογείτω.

XXXIII. Ἢν δὲ ἐξαρθρήσαντα ἕλκος ποιησάμενα ἐξίσχῃ, ἐώμενα ἀμείνω, ὥστε δὴ μὴ ἀπαιωρεῖσθαι μηδ' ἀπαναγκάζεσθαι. ἴησις δέ· πισσηρῇ ἢ σπλήνεσιν οἰνηροῖσι θερμοῖσιν—ἅπασι γὰρ τούτοισι τὸ ψυχρὸν κακόν—καὶ φύλλοισιν· χειμῶνος δέ, εἰρίοισι ῥερυπωμένοισι τῆς σκέπης εἵνεκα· μὴ καταπλάσσειν, μηδ' ἐπιδεῖν· δίαιτα λεπτή· ψῦχος, ἄχθος πολύ, πίεξις, ἀνάγκη, σχήματος τάξις· εἰδέναι μὲν οὖν ταῦτα πάντα
10 ὀλέθρια. μετρίως δὲ θεραπευθέντες, χωλοὶ αἰσχρῶς· ἢν γὰρ παρὰ πόδας γένηται, ποὺς ἀνασπᾶται, καὶ ἤν πῃ ἄλλῃ, κατὰ λόγον. ὀστέα οὐ μάλα ἀφίσταται· μικρὰ γὰρ ψιλοῦται, περιωτειλοῦται λεπτῶς. τούτων τὰ μέγιστα κινδυνωδέστατα, καὶ τὰ ἀνωτάτω. ἐλπὶς δὲ μούνη σωτηρίης, ἐὰν μὴ ἐμβάλλῃ, πλὴν τὰ κατὰ δακτύλους καὶ χεῖρα ἄκρην· ταῦτα δὲ προειπέτω[2] τοὺς κινδύνους. ἐγχειρεῖν ἐμβάλλειν ἢ τῇ πρώτῃ ἢ τῇ δευτέρῃ, ἢν δὲ μή, πρὸς τὰ δέκα· ἥκιστα
20 τεταρταῖα. ἐμβολὴ δέ, οἱ μοχλίσκοι. ἴησις δέ, ὡς κεφαλῆς ὀστέων, καὶ θερμή· ἐλλεβόρῳ δὲ καὶ αὐτίκα ἔπειτα[3] τοῖσιν ἐμβαλλομένοισι βέλτιον χρῆσθαι. τὰ δ' ἄλλα εὖ εἰδέναι δεῖ ὅτι ἐμβαλλομένων θάνατοι· τὰ μέγιστα καὶ τὰ ἀνωτάτω

[1] κηρωτῇ ῥητινώδει.
[2] προειπόντα.
[3] καὶ ἔπειτα.

Now there is more than one kind of club-foot. Here is the treatment of it: moulding, resined cerate, plenty of bandages, a sandal or sheet of lead bound in with the bandaging, not directly on the flesh; let the slinging up and attitude of the foot be in accordance.

XXXIII. If dislocated bones make a wound and project, they are best let alone, seeing, of course, that they are not left unsupported or subject to violence. Treatment with pitch cerate, or compresses soaked in warm wine (for cold is bad in all these cases), also leaves, and, in winter, crude wool as a protection; do not use a plaster application or bandaging; low diet; cold, heavy weight, constriction, violence, a forcibly ordered attitude—bear in mind that all these are pernicious. Suitably treated, they survive badly maimed; for if the lesion is near the foot, the foot is drawn up; and if anywhere else, there is a corresponding deformity. Bones do not usually come away, for only small surfaces are denuded, and a thin scar forms. In these cases there is greatest danger with the largest and proximal joints. The only hope of safety is not to reduce them, except the fingers and bones of the hand. In these cases let the surgeon explain the risks beforehand. Perform reduction on the first or second day; failing that, about the tenth; by no means on the fourth. Reduction: the small levers. Treatment: as for bones of the head; warmth; it is rather a good thing to give a dose of hellebore to the patients immediately after reduction. As to other bones, one must bear well in mind that their reduction means death, the quicker and more certain the larger and higher up they are. In the

μάλιστα καὶ τάχιστα. πούς δὲ ἐκβάς, σπασμός, γάγγραινα· καὶ γὰρ ἢν ἐμβληθέντι ἐπιγένηταί τι τούτων, ἐκβάλλοντι ἐλπίς, εἴ τις ἄρα ἐλπίς· οὐ γὰρ ἀπὸ τῶν χαλώντων οἱ σπασμοί, ἀλλ᾿ ἀπὸ
29 τῶν ἐντεινόντων.

XXXIV. Αἱ δὲ ἀποκοπαὶ ἢ ἐν ἄρθρῳ ἢ κατὰ τὰ ὀστέα, μὴ ἄνω, ἀλλ᾿ ἢ παρὰ τῷ ποδὶ ἢ παρὰ τῇ χειρὶ ἐγγὺς περιγίνονται, ἢν μὴ αὐτίκα μάλα[1] λειποθυμίῃ ἀπόλωνται. ἴησις, ὡς κεφαλῆς,
5 θερμή.

XXXV. Ἀποσφακελίσιος μέντοι σαρκῶν, καὶ ἐν τρώμασι αἱμοῤῥόοις ἀποσφιγχθέν, καὶ ἐν ὀστέων κατήγμασι πιεχθέν, καὶ ἐν δεσμοῖς ἀπομελανθέν. καὶ οἷσι μηροῦ μέρος ἀποπίπτει καὶ βραχίονος, ὀστέα τε καὶ σάρκες ἀποπίπτουσι, πολλοὶ περιγίνονται, ὡς τά γε ἄλλα εὐφορώτερα· οἷσι μὲν οὖν κατεαγέντων ὀστέων, αἱ μὲν περιῤῥήξιες ταχεῖαι, αἱ δὲ τῶν ὀστέων ἀποπτώσιες, ᾗ ἂν τὰ ὅρια τῆς ψιλώσιος ᾖ, ταύτῃ ἀποπίπτουσι,
10 βραδύτερον δέ. δεῖ[2] δὲ τὰ κατωτέρω τοῦ τρώματος προσαφαιρεῖν καὶ τοῦ σώματος τοῦ ὑγιέος —προθνήσκει γάρ—φυλασσόμενον·[3] ὀδύνῃ ἅμα γὰρ λειποθυμίῃ θνήσκουσιν. μηροῦ ὀστέον ἀπελύθη ἐκ τοιούτου ὀγδοηκοσταῖον, ἡ δὲ κνήμη ἀφῃρέθη εἰκοσταίη· κνήμης δὲ ὀστέα κατὰ μέσην ἑξηκοσταῖα ἀπελύθη. ἐκ τοιούτων ταχὺ καὶ

[1] ἅμα. [2] χρὴ Kw.

case of a (compound) dislocation of the foot, spasm and gangrene (are to be expected). If anything of this kind supervenes on reduction, there is hope from dislocation, if indeed there is hope at all; for spasms do not come from relaxation of parts, but from their tension.

XXXIV. Amputations at a joint or in the length of the bones, if not high up, but either near the foot or near the hand, usually[1] result in recovery, unless the patients perish at once from collapse. Treatment: as for the head; warmth.

XXXV. (Causes) of gangrene of the tissues are: constriction in wounds with haemorrhage, compression in fractures of bones, and mortification from bandages.[2] Even in cases where part of the thigh or arm falls off and bones and flesh come away, many survive; and in other respects this is rather well borne. In cases of fractured bones, lines of demarcation form quickly; but the falling off of the bones (it is where the limit of the denudation occurs that they fall off) occurs more slowly. One must[3] intervene to remove the parts below the lesion and the sound part of the body (for these parts die first), and be careful;[4] for patients die from pain and collapse combined. A thigh-bone separated in such a case on the eightieth day, but the leg was removed on the twentieth; leg-bones separated at the middle on the sixtieth day. In such cases the compression

[1] ἐγγὺς corresponds to τοῖς πλείστοισι, *J.* LXVIII; but it is a curious use.

[2] *J.* LXIX.

[3] "Should" (Kw.).

[4] "Avoid pain"—Kw.'s punctuation.

---

[3] φυλασσόμενον absolute: cf. *Head Wounds* XVIII. Kw. follows a conjecture of Foës and reads φυλασσόμενον ὀδύνην.

βραδέως, αἱ πιέξιες αἱ ἰητρικαί. τὰ δ' ἄλλα ὅσα ἡσυχαίως, τὰ μὲν ὀστέα οὐκ ἀποπίπτει οὐδὲ σαρκῶν ψιλοῦται, ἀλλ' ἐπιπολαιότερον.[1] προσ-
20 δέχεσθαι ταῦτα χρή· τὰ γὰρ πλεῖστα φοβερώτερα ἢ κακίω. ἡ ἴησις πραεῖα, θερμῇ διαίτῃ ἀκριβεῖ· κίνδυνος αἱμοῤῥαγιῶν, ψύχεος· σχήματα δὲ ὡς μὲν ἀνάῤῥοπα, ἔπειτα ὑποστάσιος πύου εἵνεκα ἐξ ἴσου ἢ ὅσα συμφέρει. ἐπὶ τοῖσι τοιούτοισι καὶ ἐπὶ τοῖσι μελασμοῖσιν, αἱμοῤῥαγίαι, δυσεντερίαι, περὶ κρίσιν, λαῦροι μέν, ὀλιγήμεροι δέ. οὐκ ἀπόσιτοι δὲ πάνυ οὐδὲ πυρετώδεες, οὐδέ τι
28 κενεαγγητέον.

XXXVI. Ὕβωσις, ἡ μὲν ἔσω ἐπιθάνατος, οὔρων σχέσιος, ἀποναρκώσιος·[2] τὰ δὲ ἔξω, τούτων ἀσινέα τὰ πλεῖστα, πολὺ μᾶλλον ἢ ὅσα σεισθέντα μὴ ἐξέστη. αὐτὰ μὲν ἑωυτοῖσι κρίσιν ποιησάμενα, κεῖνα δὲ ἐπὶ πλέον τῷ σώματι ἐπιδιδόντα, καὶ ἐν ἐπικαίροις ἐόντα.

Οἷον πλευραὶ κατεαγεῖσαι μέν, ὀλίγαι πυρετώδεες καὶ αἵματος πτύσιος καὶ σφακελισμοῦ, ἤν τε μία, ἤν τε πλείους μὴ καταγῇ ἔσω δέ·[3]
10 καὶ ἴησις φαύλη, μὴ κενεαγγοῦντα, ἢν ἀπύρετος ᾖ. ἐπίδεσις ὡς νόμος· ἡ δὲ πώρωσις ἐν εἴκοσιν ἡμέρῃσιν, χαῦνον γάρ. ἢν δ' ἀμφιφλασθῇ, φυματίαι, καὶ βηχώδεες, καὶ ἔμμοτοι, καὶ πλευρὰς ἐσφακέλισαν· παρὰ γὰρ πλευρὴν ἑκάστην ἀπὸ
15 πάντων τόνοι εἰσίν.

XXXVII. Τὰ δὲ ἀπὸ καταπτώσιος ἧσσον

[1] ἐπιπολαιότερα. [2] εἵνεκα understood.
[3] μὴ καταγεῖσαι δέ . . . Kw. He suspects a mutilation in the text.

---

[1] "Which have been gently constricted." Littré (Adams).

used during treatment makes it quick or slow. For the rest, in cases of mild character [1] the bones do not come away, nor are they denuded of flesh; but the mortification is more superficial. One should take on these cases, for they are most of them more terrifying than dangerous. Treatment: gentle, with warmth and strict diet; dangers: haemorrhage, chill; attitudes rather elevated; afterwards, because of collection of pus, on a level, or whatever suits. Haemorrhage supervenes in such cases, also in mortification, and dysentery at the crisis, copious, but of short duration. Patients do not lose their appetites much, nor are they feverish; and there is no reason why one should starve them.

XXXVI. Spinal curvature: inwards it is fatal, from retention of urine and loss of sensation; external curvatures are most of them without serious lesions, much more so than cases of concussion without displacement, for they make their own crisis; but the latter have a greater effect on the body and on parts of vital importance.

So, too, fractured ribs rarely give rise to fever, spitting of blood, or necrosis, where there is one or more fractured, if it is not broken inwards; [2] and the treatment is simple, without starvation diet, if there is no fever. Bandaging as customary. Callus forms in twenty days, for the bone is spongy. But if there is great contusion, tubercles, chronic coughs and suppurating wounds supervene, with necrosis of the ribs; for along each rib there are cords coming from all parts.

XXXVII. Curvatures due to a fall are less sus-

[2] Or, "if not splintered," Littré (Adams); "if they are not broken (but contused)," Kw.

δύναται ἐξιθύνεσθαι· χαλεπώτερα δὲ τὰ ἄνω φρενῶν ἐξιθύνεσθαι. οἷσι δὲ παισίν, οὐ συναύξεται, ἀλλ' ἢ σκέλη καὶ χεῖρες καὶ κεφαλή· ηὐξημένοισιν ὕβωσις, παραχρῆμα μὲν τῆς νούσου ῥύεται, ἀνὰ χρόνον δ' ἐπισημαίνεται δι'[1] ὧνπερ καὶ τοῖσι νεωτέροισιν, ἧσσον δὲ κακοήθως. εἰσὶ δὲ οἳ εὐφόρως ἤνεγκαν, οἷσιν ἂν ἐς εὔσαρκον καὶ πιμελῶδες τράπηται· ὀλίγοι δὲ τούτων περὶ
10 ἑξήκοντα ἔτεα ἐβίωσαν. ἀτὰρ καὶ ἐς τὰ πλάγια διαστρέμματα γίνεται· συναίτια δὲ καὶ τὰ σχήματα ἐν οἷσιν ἂν κατακέωνται· καὶ ἔχει προγνώσιας.

Πολλοὶ δὲ καὶ αἷμα ἔπτυσαν καὶ ἔμπυοι ἐγένοντο. ἡ δὲ μελέτη, ἴησις, ἐπίδεσις ὡς νόμος· διαίτης τὰ πρῶτα ἀτρεκέως, ἔπειτα ἁπαλύνειν· ἡσυχίῃ, σιγῇ· σχήματα, κοιλίη, ἀφροδίσια. ἀτὰρ οἷς ἄναιμα, ἐπωδυνώτερα τῶν καταγνυμένων καὶ φιλυποστροφώτερα χρόνοισιν· οἷσι δὲ καταλείπε-
20 ται μυξῶδες, ὑπομιμνήσκει ἐν πόνοισιν. ἴησις· καῦσις, τοῖσι μὲν ἀπ' ὀστέου, μέχρις[2] ὀστέου, μὴ αὐτὸ δέ· ἢν δὲ μεταξύ, μὴ πέρην, μηδὲ ἐπι πολῆς· σφακελισμός. καὶ τὰ ἔμμοτα πειρᾶσθαι· εἰρήσεται ἅπαντα τὰ ἐπεσιόντα. ὁρατά, λόγοις δ' οὐ μή· βρώματα, πόματα, θάλπος, ψῦχος, σχῆμα· ὅτι καὶ φάρμακα, τὰ μὲν ξηρά, τὰ δὲ ὑγρά, τὰ δὲ πυῤῥά, τὰ δὲ μέλανα, τὰ δὲ λευκά,
28 τὰ δὲ στρυφνά, ἐπὶ ἕλκη, οὕτω καὶ δίαιται.

XXXVIII. Νόμος ἐμβολῆς καὶ διορθώσιος· ὄνος, μοχλός, σφηνίσκος, ἶπος· ὄνος μὲν ἀνάγειν, μοχλὸς δὲ παράγειν. τὰ δὲ ἐμβλητέα ἢ διορ-

[1] ἐπισημαίνεταί τι (as in *J.* XLI). [2] μέχρι τοῦ.

ceptible to rectification; and those above the diaphragm are the more difficult to straighten. In the case of children, there is cessation of growth, except in the legs, arms, and head. Curvature in adults delivers from the disease at the moment; but in time the same symptoms appear as in younger patients, but in less malignant form. There are some who bear the affection well, those in whom there is a tendency to fulness of flesh and fat; but few of these reach sixty years. Lateral distortions also are produced, and the positions in which patients lie are accessory causes; they also serve for prognosis.

Many patients spit blood, and get an abscess.[1] Care and treatment; bandaging as usual. Diet: at first strict, then feed him up; repose and silence, position, the bowels, sexual matters. But where there is no show of blood, the parts are more painful than in fractured cases, and there is more tendency to relapse later. Where the tissue is left in a mucous state, there is a return of pains. Treatment: cautery, where bone is involved, down to the bone, but not of the bone itself; if between the ribs, not right through, yet not superficial. Necrosis: try also the treatment with tents; all that concerns this will be described. Things are to be seen—don't trust to words; food, drink, warmth, cold, attitude. As to drugs also, some are dry, some moist, some ruddy, some black, some white, some astringent, used for wounds; so too (various) diets.

XXXVIII. Usage for reduction and adjustment: windlass, lever, wedge, press; windlass for stretching, lever for bringing into place. Parts to be

[1] This passage seems out of place here, and Littré boldly joins it on to XXXVI; but we now have to do with odd notes.

θωτέα διαναγκάσαι δεῖ ἐκτείνοντα, ἐν ᾧ ἂν ἕκαστα σχήματι μέλλῃ ὑπεραιωρηθήσεσθαι· τὸ δ' ἐκβάν,[1] ὑπὲρ τούτου ὅθεν ἐξέβη. τοῦτο δέ, ἢ χερσὶν ἢ κρεμασμῷ ἢ ὄνοισιν ἤ περί τι. χερσὶ μὲν οὖν ὀρθῶς κατὰ μέρεα· καρπὸν δὲ καὶ ἀγκῶνα ἀπόχρη διαναγκάζειν, καρπὸν μὲν εἰς
10 ἰθὺ ἀγκῶνος, ἀγκῶνα δὲ ἐγγώνιον πρὸς βραχίονα ἔχοντα, οἷον παρὰ τῷ βραχίονι τὸ ὑπὸ τὴν χεῖρα ὑποτεινόμενον. ἐν οἷσι δὲ δακτύλου, ποδός, χειρός, καρποῦ, ὑβώματος τὸ ἔξω,[2] διαναγκάσαι δεῖ καὶ καταναγκάσαι, τὰ μὲν ἄλλα ὑπὸ χειρῶν αἱ διαναγκάσιες ἱκαναί, καταναγκάσαι δὲ τὰ ὑπερέχοντα ἐς ἕδρην πτέρνῃ ἢ θέναρι ἐπί τινος· ὥστε κατὰ μὲν τὸ ἐξέχον ὑποκεῖσθαι ὄγκον σύμμετρον μαλθακόν· κατὰ δὲ τὸ ἕτερον [μήστωρα] δ' ἂν[3] χρὴ ὠθεῖν ὀπίσω καὶ κάτω,
20 ἢν δὲ ἔσω ἢν δὲ ἔξω ἐκπεπτώκῃ· τὰ δὲ ἐκ πλαγίων, τὰ μὲν ἀπωθεῖν, τὰ δὲ ἀντωθεῖν ὀπίσω ἀμφότερα κατὰ τὸ ἕτερον. τὰ δὲ ὑβώματα, τὰ μὲν ἔσω, οὔτε πταρμῷ οὔτε βηχί, οὔτε φύσης ἐνέσει, οὔτε σικύῃ· δεῖ δέ τι, ἡ κατάστασις· ἡ δὲ ἀπάτη, ὅτι οἷόν τέ[4] ποτε κατεαγέντων τῶν σπονδύλων καὶ τὰ λορδώματα διὰ τὴν ὀδύνην δοκεῖ ἔσω ὠλισθηκέναι· ταῦτα δὲ ταχυφυᾶ καὶ ῥᾴδια. τὰ δὲ ἔξω, κατάτασις, τὰ μὲν ἄνω ἐπὶ πόδας, τὰ δὲ κάτω τἀναντία· κατανάγκασις δὲ
30 σὺν κατατάσει, ἢ ἕδρῃ ἢ ποδὶ ἢ σανίδι. τὰ δ'

[1] ἐμβάν Ap. [2] ἐς τὸ ἔξω Ap.
[3] μήστωρ (= "skilled assistant") δ' ἂν vulg.; μὴ στορέσαντα Lit.: μήστορα ἅμα Kw.
[4] οἴονται Kw., Littré.

---

[1] *I.e.* hand-power is strong enough.

reduced or adjusted must be separated by extension, till each comes into an attitude of sufficient elevation, the dislocated part above that from which it was dislocated; this is done with the hands, or suspension, or a windlass, or round something. Proper use of the hands varies with the part; in the case of the wrist and ankle, it suffices[1] to separate the parts, the wrist being in line with the elbow, but the elbow at right angles to the upper arm, as when the forearm is in a sling. In the case of finger or toe, foot, hand, wrist, humpback, double extension and forcing down the projection are required; in the other cases, separation by hand-power is enough, but one must force projecting parts into position with the heel or palm over something, taking care that a suitable soft pad is placed under the projection. On the other side, a skilled assistant should simultaneously press backwards and downwards, if the dislocation is either inwards or outwards; in lateral cases, press one side away and the other side back to meet it, bringing both together. As to curvatures, internal ones are not (reducible) by sneezing, coughing, injection of air, or a cupping instrument; a mode of restoration is wanting.[2] The deception people fall into when vertebrae are fractured, and incurvings due to pain simulate dislocation inwards; these heal quickly, and are not serious. Outward curvatures: extension,[3] towards the feet if the lesion is high up, if low down, the reverse; forcing into place, simultaneously with extension, by sitting on it, or by using the foot or a plank.

[2] Or "If anything, extension," reading κατάτασις, as Littré (Adams).

[3] κατάσεισις, "succussion." Littré.

ἔνθα ἢ ἔνθα, εἴ τις κατάτασις, καὶ ἔτι τὰ σχήματα ἐν τῇ διαίτῃ.

Τὰ ἄρμενα πάντα εἶναι πλατέα, προσηνέα, ἰσχυρά, εἰ δέῃ· μὴ[1] δεῖ ῥάκεσι προκατειλίχθαι. ἐσκευάσθαι πρὶν ἢ ἐν τῇσιν ἀνάγκῃσιν πάντα συμμεμετρημένως τὰ μήκεα καὶ ὕψεα καὶ εὔρεα. διάτασις, οἷον μηροῦ, τὸ παρὰ σφυρὸν δεδέσθαι καὶ ἄνω τοῦ γούνατος, ταῦτα μὲν ἐς τὸ αὐτὸ τείνοντα· παρὰ δὲ ἰξύϊ[2] καὶ περὶ μασχάλας, 40 καὶ κατὰ περίναιον καὶ μηρόν, τὰ[3] μεταξὺ τῆς ἀρχῆς, τὸ μὲν ἐπὶ στῆθος, τὸ δὲ ἐπὶ νῶτον τείνοντα, ταῦτα δ' ἐς τὸ αὐτὸ ἅπαντα[4] τείνοντα, προσδεθέντα ἢ πρὸς ὑπεροειδέα ἢ πρὸς ὄνον. ἐπὶ μὲν οὖν κλίνης ποιέοντι, τοῦτο μὲν τῶν ποδῶν πρὸς οὐδὸν χρὴ ἐρεῖσαι, πρὸς δὲ τὸ ἕτερον, ξύλον ἰσχυρὸν πλάγιον παραβεβλῆσθαι, τὰ δὲ ὕπερθεν ὑπεροειδέα πρὸς ταῦτα ἀντιστηρίζοντα διατείνειν, ἢ πλήμνας κατορύξαντα, ἢ κλίμακα διαθέντα, ἀμφοτέρωθεν ὠθεῖν. τὸ δὲ κοινόν, 50 σανὶς ἑξάπηχυς, εὖρος δίπηχυς, πάχος σπιθαμῆς, ἔχουσα ὄνους δύο ταπεινοὺς ἔνθεν καὶ ἔνθεν, ἔχουσα δὲ κατὰ μέσον στυλίσκους συμμέτρους, ἐξ[5] ὧν ὡς κλιμακτὴρ ἐπέσται ἐς τὴν ὑπόστασιν τῷ ξύλῳ, ὥσπερ τῷ κατ' ὦμον· καταγλύφους δὲ ὥσπερ ληνοὺς λείας ἔχειν, τετραδακτύλους εὖρος καὶ βάθος, καὶ διαλιπεῖν τοσοῦτον ὅσον αὐτῇ τῇ μοχλεύσει ἐς διόρθωσιν· ἐν μέσῳ δὲ τετράγωνον καταγλυφὴν ὥστε στυλίσκον ἐνεῖναι, ὃς παρὰ περίναιον ἐὼν περιῤῥέπειν τε κωλύσει ἐὼν

[1] εἰ δὲ μή, Littré's conjecture, Kw. Cf. *J.* LXXVIII.
[2] ἰξὺν.
[3] μηρῶν τὸ.
[4] ἐς τὰ ἀπεναντία.
[5] ἐφ'.

Curvatures to this side or that; one may use some extension, also postures with regimen.

The tackle should all be broad, soft, and strong, otherwise [1] they must be previously wrapped in rags; all should be suitably prepared as to length, height, and breadth before use in the reductions. In double extension of the thigh, for example, make attachments at the ankle and above the knee, drawing these in the same direction; at the loin and round the armpits; also at the perineum and between the thighs,[2] drawing one end over the chest, the other over the back, but bringing these in the opposite direction;[3] they should be fixed either to a pestle-pole or to a windlass. If one operates on a patient in bed, its legs at one end should press against the threshold, and a strong plank should be laid across the other end; then, using these as fulcra, draw back the pestle-like poles from above; or fix wheel-naves in the ground; or lay a ladder along, and apply force at both ends. For all cases: a nine-foot plank, three feet broad, a span thick, having two windlasses set low down at each end, and also having at the middle suitable props, on which is placed a sort of crossbar to act as fulcrum for the board, like that used for the shoulder.[4] It should have fossae like smooth troughs, four fingers broad and deep, with sufficient intervals between for adjustment by actual leverage. In the middle (there should be) a quadrangular excavation for a prop to fit into, which, when it is at the perineum, will prevent the patient from slipping, and when it is

[1] Reading εἰ δὲ μή. "Sufficiently strong; it should not be necessary to wrap" (Pq.'s rendering of the text).

[2] Kw.'s reading. [3] Kw.'s reading.

[4] *I.e.* the *ambê*; cf. *J.* LXXIII.

60 τε ὑποχάλαρος ὑπομοχλεύσει. χρὴ δὲ τῆς σανίδος, ἣ ἐν τῷ τοίχῳ τὸ ἄκρον καταγεγλυμμένον τι ἐχούσης, τοῦ ξύλου ὦσαι τὸ ἄκρον, ἐπὶ δὲ θάτερα καταναγκάζειν, ὑποτιθέντα μαλθακά τινα
64 σύμμετρα.

XXXIX. Οἷσιν ὀστέον ἀπὸ ὑπερῴης ἀπῆλθε, μέση ἵζει ἡ ῥὶς τούτοισιν. οἱ δὲ φλώμενοι κεφαλὰς ἄνευ ἕλκεος, ἢ πεσόντος ἢ κατάξαντος ἢ πιέσαντος, τούτων ἐνίοισι τὰ δριμέα ἔρχεται ἀπὸ κεφαλῆς κατὰ τὰς φάρυγγας, καὶ ἀπὸ τρώματος
6 ἐν τῇ κεφαλῇ καὶ ἐς τὸ ἧπαρ καὶ ἐς τὸν μηρόν.

XL. Σημεῖα παραλλαγμάτων καὶ ἐκπτωμάτων· καὶ ᾗ καὶ ὅπως καὶ ὅσον διαφέρει ταῦτα πρὸς ἄλληλα· καὶ οἷσιν ἡ κοτύλη παρέαγε, καὶ οἷσι νευρίον ἀπεσπάσθη, καὶ οἷσι ἐπίφυσις ἀπέαγε, καὶ οἷσι καὶ ὡς, καὶ ἓν ἢ δύο, ὧν δύο ἐστίν· ἐπὶ τούτοισι κίνδυνοι, ἐλπίδες οἷσι κακαί, καὶ ὅτε κακώσιες θανάτου, ὑγιείης, ἀσφαλείης. καὶ ἃ ἐμβλητέα ἢ χειριστέα καὶ ὅτε, καὶ ἃ οὒ ἢ ὅτε οὔ· ἐπὶ τούτοισιν ἐλπίδες, κίνδυνοι· οἷα καὶ ὅτε χει-
10 ριστέα, καὶ τὰ ἐκ γενεῆς ἔξαρθρα, τὰ αὐξανόμενα, τὰ ηὐξημένα, καὶ ὅ τι θᾶσσον, καὶ ὅ τι βραδύτερον, καὶ ὅ τι χωλόν, καὶ ὡς καὶ οὔ· καὶ διότι καὶ ὅ τι μινυθήσει, καὶ ᾗ καὶ ὡς καὶ οἷσιν ἧσσον καὶ ὅτι τὰ καταγέντα θᾶσσον καὶ βραδύτερον φυόμενα, ᾗ αἱ διαστροφαὶ καὶ ἐπιπωρώσιες γίνονται, καὶ ἀκὴ τούτων. οἷσιν ἕλκεα αὐτίκα

[1] This is condensed from *J.* XLVII and LXXV, on pressing down a hump by bringing a plank across it, one end being in a groove in a post or wall. The translation makes the epitomiser say this; but in the Greek he seems to confuse the plank with the *ambê*, which had a sort of excavation at its end. Littré omits ἣ and the first τὸ ἄκρον.

rather loose will serve as a lever. Use of the plank: one should push it in at one end; the end should occupy an excavation in a post or in a wall;[1] press down at the other end, putting some suitable soft substance underneath.

XXXIX. In cases where a bone comes away from the roof of the mouth, the nose falls in in the middle.[2] Patients with contused heads without a wound, due to a fall, fracture, or compression; some of them have a flow of acrid humour from the head down to the fauces, and from the lesion in the head to both liver and thigh.[3]

XL. Symptoms of subluxations and dislocations: their difference from one another in position, nature, and extent, where the socket is fractured, where a small ligament is torn away, where the epiphysis is broken off. In what cases and how either one or two bones (are broken), when there are two; dangers and expectations in these cases; in which cases they are bad, and when injuries are mortal, or when there is more hope of recovery. Also what cases are to be reduced or treated surgically, and when, and which not, and when not; the expectations and dangers in these cases. In what cases and at what time one should treat congenital dislocations or those occurring during and after adolescence. Which case is quicker and which slower to recover where a patient is (permanently) lame, and how, and when not; and why, and in what cases, there is atrophy; on which side, and how, and the cases in which it is less; and that fractured bones are quicker or slower to consolidate, where distortions and accumulation of callus occur, and the cure for these. Cases

[2] *Epid.* IV. 1. 9, VI. 1. 3. [3] *Epid.* II. 5. 4.

ἢ ὕστερον γίνονται· οἷσι καὶ ὀστέα καταγεῖσι μείω, οἷσιν οὔ· οἷσι καταγέντα ἐξέσχεν, καὶ ᾗ ἐξίσχει μᾶλλον· οἷσιν ἐκβάντα ἢ ἄρθρα ἐξίσχια·
20 ἀπατῶνται[1] καὶ δι' ἅ, ἐν οἷσιν ὁρῶσιν, ἐν οἷσιν διανοεῦνται, ἀμφὶ τὰ παθήματα, ἀμφὶ τὰ θερα-
22 πεύματα.

XLI. Νόμοισι τοῖσι νομίμοισι περὶ ἐπιδέσιος· παρασκευή, πάρεξις, κατάτασις, διόρθωσις, ἀνάτριψις, ἐπίδεσις, ἀνάληψις, θέσις, σχῆμα, χρόνοι, δίαιται. τὰ χαυνότατα τάχιστα φύεται, τὰ δὲ ἐναντία, ἐναντίως· διαστροφαί, ᾗ κυρτοί· ἄσαρκοι, ἄνευροι. τὸ ἐμπεσὸν ὡς προσωτάτω[2] ἢ τὸ ἐκπεσὸν ἔσται τοῦ χωρίου οὗ ἐξέπεσεν.[3] νεύρων, τὰ μὲν ἐν κινήσει καὶ ἐν πλάδῳ, ἐπιδοτικά· τὰ δὲ μή, ἧσσον· ἄριστον ᾗ ἂν ἐκπέσῃ,
10 εἰ ἐμπέσοι τάχιστα·[4] πυρεταίνοντι μὴ ἐμβάλλειν, μηδὲ τεταρταῖα, πεμπταῖα, ἥκιστα ἀγκῶνα. καὶ τὰ ναρκώδεα πάντα, ὡς τάχιστα ἄριστα, ἢ τὴν φλεγμονὴν παρέντα. τὰ ἀποσπώμενα, ἢ νεῦρα ἢ χόνδρια ἢ ἐπιφύσιες, ἢ διϊστάμενα κατὰ συμφύσιας, ἀδύνατα ὁμοιωθῆναι· διαπωροῦται ταχέως τοῖσι πλείστοισιν· ἡ δὲ χρῆσις σώζεται. ἐκβάντων, τὰ ἔσχατα, ῥᾷον· τὰ ῥᾷστα ἐκπεσόντα ἥκιστα φλεγμαίνει· τὰ δὲ ἥκιστα θερμαίνοντα, καὶ μὴ ἐπιθεραπευθέντα, μάλιστα αὖθις ἐκπί-
20 πτει. κατατείνειν ἐν σχήματι τοιούτῳ, ἐν ᾧ

[1] ἃ ἀπατῶνται Kw. [2] ἑκαστάτω.
[3] Obscure ; seems to be taken from *J.* IX.
[4] Cf. *J.* LXXIX.

---

[1] Apparently "intervals" between changes of dressing and the like.

where wounds occur at once or later; where the fractured bones are shortened, and where they are not. In what cases fractured bones project, and at what part they chiefly do this. The confusion between dislocations and prominent joints, causes of deception in what men see, and conjecture concerning maladies and treatments.

XLI. Recognised usages as regards bandaging: preparation, presentation, extension, adjustment, friction, bandaging, suspension, putting up, attitude, periods,[1] diets. The most spongy bones consolidate quickest, and vice versa; distortions on the side towards which they curve; atrophy of flesh and sinews. The reduced bone shall be (kept) as far as possible from the place where it was dislocated.[2] Of ligaments, those in mobile and moist parts are yielding; those which are not are less so. Wherever a dislocation may be, prompt reduction is best. Do not reduce when a patient has fever, or on the fourth or fifth days, least of all in an elbow case. All cases with loss of sensation, the quicker the better; or wait till inflammation has subsided. Parts torn away: ligaments, cartilages, epiphyses or separations at symphyses cannot be made the same as before; in most cases there is rapid ankylosis, but the use of the limb is preserved. Of dislocated joints, the most distal are the more easily (put out?);[3] those most easily put out suffer least inflammation; but where there is least heat and no after-treatment, there is greatest liability to another dislocation. Make extension in such a posture that

[2] "Force used in reduction to be applied at as great a distance as possible" (Adams).

[3] Or "treated"; but it seems best to follow the context.

μάλιστα ὑπεραιωρηθήσεται, σκεπτόμενον ἐς τὴν φύσιν καὶ τὸν τόπον ᾗ ἐξέβη. διόρθωσις· ὀπίσω ἐς ὀρθὸν καὶ ἐς πλάγιον παρωθεῖν· τὰ δὲ ταχέως ἀντισπάσαντα ἀντισπάσαι ταχέως ἢ δὴ ἐκ περιαγωγῆς· τὰ δὲ πλειστάκις ἐκπίπτοντα ῥᾷον ἐμπίπτει· αἴτιον νεῦσις ἢ νεύρων ἢ ὀστέων. νεύρων μὲν μῆκος ἢ ἐπίδοσις· ὀστέων δέ, κοτύλης ὁμαλότης, κεφαλῆς φαλακρότης· τὸ ἔθος τρίβον ποιεῖ· αἰτίη καὶ σχέσις καὶ ἕξις καὶ ἡλικίη. τὸ
30 ὑπόμυξον ἀφλέγμαντον.

XLII. Οἷσιν ἕλκεα ἐγένετο, ἢ αὐτίκα ἢ ὀστέων ἐξισχόντων, ἢ ἔπειτα, ἢ κνησμῶν ἢ τρηχυσμῶν, ταῦτα μὲν ἢν αἰσθῇ, εὐθέως λύσας, πισσηρὴν ἐπὶ τὸ ἕλκος ἐπιθείς, ἐπιδεῖν ὡς ἐπὶ τὸ ἕλκος πρῶτον τὴν ἀρχὴν βαλλόμενος, καὶ τἄλλα ὡς οὐ ταύτῃ τοῦ σίνεος ἐόντος· οὕτω γὰρ αὐτό τε ἰσχνότατον καὶ ἐκπυήσει τάχιστα καὶ περιῤῥήξεται, καὶ καθαρθέντα τάχιστα φύσεται. νάρθηκας δὲ μήτε κατ' αὐτὸ τοῦτο προσάγειν μήτε
10 πιέζειν· καὶ ὧν ὀστέα μὴ μεγάλα ἄπεισιν, ὧν δὲ μεγάλα, οὕτω ποιεῖν·[1] πολλὴ γὰρ ἐμπύησις καὶ ταῦτ' οὐκ ἔτι οὕτως, ἀλλ' ἀνέψυκται τῶν ὑποστασίων εἵνεκα. τὰ δὲ τοιαῦτα ὁπόσα ἐξέσχε, καὶ εἴ τε ἐμβληθῇ εἴ τε μή, ἐπίδεσις μὲν οὐκ ἐπιτήδειον, διάτασις δέ. σφαῖραι ποιηθεῖσαι οἷαι πέδαις, ἡ μὲν παρὰ σφυρόν, ἡ δὲ

[1] Littré joins οὕτω ποιεῖν to ἄπεισιν and adds οὐ after μεγάλα, de suo: ἄπεισιν ὡσαύτως· ὧν δὲ μεγάλα δῆλον, Kw. M.

---

[1] Second ἢ perhaps added for sake of symmetry; there are only two classes of wounds, "immediate" and "later."

[2] Adopting Kw.'s reading, which has some support from the MSS.

the (dislocated bone) will be best lifted above (the socket), having regard to its conformation and the place where it is dislocated. Adjustment: push backwards, either straight or obliquely; where there has been a rapid twist, make a rapid twist (backwards), or at any rate by circumduction. Often repeated dislocations are more easily reduced; they are due to the disposition of the ligaments or bones—in the former, to length or yielding character; in the latter, to flatness of the socket and rounded shape of the head. Use makes a friction-joint; it depends on the state of the patient, his constitution and age. Rather mucous tissue does not get inflamed.

XLII. In cases where wounds occur either at once, with projection of the bones,[1] or afterwards, from irritation or roughnesses, when you recognise these latter, at once remove the dressing, and apply pitch cerate to the wound. Bandage, putting the beginning of the roll first on the wound, and the rest as though there were no lesion there, for so there will be least swelling at the part; suppuration and separation will be most prompt, and the cleansed parts heal up most rapidly. As to splints, do not apply them to this part, and do not make pressure. This treatment applies to cases where small pieces of bone come away; when large it is clear[2] (what to do), for there is much pus formation, and this treatment is no longer suitable, but the wound is left open because of the accumulations. But in all such cases as have bones projecting, whether they are reduced or not, bandaging is not suitable; what is required is stretching. Rounds are made like fetters, one at the ankle, the other

παρὰ γόνυ, ἐς κνήμην πλατεῖαι, προσηνέες, ἰσχυραί, κρίκους ἔχουσαι· ῥάβδοι τε σύμμετροι κρανίης καὶ μῆκος καὶ πάχος, ὥστε διατείνειν·
20 ἱμάντια δὲ ἐξ ἄκρων ἀμφοτέρωθεν ἔχοντα ἐς τοὺς κρίκους ἐνδεδέσθαι, ὡς τὰ ἄκρα ἐς τὰς σφαίρας ἐνστηριζόμενα διαναγκάζῃ. ἴησις δέ, πισσηρὴ θερμή·[1] σχήματα καὶ ποδὸς θέσις καὶ ἰσχίου· δίαιτα ἀτρεκής. ἐμβάλλειν τὰ ὀστέα τὰ ὑπερίσχοντα αὐθήμερα ἢ δευτεραῖα· τεταρταῖα δὲ ἢ πεμπταῖα, μή, ἀλλ' ἐπὴν ἰσχνὰ ᾖ. ἡ δὲ ἐμβολὴ τοῖσι μοχλικοῖσιν· ἢ τὸ ἐμβαλλόμενον τοῦ ὀστέου, ἢν μὴ ἔχῃ ἀποστήριξιν, ἀποπρῖσαι τῶν κωλυόντων· ἀτὰρ καὶ ὡς τὰ ψιλωθέντα ἀπο-
30 πεσεῖται, καὶ βραχύτερα τὰ μέλεα.

XLIII. Τὰ δὲ ἄρθρα, τὰ μὲν πλέον, τὰ δὲ μεῖον ὀλισθάνει· καὶ τὰ μὲν μεῖον ἐμβάλλειν ῥᾴδιον· τὰ δὲ μέζους ποιεῖ τὰς κακώσιας καὶ ὀστέων καὶ νεύρων καὶ ἄρθρων καὶ σαρκῶν καὶ σχημάτων. μηρὸς δὲ καὶ βραχίων ὁμοιότατα
6 ἐκπίπτουσιν.

[1] πισσηρῇ θερμῇ.

at the knee, flattened on the leg side, soft and strong, provided with rings; rods of cornel-wood, suitable in length and thickness, to keep the limb stretched; leather thongs adapted at each end to the extremities (of the rods) are fastened to the rings, so that the ends of the rods, being fixed to the rounds, make extension both ways. Treatment: warm pitch cerate, attitude, position of foot and hip, strict diet. Reduce projecting bones on the first or second day, not on the fourth or fifth, but when swelling has gone down. The reduction with small levers: if the fragment to be reduced does not afford a fulcrum, saw off what is in the way. For the rest, shortening of the limbs is proportional to the denuded bone which comes away.

XLIII. Joints are dislocated, some to a greater, some to a less extent; and the less are easy to reduce, but the greater produce more serious lesions of bones, ligaments, joints, flesh, and attitudes. The thigh and upper arm are very similar in their manner of dislocation.[1]

[1] *I.e.* completely, or not at all. See *J.* LXI.

APPENDIX

NOTES ON JOINTS LXXX

We have seen that, according to Galen, Chapter LXXVIII is the ὕστατος λόγος, or "final discourse," of *Joints*. His commentary ends rather abruptly in the middle of it, but he has already intimated that he is not going to say much, and he can hardly have gone beyond, though some manuscripts contain the rest of the Hippocratic treatise. Of this appendix the most interesting part is Chapter LXXX. It looks like, and has always been considered, the original Hippocratic account of finger-joint dislocation, which somehow got displaced and replaced by the very poor substitute, Chapter XXIX, identical with *Mochlicon* XIX.

But there are difficulties in this view. No ancient writer, till we get back to Diocles, early in the fourth century B.C., seems aware of its existence. Galen excludes it from *Joints*, but had he known that Hippocrates anywhere mentioned "lizards" as surgical instruments he would surely not have left them to puzzle succeeding generations till Diels happened to visit a toy shop. He would have explained it in his Hippocratic Glossary. Even Erotian, who tells us twice over that σειρά in Hippocrates means ἱμάς (strap), would hardly have left σαύρα unexplained. The analogous but less peculiar use of τύρσις (*Joints* XLIII) is explained twice over both by Erotian and Galen.

Apollonius obviously knew nothing about it. He apologises for the poverty of XXIX, and supplements it by an extract from Diocles, but seems quite unaware that this extract is an abbreviation of the genuine Hippocratic account. Apollonius was the chief Alexandrian surgeon of his day (first century B.C.), so we may safely conclude that the chapter was not in the Alexandrian edition of Hippocrates.

APPENDIX

One would hardly add a poor account of a matter to a treatise which already contained a good one; it is therefore improbable that *Joints* contained Chapter LXXX when it got separated from *Fractures*, and had its more glaring omissions made up by insertions from *Mochlicon*. We thus get back to the author of *Mochlicon*. Did he abbreviate his Chapter XIX (XXIX *J.*) from LXXX? Able editors such as Littré, Adams, Petrequin say he did. I venture to think that the reader will find no evidence of this, but will discover without much trouble that XXIX is practically made up of stock phrases taken from the three previous chapters, one of them ("the flesh wastes chiefly on the side opposite to the dislocation") being dragged in rather absurdly. Unusual words, εὔσημον ἀντωθεῖν ἔκπτωμα ἐπιπωροῦται, are all absent from LXXX, but have been just used or seen by the epitomist (ἐπιπωροῦται *F.* XXXVIII which he has just abridged), while the peculiar words and expressions of LXXX are all absent.

Coming to the Diocles quotation we find a great contrast. The correspondence of words and phrases is so close, that, though the hand is looked at from a different position, it seems almost certain that the two passages are connected. The natural view is that Diocles is copying Hippocrates, and this seems confirmed by Galen's assertion that he paraphrased other parts of *Joints*. On the other side there is the ignorance of Apollonius; the difficulty in believing that Chapter LXXX could have been so entirely lost and so entirely recovered after many centuries, and another fact which perhaps turns the balance against the accepted theory. Besides σαύρα the writer uses another word in a peculiar sense, χώρα = "joint socket." This occurs no less than six times in the two chapters LXXIX–LXXX, which is strong evidence that they are by the same author, and against the view that he is identical with the author of *Fractures-Joints*; for though the old writer uses χώρα[1] occasionally, it always has its natural sense of "place," whereas in LXXIX–LXXX the "natural" and sometimes necessary sense is "socket." The remaining Chapter (LXXXI) is made up largely of passages taken from the two previous

[1] Usually with ἑωυτοῦ, cf. *F.* IX, XIV. In *J.* LXXIX–LXXX this word is omitted in all six cases.

ones, with the highly un-Hippocratic addition that all dislocation patients should be starved for seven days (!). Even if we soften this down by inserting καὶ ("even for seven days") as do some manuscripts, it is still inconsistent with the rules given by the author of *Fractures-Joints.* We conclude therefore that these three chapters are probably a late addition. Perhaps a surgeon who had read the apology and supplement of Apollonius, and believed, as we do, that the latter is really taken from Hippocrates, thought it no forgery to try to rewrite the latter in an expanded form and in Hippocratic style. While he was about it, he might also wish to remedy another defect in *Joints,* which, as he justly observes, should first tell us what joints are. He therefore composed Chapters LXXIX–LXXX and probably LXXXI which became firmly attached to the end of the treatise.

THE DIOCLES SUPPLEMENT TO XXIX

Δακτύλου μὲν ἄρθρον ἄν τε ποδὸς ἄν τε χειρὸς ἐκπέσῃ, τετραχῶς ἐκπίπτει, ἢ ἐντὸς ἢ ἐκτὸς ἢ εἰς τὰ πλάγια. ὅπως δ' ἂν ἐκπέσῃ, ῥᾴδιον γνῶναι πρὸς τὸ ὁμώνυμον καὶ τὸ ὑγιὲς θεωροῦντα. ἐμβάλλειν δὲ κατατείνοντα εὐθὺ ἀπὸ χειρῶν, περιελίξαι δὲ ὅπως μὴ ἐξολισθάνῃ. ἀστεῖον δὲ καὶ τὰς σαύρας, ἃς οἱ παῖδες πλέκουσι, περιθέντα περὶ ἄκρον τὸν δάκτυλον κατατείνειν, ἐκ δὲ τοῦ ἐπὶ θάτερα ταῖς χερσίν.

A joint either of a toe or finger may be put out. It is put out in four ways, inwards, outwards, or to the sides. The way it is put out is easy to distinguish by comparing it with the sound and corresponding joint. Put it in by making extension in a straight line with the hands, but wrap a band round it that it may not slip away. It is also ingenious to put the lizards, which children plait, round the end of the finger and make extension, pulling in the opposite direction with the hands.

THE HIPPOCRATIC BENCH

Though we have three complete accounts of the Hippocratic Bench, by "Hippocrates,"[1] Rufus (or Heliodorus),[2]

[1] *Joints* LXXII–LXXIII. [2] Oribasius XLIX. 26 ff.

and Paulus Ægineta[1] respectively, attempts at restoration have been unfortunate. Till the time of Littré they were based on that of Vidus Vidius (1544), who read *μικράς* for *μακράς* in *Joints* LXXII and produced a bench with a row of square holes down the middle. He represented the perineal peg as angular and pointed, and made the corner supports so high that the patient would be lifted as well as stretched.

Littré pointed out that the *κάπετοι* were long grooves parallel to one another. He also reduced the height of the corner posts, and was on the point of making them project horizontally lengthways, so sunk into the bench that the axles would come below its surface.[2] This view, which seems admitted as an alternative in *Joints* XLVII, is still supported by Schöne.

On the whole, however, Littré's figure, including the uncomfortable form of perineal peg which he retained, is still generally accepted: but there are serious doubts as to the intermediate supports. Littré like his predecessors represented them as fixtures at the sides of the bench, though Scultetus had suggested that they were movable, a view adopted by Petrequin, who, however, still keeps them well to the sides. The chief object of this note is to suggest that they were not only movable, but were inserted when required into the grooves not more than a foot apart.

Paulus in his renovated text is clear as to the first point.[3] "As a last resort in internal dislocation of the thigh, let the perineal peg be removed and let two other pieces of wood be inserted on either side of its position"—*ἐκ πλαγίου τῆς τούτου θέσεως ἑκατέρωθεν ἕτερα δύο ξύλα πεπήχθω.* This seems intended for a paraphrase of the Hippocratic *κατὰ μέσον καὶ ἐκ πλαγίων,*[4] for *κατὰ μέσον* has just been used to describe the position of the peg. A cross-piece is then inserted "so that the shape of the three resembles the letter pi (Π), or eta (Η) if the cross piece is a little below the top. Then, with the patient lying on his sound side, we may bring (*ἀγάγωμεν*) the sound leg between these supports."

In Rufus the apparatus is apparently in one piece, a pi-shaped prop.[5] It is noticed first merely as "another

[1] VI. 118. [2] IV. 46. [3] VI. 118. 5.
[4] LXXIII. [5] *πιοειδὴς φλιά.*

THE HIPPOCRATIC BENCH OR SCAMNUM

i. According to Vidius. 1544

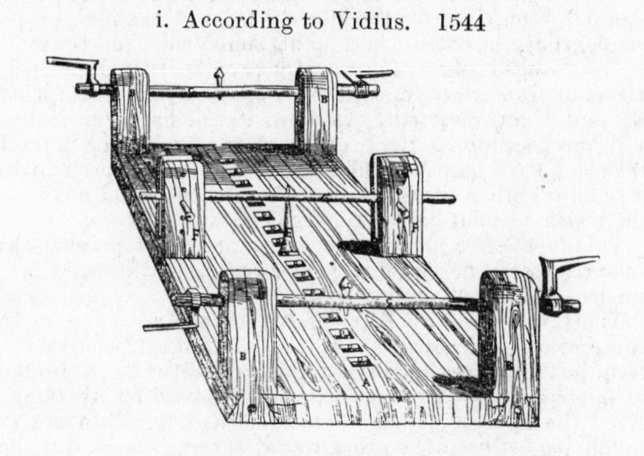

ii. According to Littré. 1844

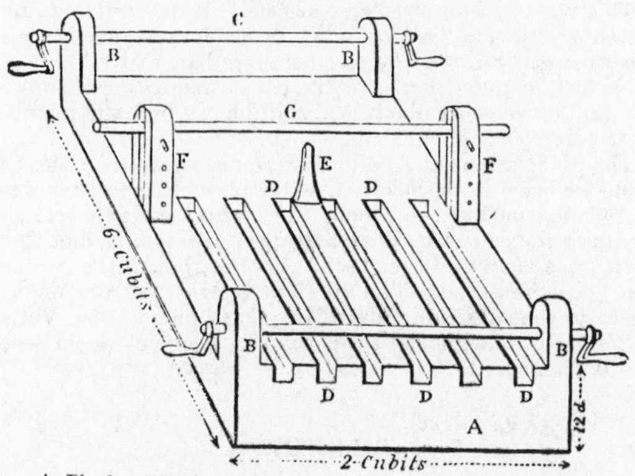

A. Plank. BB. Corner Supports. C. Axle. DD. Grooves.
E. Perineal Peg. FF. Intermediate Supports. G. Crossbar.

central contrivance besides the perineal peg."[1] In describing the use of the bench for thigh dislocation he adds that it was especially contrived for the internal form; "the perineal peg is taken out, the patient laid on his sound side, and the sound leg is arranged (τάσσεται) under the prop." It is also called a πῆγμα or framework, and perhaps could stand on the bench without being inserted. Anyhow, it can hardly have been a fixture occupying the breadth of the bench, for it would then not have been very pi-shaped, would have been in the way on all other occasions, and the patient could not lie on the bench without having his legs beneath it.

This fact seems alone sufficient to prove our points—that the props were not only movable, but, when inserted, were so close as just to admit one leg.

The terms used by Hippocrates are the strongest of the three, whether we read διέρσαι μεσηγύ ("insert between"), a term just employed for inserting an arm between the thighs,[2] or ἐρείσειε μεσηγύ ("press between"), as read by Apollonius. Even the mildest of the expressions used for bringing the sound leg between the props would surely be absurd if they were so far apart that the patient could not lie on the bench without having it there already!

This view enables us to give ποδιαίας[3] its natural meaning: the supports were "a foot long" in order to stand firmly in the grooves. So, too, the wooden cross-bar, instead of being three feet long and expected to resist immense pressure at its middle, was only about a foot in length and the pressure distributed throughout.

The illustrations of Apollonius are disappointing; the one thing we learn from them is that the grooves sometimes went the whole length of the bench. The wheel and axle arrangements at the ends are apparently separate from it, and there is no trace of any intermediate supports, though the perineal peg is represented. The Wellman Museum of Medical History contains an interesting example of the Vidian restoration, though the supports had been cut down when it was discovered.

[1] πριαπισκός. [2] LXXI. [3] LXXIII.